ENCYCLOPAEDIA OF
SCIENCE

ENCYCLOPAEDIA OF SCIENCE

Anubhav Ahuja

ANMOL PUBLICATIONS PVT. LTD.
NEW DELHI - 110 002 (INDIA)

ANMOL PUBLICATIONS PVT. LTD.
H.O.: 4374/4B, Ansari Road, Darya Ganj,
New Delhi-110 002 (India)
Ph.: 23278000, 23261597

B.O.: No. 1015, Ist Main Road, BSK IIIrd Stage
IIIrd Phase, IIIrd Block,
Bangalore - 560 085 (India)
Visit us at: www.anmolpublications.com

Encyclopaedia of Science

First Published, 2006

ISBN 81-261-3030-X

PRINTED IN INDIA

Printed at Mehra Offset Press, Delhi.

Preface

The aim of this encyclopaedia is to give, in the light of present knowledge and opinion, definitions of terms that are of importance in Science. It is an encyclopaedia of Science terms, written by specialists, partly for other specialists but more particularly for the man-in-the-street, and for students and interested workers of all kinds and ages: indeed, for all who wish to understand what science is.

The terms in this book is taken from all fields of Science such as Physics, Chemistry, Botany and Zoology. All sub fields are covered in detail. Latest developments, research and new emerging fields are taken into consideration. The terms defined in this book include laws, relationships, equations, basic principles and concepts, as well as the most widely used instruments and apparatus and are discussed in detail with diagrams. Various latest technological developments in the fields such as Nanotechnology, Biotechnology, genetics, Bioinformatics, Computer Science, information technology and much more are dealt in detail. In short, this treatment comprises the terms both of pure science and of its applications. A considerable number of terms have been included from the subject-areas bordering on Science, not only mathematics, but physical chemistry, applied electronics, applied electricity, etc. Obviously, all the terms in pure and applied Science cannot be defined adequately in a book, or for that matter in a library. The objective of editors and contributors has been the more modest one of providing a book useful as a general reference in Science, helpful even to the specialist in regions outside of his domain of specialized knowledge

The aim of this volume is to provide a concise repository of number of major concepts in this field. In preparation of this book, the author has freely consulted large number of books and journals so no authenticity is claimed. Author is especially thankful to M/S Anmol Publications Pvt. Ltd. for shaping this book in its final form. Suggestions for further improvement of this book are not only welcome but also greatly appreciated.

Author

Preface

A

A-amylase An enzyme that converts starch to sugars.

Abdomen The portion of a tetrapod's body between the thorax and pelvic girdle.The region of an arthropod's body behind the thorax. It contains the visceral organs.

Abductor A muscle that performs the action of abduction.

Abelian This describes an algebraic operation (multiplication, addition, division, subtraction, etc.). If the operation is Abelian, then the elements commute. For example regular multiplication commutes: 3x4 = 4x3 = 12. Addition also commutes: 3+4 = 4+3 = 7. There are mathematical elements (matrices) which do not commute under multiplication. (If A and B are matrices, then A times B does not equal B times A.) Operations which are different when the elements are exchanged (which do not commute) are called non-Abelian. These ideas are named for the developer, Niels Abel.

Aberration The apparent change in position of a light-emitting object due to the constancy of the speed of light and the motion of the observer relative to the emitter. The effect is nonrelativistic; that is, special relativity is not required to derive it: all that is needed is newtonian mechanics and the assumption of the constancy of the speed of light. The effect is observable in the apparent change of position of stars due to earth's relative motion, and is responsible for the "tunnel vision" effect of travelling at relativistic speeds.

Abiotic A non-living (physical or chemical) component of the environment.

Abrasion The mechanical process of gradually breaking down a hard layer, as in a seed coat.

Abscess Dead cells and tissue fluid confined in a localized area, causing swelling.

Abscisic acid A plant hormone associated with dormancy, abscission of organs, and water stress.

Abscissa The value corresponding to the horizontal distance of a point on a graph from the Y axis.

Absolute uncertainty The uncertainty in a measured quantity is due to inherent variations in the measurement process itself. The uncertainty in a result is due to the combined and accumulated effects of these measurement uncertainties which were used in the calculation of that result. When these uncertainties are expressed in the same units as the quantity itself they are called absolute uncertainties. Uncertainty values are usually attached to the quoted value of an experimental measurement or result, one common format being: (quantity) ± (absolute uncertainty in that quantity). Compare: relative uncertainty.

Absolute zero The temperature of a body at which the kinetic energy of its molecules is at a minimum; O°K or -273.16°C.

Absorption spectrum A continuous spectrum interrupted by dark lines or bands that are characteristic of the medium through which the radiation has passed.

Absorptive feeding A means of taking up dissolved foodmaterial through specialized organs or across the body wall.

Absorptive heterotroph Mode of nutrition in which organisms secrete digestive enzymes into a substrate and then absorb the products of digestion; nutritional mode of fungi.

Abstraction A chemical reaction or transformation, the main feature of which is the bimolecular removal of an atom (neutral or charged) from a molecular entity. For example:

$CH_3COCH_3 + (i\text{-}C_3H_7)_2n^- \longrightarrow (CH_3COCH_2)^- + (i\text{-}C_3H_7)_2nH$

(proton abstraction from acetone)

$CH_4 + Cl \longrightarrow H_3C^- + HCl$

(hydrogen abstraction from methane).

Abyssal plains Flat, sediment-covered areas in the ocean basin usually 3,000 to 5,000 m deep.

Acceleration Time rate of change of velocity.

Acceptor An element with three valence electrons per atom which when added to a semiconductor crystal provides electron "holes" in the lattice structure of the crystal.

Accessory buds Those buds adjacent to a primary bud and usually smaller in size.

Accessory fruit Most or part of the fruit is derived from tissue other than the ovary of a flower.

Accessory pigment One of several nongreen photosynthetic pigments found in marine plants that absorbs light energy from the center of the visible light spectrum and transfers it to the green pigment chlorophyll.

Acclimation The change in tolerance of an animal for a condition in its environment.

Acclimatization Gradual physiological adaptation in response to relatively long-lasting environmental changes.

Accommodation The adjustment of the eye for various distances.

Accuracy Closeness of a measurement to the accepted value for a specific physical quantity; expressed in terms of error.

Acetabulum True sucker, especially in flukes and leeches; the socket in the hip bone that receives the thigh bone.

Acetospora The protozoan phylum characterized by members having multicellular spores; all parasitic in invertebrates. Examples: acetosporans (Paramyxa, Halosporidium).

Acetylcholine A neurotransmitter liberated by certain neurons. It is excitatory at neuromuscular junctions and inhibitory at other synapses.

Achene A small, dry, one-seeded indehiscent fruit; the pericarp is easily separated from the seed coat.

Acid A molecular entity or chemical species capable of donating a hydron (proton) or capable of forming a covalent bond with an electron pair.

Acid rain The combination of sulfur dioxide and nitrogen oxides with water in the atmosphere. This combination produces acidicprecipitation called acid rain. The burning of fossil fuels is a major contributor to acid rain.

Acidic Possessing a relatively large number of hydrogen ions; having a ph less than 7.0.

Acidity 1. Of a compound:

For Brønsted acids it means the tendency of a compound to act as a hydron donor. It can be quantitatively expressed by the acid dissociation constant of the compound in water or some other specified medium. For Lewis acids it relates to the association constants of Lewis adducts and π-adducts.

2. Of a medium:

The use of the term is mainly restricted to a medium containing Brønsted acids, where it means the tendency of the medium to hydronate a specific reference base. It is quantitatively expressed by the appropriate acidity function.

Acidity function Any function that measures the thermodynamic hydron-donating or -accepting ability of a solvent system, or a closely related thermodynamic property, such as the tendency of the lyate ion of the solvent system to form Lewis adducts. Acidity functions are not unique properties of the solvent system alone, but depend on the solute (or family of closely related solutes) with respect to which the thermodynamic tendency is measured.

Commonly used acidity functions refer to concentrated acidic or basic solutions. Acidity functions are usually established over a range of composition of such a system by UV/VIS spectrophotometric or NMR measurements of the degree of hydronation (protonation or Lewis adduct formation) for the members of a series of structurally similar indicator bases (or acids) of different strength: the best known of these functions is the Hammett acidity function Ho (for uncharged indicator bases that are primary aromatic amines).

Acquired immunity Immunity arising from a specific immune response, stimulated by antigen in the host's body (active) or in the body of another individual with the antibodies or lymphocytes transferred to the host (passive).

Acquisition Gaining a response in a learning paradigm.

Acridine dyes Organic pigment molecules that are capable of causing permanent genetic changes (mutations).

Acrocentric Chromosome with centromere near the end.

Actin A protein in a muscle fiber that, together with myosin, is responsible for contraction and relaxation.

Actinomorphic Regular or radially symmetric flowers.

Actinotroch Larval form found in Phoronida.

Action This technical term is a historic relic of the 17th century, before energy and momentum were understood. In modern terminology, action has the dimensions of energy×time. Planck's constant has those dimensions, and is therefore sometimes called Planck's quantum of action. Pairs of measurable quantities whose product has dimensions of energy×time are called conjugate quantities in quantum mechanics, and have a special relation to each other, expressed in Heisenberg's uncertainty principle.

Action potential The sequence of electrical changes occurring when a nerve cell membrane is exposed to a stimulus that exceeds its threshold.

Activated complex An activated complex, often characterized by the superscript, is defined as that assembly of atoms which corresponds to an arbitrary infinitesimally small region at or near the col (saddle point) of a potential energy surface.

Active transport A process that requires an expenditure of ATP energy to move molecules across a cell membrane; usually moved against the concentration gradient with the aid of specific transport proteins.

Adaptation An anatomical structure, physiological process, or behavioral trait that evolved by natural selection and improves an organism's ability to survive and leave descendants.

Adaptive behavior Behavior patterns that make an organism more fit to survive and reproduce in comparison with other members of the same species.

Adaptive radiation Evolutionary diversification that produces numerous ecologically disparate lineages from a single ancestral one, especially when this diversification occurs within a short interval of geological time.

Addictive drug Substance that induces an addiction by causing physiological dependence, psychological dependence, and/or tolerance.

Addition (alpha-addition) A chemical reaction resulting in a single reaction product from two or three reacting chemical species, with formation of two new chemical bonds to the same atom in one of the reactant molecular entities. The synonymous term 1/1/addition is also used. For example:

$C_{12}C: + CH_3OH \longrightarrow C_{12}CHOCH_3$

Addition reaction A chemical reaction of two or more reacting molecular entities, resulting in a single reaction product containing all atoms of all components, with formation of two chemical bonds and a net reduction in bond multiplicity in at least one of the reactants. The reverse process is called an elimination reaction. The addition may occur at only one site (-addition, 1/1/addition), at two adjacent sites (1/2/addition) or at two non-adjacent sites (1/3/- or 1/4/addition, etc.). For example

(a) $H^+ + Br^- + (CH_3)_2C{=}CH_2 \longrightarrow (CH_3)_2CBr{-}CH_3$ (1/2/addition)

(b) $Br_2 + CH_2{=}CH{-}CH{=}CH_2 \longrightarrow BrCH_2{-}CH{=}CH{-}CH_2Br$ (1/4/addition) and $BrCH_2{-}CH(Br){-}CH{=}CH_2$ (1/2/addition)

If the reagent or the source of the addends of an addition are not specified, then it is called an addition transformation.

Additivity principle The hypothesis that each of several structural features of a molecular entity makes a separate and additive contribution to a property of the substance concerned. More specifically, it is the hypothesis that each of the several substituent groups in a parent molecule makes a separate and additive contribution to the standard Gibbs energy change (or Gibbs energy of activation) corresponding to a particular equilibrium (or rate of reaction).

Adduct A new chemical species AB, each molecular entity of which is formed by direct combination of two separate molecular entities A and B in such a way that there is change in connectivity, but no loss, of atoms within the moieties A and B. Stoichiometries other than 1:1 are also possible, e.g. a bis-adduct (2:1). An "intramolecular adduct" can be formed when A and B are groups contained within the same molecular entity.

This is a general term which, whenever appropriate, should be used in preference to the less explicit term complex. It is also used specifically for products of an addition reaction.

Adductor A muscle that draws a part toward a median axis, or a muscle that draws the two valves of a mollusc shell together.

Adenine A purine base; component of nucleotides and nucleic acids. A nitrogen base found in both DNA and RNA.

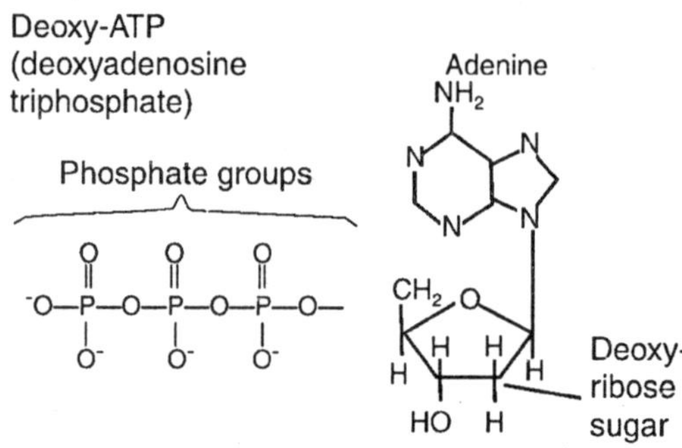

Fig. Adenine

Adenosine A nucleotide composed of adenine, ribose sugar, and two (ADP) or three (ATP) phosphate units; ATP is an energy-rich compound that, with ADP, serves as a phosphate bond-energy transfer system in cells.

Adenosine diphosphate (ADP) The building block for ATP; by adding a terminal phosphate group and a large amount of energy, ATP can be formed.

Adenosine monophosphate AMP molecule; created when the terminal phosphate is lost from a molecule of adenosine diphosphate.

Adenosinetriphosphate (ATP) A complex organic compound composed of the molecule adenosine and three phosphates, which serves in short-term energy storage and conversion in all organisms.

Adhesio Interthalamica Part of the brain.

Adhesion Attraction between unlike molecules. The polar nature of water molecules causes them to adhere to a surface.

Adiabatic process A thermal process in which no heat is added to or removed from a system.

Adipose Fatty tissue; fatty.

Adipose tissue Fat-storing tissue.

Adoptive immunity Immune state conferred by inoculation of lymphocytes, not antibodies, from an immune animal rather than by exposure to the antigen itself.

Adrenal glands Paired endocrine glands, located next to the kidneys in the abdomen. The adrenal cortex produces steroid hormones involved in water balance, glucose metabolism, and electrolyte balance. The adrenal medulla produces adrenaline and noradrenaline, which are involved in glucose metabolism, heart rate, and blood pressure.

Adrenaline A hormone produced by the adrenal, or suprarenal, gland; epinephrine.

Adrenocorticotrophic A hormone secreted by the pituitary.

Adsorption The adhesion of molecules to solid bodies.

Adventitious A structure arising at some location not usually expected, such as on a stem.

Aerial Pertaining to being in the air, such as a root projecting from an aboveground stem.

Aerobic Having molecular oxygen present; an oxygen-dependent form of respiration.

Aerobic respiration Respiration in the presence of oxygen.

Aestivation The condition of dormancy or torpidity during the hot summer months.

Aflatoxin A complex of four mycotoxins produced by Aspergillus flavus; often found in peanut products.

Agar Gelatinous product extracted from the walls of some red algae; used in growth media for microorganisms and in tissue culture.

Agent orange A herbicide used as a defoliant in the Vietnam War; composed of 2,4-D and 2,4,5-T.

Aggregated Describes the concentration of most parasites of a single species in a minority of hosts.Also called overdispersed.

Aggression Behavior that appears to be intended to inflict noxious stimulation or destruction on another organism.

Aggressive mimicry A technique for capturing prey in which the predator uses lures or other means to misinform the prey.

Agnatha A superclass of vertebrates whose members lack jaws and paired appendages and possess a cartilaginous skeleton and a persistent notochord. Lampreys and hagfishes.

Agonist The "prime mover" or muscle of reference.

Agostic The term designates structures in which a hydrogen atom is bonded to both a carbon atom and a metal atom. The term is also used to characterize the interaction between a CH bond and an unsaturated metal centre, and to describe similar bonding of a transition metal with Si-H compounds. The expression "-hydrido-bridged" is also used to describe the bridging hydrogen.

Agonistic behavior A suite of behavior patterns used during conflict with a conspecific, usually indicating whether an individual is going to submit to the other animal or fight if the other does not submit.

Agregate fruit A fruit derived from a single flower with several separate ovaries; example: blackberry.

Agricultural infrastructure The complexities of processing, storing, packaging, and transporting farm products.

Ahermatypic coral A coral that does not build reefs.

Air sacs Lateral branches of the nasal passages of smaller toothed whales; sources of echolocation sounds.

Airfoil A surface, such as a wing, that provides lift by using currents of air it moves through.

Ala Term often applied to winglike structure on plants or animals: the lateral winglike expansions of the branchiuran carapace to form respiratory alae, cuticular winglike expansions of nematodes, and others.

Alare Usually a bone process shaped like a wing.

Alate Winged.

Albedo A description of a surface's reflective properties.

Albumin Any of a large class of simple proteins that are important constituents of vertebrate blood plasma and tissue fluids and also present in milk, whites of eggs, and other animal substances.

Alcoholic fermentation A form of anaerobic respiration in which sugar is degraded to alcohol and CO_2 and energy is released.

Aleurone layer A group of cells rich in protein granules and located as the outer layer of the endosperm of many grain seeds.

Alga A photosynthetic protist containing plastids. Any of several groups of autotrophs that lack the structural features (true leaves, roots, and stems) of the higher plants.

Algal bloom A proliferation of algae due to a nutrient-rich medium, usually resulting in a green scum on the water surface.

Algal grazer An animal that consumes either large algae or thin algal films.

Algal ridge A ridge of coralline algae that is found on the outer edge of some coral reefs.

Algal turf A dense growth of often filamentous algae.

Algin A polysaccharide derived from brown algae and used for many industrial processes.

Alginic acid Gelatinous material extracted from the wall of certain brown algae; used in a variety of commercial and industrial products.

Alien species (exotic species)A species introduced by humans.

Alimentary Having to do with nutrition or nourishment.

Alkaline Denoting substances that release hydroxyl (OH^-) ions into solution.

Alkaloid A group of nitrogen-containing compounds having diverse structures; many alkaloids have medicinal, hallucinogenic, or toxic properties.

Allantois One of the extraembryonic membranes of the amniotes that functions in respiration and excretion in birds and reptiles and plays an important role in the development of the placenta in most mammals.

Allee effect The unfavorable consequences of undercrowding, as when populations fail to breed when numbers are below some critical density.

Allele Alternative forms of genes coding for the same trait; situated at the same locus in homologous chromosomes.

Allelopathy The release of chemicals by certain plants that inhibit the growth of competing plants.

Allergens Substances that activate the immune system and cause an allergic response; may not be directly antigenic themselves but may make other materials antigenic.

Allograft A piece of tissue or an organ transferred from one individual to another individual of the same species, not identical twins; homograft.

Allometry Relative growth of a part in relation to the whole organism.

Allopatric In separate and mutually exclusive geographical regions.

Allopatric speciation The formation of new species by a process involving geographic barriers.

Allopatry Populations or species with nonoverlapping geographic distributions. Contrast with sympatry.

All-or-none law The phenomenon in which a muscle fiber contracts completely when it is exposed to a stimulus of threshold strength. Also, the principle that states a neuron will "fire" at full power or not at all.

Allylic substitution reaction A substitution reaction occurring at position 1/ of an allylic system, the double bond being between positions 2/ and 3/. The incoming group may be attached to the same atom 1/ as the leaving group, or the incoming group becomes attached at the relative position 3/, with movement of the double bond from 2/3 to 1/2. For example:

$CH_3CH{=}CHCH_2Br \longrightarrow CH_3CH{=}CHCH_2OAc$

or

$CH_3CH{=}CHCH_2Br \longrightarrow CH_3CH(OAc)CH{=}CH_2$

Alpha particle A helium-4 nucleus, especially when emitted from the nucleus of a radioactive atom.

Alpha-helix Literally the first spiral arrangement of the genetic DNA molecule; regular coiled arrangement of polypeptide chain in proteins; secondary structure of proteins.

Alpine The high, treeless biogeographic zone of mountains that consists of slopes above the timberline.

Alternant A conjugated system of π electrons is termed alternant if its atoms can be divided into two sets so that no atom of one set is directly linked to any other atom of the same set.

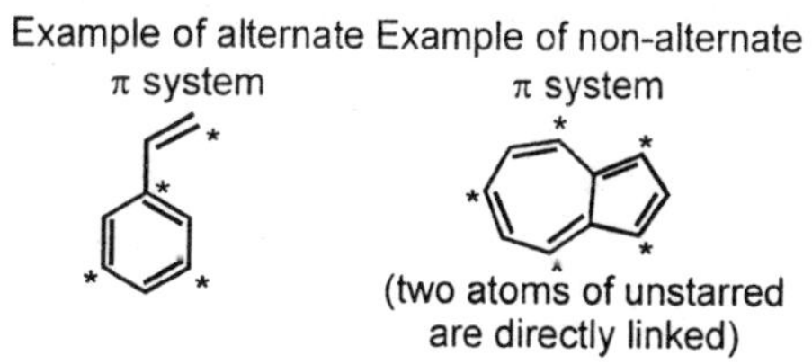

Alternate host The alternate plant required to complete the life cycle of some microorganisms. For example, for Puccinia graminis tritice, wheat is the primary host and Berberis vulgaris is the alternate host.

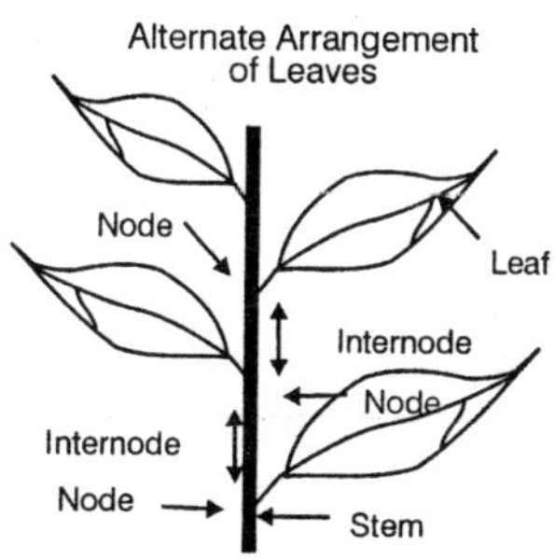

Fig. Alternate

Alternating current An electric current that has one direction during one part of a generating cycle and the opposite direction during the remainder of the cycle.

Alternation of generations A reproductive cycle in which a sexual stage alternates with an asexual one, as in the case of a gametophyte alternating with a sporophyte.

Altricial An animal that is helpless at hatching or birth.

Ambident A description applied to a chemical species whose molecular entities each possess two alternative and strongly interacting distinguishable reactive centres, to either of which a bond may be made in a reaction: the centres must be connected in such a way that reaction at either site stops or greatly retards subsequent attack at the second site. The term is most commonly applied to conjugated nucleophiles, or -pyridones, and also to the vicinally ambident cyanide ion, cyanate ion, thiocyanate ion, sulfinate ion, nitrite ion, and unsymmetrical hydrazines. Ambident electrophiles are exemplified by carboxylic esters $RC(O)OCR_3$ which react with nucleophiles either at the carbonyl carbon or the alkoxy carbon.

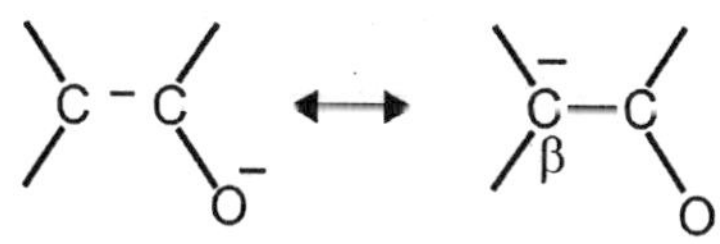

Molecular entities, such as dianions of dicarboxylic acids, containing two non-interacting (or feebly interacting) reactive centres, are not generally considered to be ambident and are better described as "bifunctional".

The Latin root of the word implies two reactive centres, but the term has in the past also incorrectly been applied to chemical species with more than two reactive centres. For such species the existing term "polydent" (or, better, "multident") is more appropriate.

Amebocyte Cell in metazoan invertebrate, often functioning in defense against invading particles.

Ameboid Ameba-like in putting forth pseudopodia.

Ameboma Granuloma containing active trophozoites, occasionally resulting from a chronic amebic ulcer; rare except in Central and South America.

Amebula Daughter cell resulting from mitosis and cytokinesis of an encysted ameba.

Ametabolous metamorphosis Development in which the number of molts is variable; immature stages resemble adults, and molting continues into adulthood.

Amictic Pertaining to female rotifers, which produce only diploid eggs that cannot be fertilized, or to the eggs produced by such females. Compare with mictic.

Amino acid An organic molecule including one or more amino (-NH_2) and acid (-COOH) groups; one of the 20 nitrogen-containing molecules that make up proteins.

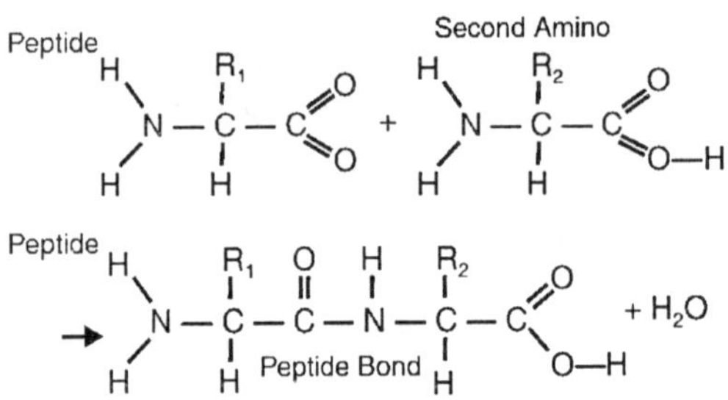

Fig. Amino acid

Amino group (-NH_2) A chemical part of a molecule that imparts basic properties to an amino acid.

Amitosis A form of cell division in which mitotic nuclear changes do not occur; cleavage without separation of daughter chromosomes.

Ammeter An electric meter designed to measure current.

Ammonotelic excretion Having ammonia as the chief excretory product of nitrogen metabolism; occurs in freshwater fishes.

Amniocentesis Procedure for withdrawing a sample of fluid around the developing embryo for examination of chromosomes in the embryonic cells and other tests.

Amnion One of the extraembryonic membranes of the embryos of reptiles, birds, and mammals. The amnion encloses the embryo in a fluid-filled sac.

Amniotic egg The egg of reptiles, birds, and mammals. It possesses a series of extraembryonic membranes that help prevent desiccation, store wastes, and promote gas exchange. These adaptations allowed vertebrates to invade terrestrial habitats.

Amoeboid movement A form of movement similar to that found in amoebae. Fluid endoplasm (plasmasol) flows forward inside the cell and changes state to viscous ectoplasm (plasmagel) on reaching the tip of a pseudopodium. At the opposite end of the cell, ectoplasm is converted into endoplasm.

Ampere The unit of electric current; one coulomb per second.

Ampere's law, modified form The line integral of the magnetic field around a closed curve is proportional to the sum of two terms: first, the algebraic sum of electric currents flowing through that closed curve; and second, the instantaneous time rate of change of the electric flux through a surface bounded by that closed curve; in differential form,

Curl h = j + dd/dt,

Where d/dt here represents partial differentiation.

In addition to describing electromagnetism, his equations also predict that waves can propagate through the electromagnetic field, and would always propagate at the same speed — these are electromagnetic waves.

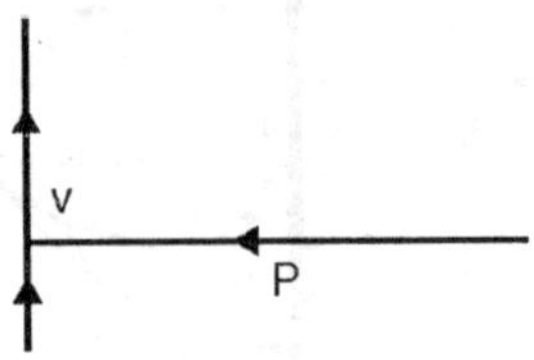

Fig. Ampere's law

Amphibia The class of vertebrates whose members are characterized by skin with mucoid secretions, which serves as a respiratory organ. Developmental stages are aquatic and are usually followed by metamorphosis to an amphibious adult. Frogs, toads, and salamanders.

Amphiblastula Free-swimming larval stage of certain marine sponges; blastula-like but with only the cells of the animal pole flagellated; those of the vegetal pole unflagellated.

Amphid One of a pair of chemosensory organs found on the anterior end of certain nematodes.

Amphipathic A molecule with one part soluble in water (polar) and another part insoluble in water (nonpolar).

Amphiphilic A compound containing a large organic cation or anion which possesses a long unbranched hydrocarbon chain, e.g.

$H_3C(CH_2)_nCO_2\text{-}M^+H_3C(CH_2)_nSO_3\text{-}M^+H_3C(CH_2)_nN(CH_3)_3{}^+X^-$

(n > 7).

The existence of distinct polar (hydrophilic) and non polar (hydrophobic) regions in the molecule promotes the formation of micelles in dilute aqueous solution.

Amphipods A group of small, laterally compressed crustaceans that includes beach hoppers and others.

Amphiprotic (solvent) Self-ionizing solvent possessing both characteristics of Brønsted acids and bases, for example H_2O and CH_3OH, in contrast to aprotic solvent.

Amphistome Fluke with the ventral sucker located at the posterior end.

Amphoteric A chemical species that behaves both as an acid and as a base is called amphoteric. This property depends upon the medium in which the species is investigated: H_2SO_4 is an acid when studied in water, but becomes amphoteric in superacids.

Amplexus The positioning of a male amphibian dorsal to female amphibian, his forelimbs around her waist. During amplexus, the male releases sperm as the female releases eggs.

Amplifier A device consisting of one or more vacuum tubes (or transistors) and associated circuits, used to increase the strength of a signal

Amplitude The maximum absolute value of a periodically varying quantity, as the peak in activity in a circadian or circannual cycle.

Ampulla Membranous vesicle; dilation at one end of each semicircular canal containing sensory epithelium; muscular vesicle above tube foot in water-vascular system of echinoderms.

Ampulla of Lorenzini One of several sensory structures in the head of sharks that detect weak electric fields.

Ampullary organ A receptor that can detect electrical currents. These electroreceptors are found in most fishes, some amphibians, and the platypus.

Amygdaloid A part of the brain.

Amylase An enzyme that breaks down starch into smaller units.

Amylopectin Component of starch consisting of highly branched chains of repeating glucose units; insoluble in water.

Amyloplast Starch-storing plastid.

Amylose Component of starch consisting of unbranched chains of repeating glucose units; soluble in water.

Anabolic Chemical reactions that synthesize and require energy.

Anabolism Any constructive metabolic process by which organisms convert substances into other components of the organism's chemical architecture. System of biosynthetic reactions in a cell by which large molecules are made from smaller ones.

Anadromous Refers to fishes that migrate up streams from the sea to spawn.

Anaerobic Not dependent on oxygen for respiration.

Anaerobic bacteria Bacteria that do not need oxygen.

Anaerobic respiration The incomplete intracellular breakdown of sugar or other organic compounds in the absence of oxygen that releases some energy and produces organic acids and/or alcohol.

Anal fin Each of the last pair of ventral fins of fishes.

Analgesic Pain-relieving remedy.

Analogous Structures that have similar functions in two organisms but have not evolved from a common ancestral form.

Analogy Similarity of function but not of origin.

Anamnestic response Immune response to a challenge or secondary antigen inoculation, marked by more rapid and stronger manifestation of the immune reaction (specifically, antibody titer) than after the primary immunizing dose.

Anamorphic The asexual or imperfect stage in a fungus life cycle.

Anaphase The stage in mitosis and meiosis, following metaphase, in which the centromeres divide and the chromatids, lined up on the mitotic spindle, begin to move apart toward the poles of the spindle to form the daughter chromosomes.

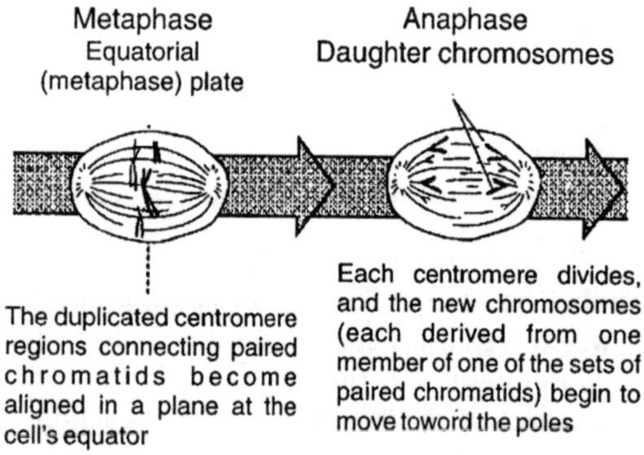

Fig. Anaphase

Anaphylaxis A systemic (whole body) immediate hypersensitivity reaction.

Anapolysis Detachment of a senile proglottid after it has shed its eggs.

Anapsid Amniotes in which the skull lacks temporal openings, with turtles the only living representatives.

Anastomosis A union of two or more blood vessels, fibers, or other structures to form a branching network.

Anatomy The study of the structure of an organism and its parts.

Anautogeny In some Diptera the necessity of a blood meal before eggs can develop within the female.

Androecium A collective term referring to the stamens within a flower.

Androgen Any substance that contributes to masculinization, such as the hormone testosterone.

Androgenic gland Gland located near the vas deferens in many Crustacea. Its secretions are responsible for development of male secondary sexual characteristics.

Androsterone An animal hormone not synthesized by plants.

Anecdysis Ecdysis in which successive molts are separated by quite long intermolt phases; referred to as terminal anecdysis when maximal size is reached and no more ecdyses occur.

Anemia Low levels of hemoglobin due to iron deficiency or lack of red blood cells.

Aneuploidy Loss or gain of a chromosome, cells of the organism have one fewer than normal chromosome number, or one extra chromosome, for example, trisomy 21 (Down syndrome).

Angiosperm The group of plants characterized by having flowers as their sexual reproductive structures.

Angiotensin Blood protein formed from the interaction of renin and a liver protein, causing increased blood pressure and stimulating release of aldosterone and ADH.

Angle of refraction The angle between the refracted ray and the normal drawn to the point of refraction.

Angstrom A unit of one ten-millionth of a millimeter (one ten-thousandth of a micrometer); it is represented by the symbol Å.

Angular acceleration The time rate of change of angular velocity.

Angular impulse The product of a torque and the time interval during which it acts.

Angular Magnification is the ratio of the angular size of the object as seen through the instrument to the angular size of the object as seen with the 'naked eye'. The 'naked eye' view is without use of the optical instrument, but under optimal viewing conditions. Some texts express angular magnification as the ratio of the angles, some express it as the ratio of the tangents of the angles.

Angular momentum The product of the rotational inertia of a body and its angular velocity.

Angular velocity The time rate of change of angular displacement.

Anhydrase An enzyme involved in the removal of water from a compound. Carbonic anhydrase promotes the conversion of carbonic acid into water and carbon dioxide.

Animal behavior Activities animals perform during their lifetime.

Animal pole The region of a fertilized egg where meiosis is completed. It contains less yolk and is more metabolically active than the opposite vegetal pole.

Animalia The kingdom of organisms whose members are multicellular, eukaryotic, and heterotrophic. The animals.

Animals Members of the kingdom Animalia, which consists of heterotrophic, eukaryotic, and multicellular organisms.

Anionotropic rearrangement (or anionotropy) A rearrangement in which the migrating group moves with its electron pair from one atom to another.

Anisogametes Outwardly dissimilar male and female gametes. Anisogamy is the condition of having dissimilar male and female gametes.

Anisogamy The condition in which the female gamete (ovum) is larger than the male gamete (sperm).

Anlage Rudimentary form; primordium.

Annelation Alternative, but less desirable term for annulation. The term is widely used in German and French language.

Annelida The phylum of triploblastic, coelomate animals whose members are metameric (segmented) and wormlike. Annelids have a complete digestive tract and a ventral nerve cord.

Annual A plant that lives for a single growing season.

Annual ring Ring of xylem in woody stem composed of springwood and summerwood that corresponds in temperate regions to a chronological year.

Annulation A transformation involving fusion of a new ring to a molecule via two new bonds.

Some authors use the term "annelation" for the fusion of an additional ring to an already existing one, and "annulation" for the formation of a ring from one or several acyclic precursors, but this distinction is not made generally.

Annulene Mancude (i.e. having formally the maximum number of noncumulative double bonds) monocyclic hydrocarbon without side chains of the general formula CnHn (n is an even number) or $CnHn_{+1}$ (n is an odd number).

Annuli Rings on the body of a parasite; not necessarily indicative of internal segmentation.

Annulus In fern sporangia, a row of cells with both thin and thickened cell walls which facilitate the opening of the sporangia and the release of spores because of the unequal expansion and tension produced as they are moistened and dried.

Anode When two metal plates are placed facing each other in an electrical circuit the positively charged plate is called the anode (the negatively charged plate is called the cathode).

Anoxic Lacking oxygen.

Antagonism The condition of being an opposing principle, force, or factor, as when two hormones have opposite effects on target tissues.

Antagonist The muscle or muscles opposing the agonist.

Antarafacial, suprafacial When a part of a molecule ("molecular fragment") undergoes two changes in bonding (bond-making or bond-breaking), either to a common centre or to two related centres, external to itself, these bonding changes may be related in one of two spatially different ways. These are designated as "antarafacial" if opposite faces of the molecular fragment are involved, and "suprafacial" if both changes occur at the same face. The concept of "face" is clear from the diagrams in the cases of planar (or approximately planar) frameworks with isolated or interacting pi orbitals.

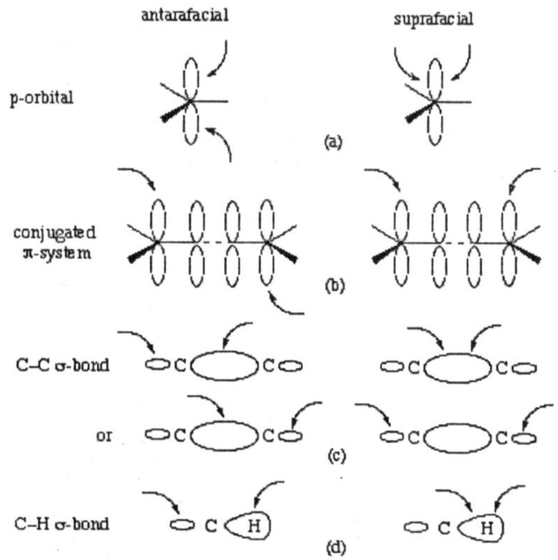

The terms antarafacial and suprafacial are, however, also employed in cases in which the essential part of the molecular fragment undergoing changes in bonding comprises two atoms linked only by a sigma bond. In these cases it is customary to refer to the phases of the local sigma-bonding orbital: occurrence of the two bonding changes at sites of like orbital phase is regarded as suprafacial, whereas that at two sites of opposite phase is antarafacial. There may be two distinct and alternative stereochemical outcomes of a suprafacial process involving a sigma bond between saturated carbon atoms, i.e. either retention or inversion at both centres. The antarafacial process results in inversion at one centre and retention at the second.

Antennae Second pair of appendages in Crustacea, with bases usually immediately posterior to antennules; primarily sensory but sometimes adapted for other functions; derived from appendages on primitive third preoral somite; no homologous appendage in insects.

Antennal gland The excretory organ in some crustaceans (crayfish). Called antennal glands because of their location near the base of each second antenna and their green color; also called green glands.

Antennules (first antennae) Anteriormost pair of appendages of Crustacea; primarily sensory but often adapted for additional or other functions in particular species; derived from appendages on primitive second preoral somite; homologous to antennae of insects.

Anterior station Development of a protozoan in the middle or anterior intestinal portions of its insect host, such as the section Salivaria of Trypanosomatidae.

Anther The male reproductive organ enclosing and containing the pollen grains.

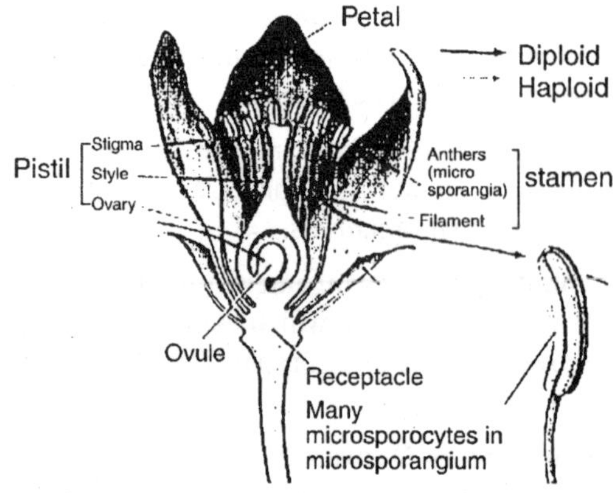

Fig. Anther

Antheridiophore in some liverworts, a stalk that bears antheridia embedded on an elevated and expanded tip.

Antheridium The multicellular male sex structure of plants other than seed plants.

Anthocyanins A group of water-soluble red to blue flavonoid pigments found in certain plants; especially important pigmentation in flower petals.

Anthozoa The class of Cnidarians whose members are solitary or colonial polyps. Medusae absent; gametes originate in the gastrodermis; mesenteries divide the gastrovascular cavity.

Anthracosaurs A group of Paleozoic labyrinthodont amphibians.

Anthropic principle 1. **Weak anthropic principle** The conditions necessary for the development of intelligent life will be met only in certain regions that are limited in space and time. That is, the region of the universe in which we live is not necessarily representative of a purely random set of initial conditions; only those favorable to intelligent life would actually develop creatures who wonder what the initial conditions of the universe were, and this process can only happen at certain times through the evolution of any given universe.

2. Strong anthropic principle A more forceful argument than the weak principle: it implies that if the laws of the universe were not conducive to the development of intelligent creatures to ask about the initial conditions of the universe, intelligent life would never have evolved to ask the question in the first place. In other words, the laws of the universe are the way they are because if they weren't, no intelligent beings would be able to consider the laws of the universe at all.

Anthropocentric The belief that humans hold a special place in nature; being centered primarily on humans and human affairs.

Anthropomorphism The attribution of human characteristics to nonhuman beings and objects.

Anti In the representation of stereochemical relationships "anti" means "on opposite sides" of a reference plane, in contrast to "syn" which means "on the same side", as in the following examples. (A) Two substituents attached to atoms joined by a single bond are anti if the torsion angle (dihedral angle) between the bonds to the substituents is greater than 90°, or syn if it is less than 90°. (B) In the older literature the terms anti and syn were used to designate stereoisomers of oximes and related compounds. That usage was superseded by the terms "trans" and "cis" or E and Z, respectively. (C) When the terms are used in the context of chemical reactions or transformations, they designate the relative orientation of substituents in the substrate or product: 1. Addition to a carbon-carbon double bond:

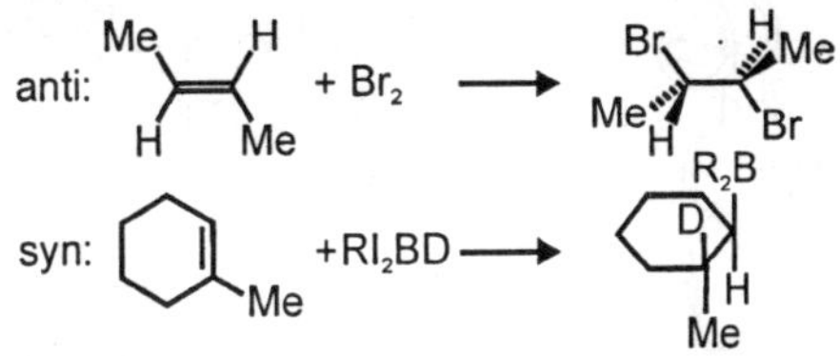

2. Alkene-forming elimination: IUPAC organic rules (1979).

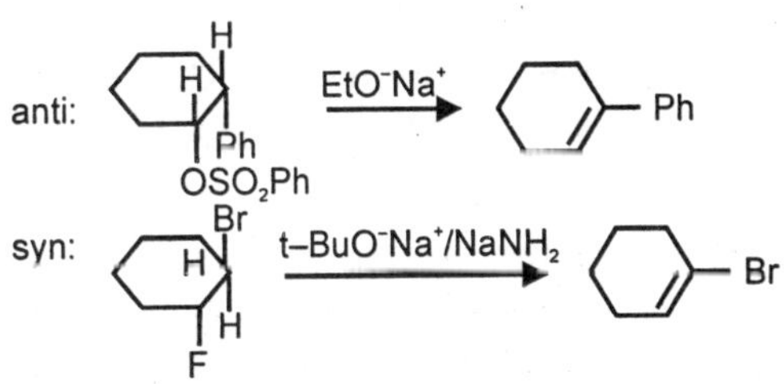

In the examples described under (1) and (2) anti processes are always antarafacial, and syn processes are suprafacial.

Antibiosis The inhibition of growth of a microorganism by a substance produced by another microorganism.

Antibiotic Substance that inhibits the growth of microorganisms, antibodiesproteins (immunoglobulins) in cell surfaces and dissolved in blood, capable of combining with the antigens that stimulated their production.

Antibody Immunoglobulin protein, produced by B cells (or plasma cells derived from B cells), that binds with a specific antigen.

Antibody titer Measure of the amount of antibody present, usually given in units per milliliter of serum.

Anticoagulant Substance that prevents blood clotting.

Anticodon A sequence of three bases on transfer RNA that pairs with codons of messenger RNA to position amino acids during protein synthesis.

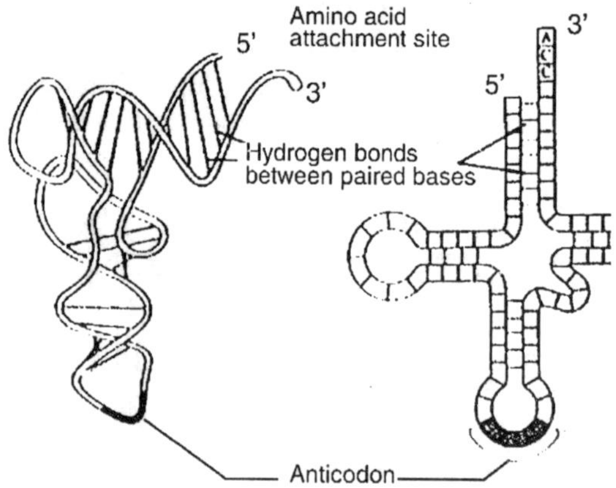

Fig. Anticodon

Anticodon loop The portion of a tRNA molecule responsible for the anticodon triplet, which pairs with the codon of mrna.

Antigen A foreign (nonself) substance (such as a protein, nucleoprotein, polysaccharide, and some glycolipids) to which lymphocytes respond; also known as an immunogen because it induces the immune response.

Antigen challenge Dose or inoculation with an antigen given to an animal some time after primary immunization with that antigen has been achieved.

Antigenic determinant Area on an antigen molecule that binds with antibody or specific receptor sites on the sensitized lymphocyte; it "determines" the specificity of the antibody or lymphocyte.

Antimatter A substance composed of antiparticles.

Antiparallel Refers to opposing strands of DNA that are oriented in opposite directions.

Antiparticle A counterpart of a subatomic particle having opposite properties (except for equal mass).

Antipodal cell The three haploid cells at the end of the embryo sac away from the micropyle.

Antipodal One of several cells (usually three) of the embryo sac (female gametophyte) of angiosperms. They are located opposite the egg and synergids.

Aorta aperture An opening; the opening into the first whorl of a gastropod shell.

Aperture Any opening through which radiation may pass The diameter of an opening that admits light to a lens or

Apex Highest or uppermost point; the lower pointed end of the heart. The tip of a structure; a leaf apex, for example, is the tip of the leaf.

Aphotic zone The portion of the ocean where the absence of sunlight prohibits plant growth.

Apical complex Dense ring and conelike structure, along with associated microtubules, micronemes, and rhoptries, at the anterior end of an apicomplexan sporozoite. (apical = at the apex.)A certain combination of organelles found in the protozoan phylum Apicomplexa.

Apical dominance The phenomenon leading to controlled growth of lateral shoots; growth occurs primarily at the top of the plant.

Apical meristem A group of cells at the tip of the stem and root that give rise by cell division to the primary tissues and are ultimately responsible for the structural organization of the entire primary plant body.

Apical organ Organ of unknown function at the apex of a cestode's scolex.

Apical Pertaining to the tip or apex.

Apicomplexa The protozoan phylum characterized by members having an apical complex used for penetrating host cells; cilia and flagella lacking, except in certain reproductive stages. Examples include the gregarines (Monocystis), coccidians (Eimeria, Isospora, Sarcocystis, Toxoplasma), Pneumocystis, and Plasmodium.

Aplacophora The class of molluscs whose members lack a shell, mantle, and foot. Wormlike, burrowing animals with head poorly developed. Some authors divide this group into two classes: Caudofoveata and Solengasters.

Apneustic breathing Breathing pattern exhibited by marine mammals in which several rapid breaths alternate with a prolonged cessation of breathing.

Apocrine Applies to a type of mammalian sweat gland that produces a viscous secretion by breaking off a part of the cytoplasm of secreting cells.

Apodeme Spinelike inward projection of the cuticle in arthropods on which a muscle inserts; a ridgelike projection is an apophysis.

Apodous larva Larva with no legs and with reduced head; usual in Hymenoptera, Diptera, some Coleoptera; requires maternal care or deposition in or on food source.

Apolysis Disintegration or detachment of a gravid tapeworm segment; also, the detachment of the hypodermis from the old procuticle in arthropods before molting.

Aponeurosis Dense connective tissue forming a broad tendon.

Apoplast Pathway of water entering a root that flows along intercellular spaces and cell walls.

Apoplastic Pertaining to the movement of water in the free space of tissue; free space includes cell walls and intercellular spaces.

Apoptosis Genetically determined cell death, "programmed" cell death.

Apopyle In sponges, opening of the radial canal into the spongocoel.

Aposematic coloration Sharply contrasting colors of an animal that warn other animals of unpleasant or dangerous effects.

Apparent power The product of the effective values of alternating voltage and current.

Appendicular Pertaining to appendages; pertaining to vermiform appendix.

Appendicular skeleton The bones of the upper and lower extremities; includes the shoulder and pelvic girdles.

Appendix Refers to the appendix vermiformis of the colon.

Appetitive behavior The flexible introductory phase of a behavior sequence during which the organism is searching to obtain something to meet a need, as in seeking food, a mate, stimulation of a specific type, etc.

Applied animal behavior A subdiscipline concerned with all aspects of the behavioral biology of pets, domestic animals, animals in zoos and circuses, and animals in aquaria.

Applied research Research conducted to address a specific problem for the purpose of application for productivity or commercial gain.

Aprotic (solvent) Non-protogenic (in a given situation). (With extremely strong Brønsted acids or bases, solvents that are normally aprotic may accept or lose a proton. For example, acetonitrile is in most instances an aprotic solvent, but it is protophilic in the presence of concentrated sulfuric acid and protogenic in the presence of potassium tert-butoxide. Similar considerations apply to benzene, trichloromethane, etc.)

Aquaculture The farming of marine and freshwater organisms.

Aquation The incorporation of one or more integral molecules of water into another species with or without displacement of one or more other atoms or groups. For example the incorporation of water into the inner ligand sphere of an inorganic complex is an aquation reaction.

Aquifers Porous, water-bearing layers of sand, gravel, and rock below the earth's surface; reservoirs for groundwater.

Arachnida The class of chelicerate arthropods whose members are mostly terrestrial, possess book lungs, or tracheae, and usually have four pairs of walking legs as adults. Spiders, scorpions, ticks, mites, and harvestmen.

Arachnoid The weblike middle covering (meninge) of the central nervous system.

Arago spot A bright spot that appears in the shadow of a uniform disc being backlit by monochromatic light emanating from a point source.

Arboreal Living in trees.

Arc tangent The inverse function to the tangent Symbol: arctan or tan-l Interpretation.

Archaea Early prokaryotes, differing from bacteria in having unusual types of metabolism, membrane lipids, and amino acid and DNA and RNA base sequences. Inhabit extreme environments.

Archaeocytes Ameboid cells of varied function in sponges.

Archegonium Female gametangium in which the female gamete or egg is produced and housed.

Archenteron The main cavity of an embryo in the gastrula stage; it is lined with endoderm and represents the future digestive cavity.

Archimedes' principle A body that is submerged in a fluid is buoyed up by a force equal in magnitude to the weight of the fluid that is displaced, and directed upward along a line through the center of gravity of the displaced fluid.

Archinephros Ancestral vertebrate kidney, existing today only in the embryo of hagfishes.

Archosaur Advanced diapsid vertebrates, a group that includes the living crocodiles and the extinct pterosaurs and dinosaurs.

Areolar A small area, such as spaces between fibers of connective tissue.

Areolus A unit of diatom frustules.

Arginine phosphate Phosphate storage compound (phosphagen) found in many invertebrates and used to regenerate stores of ATP.

Aril Thick, fleshy seed covering around some seeds.

Arista Flagellumlike appendage on the antenna of a fly of the suborder Brachycera and some members of the Nematocera.

Aristotle's lantern The set of jaws and associated muscles used by sea urchins to bite off food.

Arithmetic growth A pattern of growth that increases at a constant amount per unit time, such as 1, 2, 3, 4 or 1, 3, 5, 7.

Armature A coil of wire formed around an iron or steel core that rotates in the magnetic field of a generator or motor

Aromatic, aromaticity 1. In the traditional sense, "having a chemistry typified by benzene".

2. A cyclically conjugated molecular entity with a stability (due to delocalization) significantly greater than that of a hypothetical localized structure (e.g. Kekulé structure) is said to possess aromatic character. If the structure is of higher energy (less stable) than such a hypothetical classical structure, the molecular entity is "antiaromatic".

3. The terms aromatic and antiaromatic have been extended to describe the stabilization or destabilization of transition states of pericyclic reactions. The hypothetical reference structure is here less clearly defined, and use of the term is based on application of the Hückel (4n + 2) rule and on consideration of the topology of orbital overlap in the transition state. Reactions of molecules in the ground state involving antiaromatic transition states proceed, if at all, much less easily than those involving aromatic transition states.

Arrhenotoky Parthenogenetic production of males.

Arrhythmic activity Activity that does not exhibit any clear cyclical pattern.

Arteriole A small artery that delivers blood to capillaries.

Artery A blood vessel that carries blood away from the heart and toward a peripheral cavity.

Artesian well The result of a pressurized aquifer intersecting the surface or being penetrated by a pipe or conduit, from which water gushes without being pumped; also called a spring.

Arthropoda The phylum of animals whose members possess metamerism with tagmatization, a jointed exoskeleton, and a ventral nervous system. Insects, crustaceans, spiders, and related animals.

Arthropodization Evolutionary development of the combination of characteristics associated with the Arthropoda, including a firm cuticular exoskeleton containing chitin.

Arthropods (phylum Arthropoda) Invertebrates that have jointed appendages and a chitinous, segmented exoskeleton.

Articulation A joint or union of two bones.

Artificial selection Selective breeding as practiced by humans on domesticated plants and animals.

Artificially acquired passive immunity The type of immunity that results from introducing antibodies that have been produced either in another animal or by specific in vitro methods into an animal.

Artiodactyl One of an order of mammals with two or four digits on each foot.

Aryne A hydrocarbon derived from an arene by abstraction of two hydrogen atoms from adjacent carbon atoms; thus 1,2-didehydroarene. Arynes are commonly represented with a formal triple bond. The analogous heterocyclic compounds are called heteroarynes or hetarynes.

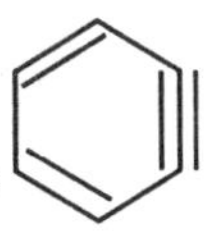
Benzyne

Arynes are usually transient species.

Arytenoid Cartilages in the larynx.

Ascaridine Protein of unknown function in the sperm of Ascaris.

Ascaroside Glycoside found in Ascaris, made of the sugar ascarylose and a series of secondary monol and diol alcohols.

Aschelminth Any animal in the phyla Gastrotricha, Rotifera, Kinorhyncha, Nematoda, Nematomorpha, Acanthocephala, Loricitera, Priapulida, or Entoprocta.

Aschoff's Rule Pertaining to circadian rhythms, the rule stating that when nocturnal animals are held in constant darkness, their free-running period becomes slightly shorter each day and when a diurnal animal is held in constant darkness, the free-running period becomes slightly longer.

Ascidiacea A class of urochordates whose members are sessile as adults, and solitary or colonial.

Ascites Edema, or accumulation of tissue fluid, in the mesenteries and abdominal cavity.

Ascocarp The dense, compact mass of hyphae constituting the fruiting body of the Ascomycetes.

Ascogonium A female reproductive structure in some ascomycetes.

Ascomycetes A group of fungi whose spores are borne in an ascus or bag.

Ascon The simplest of the three sponge body forms. Asconoid sponges are vaselike, with choanocytes directly lining the spongocoel.

Asconoid Simplest form of sponges, with canals leading directly from the outside to the interior.

Ascospore Sexual spore of an ascomycete fungus that is produced by meiosis, followed by mitosis.

Ascus Saclike reproductive structure of ascomycetes in which meiosis, followed by mitosis, produces eight haploid cells called ascospores.

Asexual (vegetative) reproduction The type of reproduction that takes place without the formation of gametes. Compare sexual reproduction.

Aspartate Four-carbon amino acid found with malate to be the first products of the C4.

Aspect ratio Index of propulsive efficiency obtained by dividing the square of a fish's fin height by the fin area.

Asphyxiants Chemicals that exclude oxygen or actively interfere with oxygen uptake and distribution; includes inert chemicals, such as nitrogen gas or halothane, that can displace oxygen and fill enclosed spaces.

Assimilation Absorption and building up of digested nutriments into complex organic protoplasmic materials.

Association The assembling of separate molecular entities into any aggregate, especially of oppositely charged free ions into ion pairs or larger and not necessarily well-defined clusters of ions held together by electrostatic attraction. The term signifies the reverse of dissociation, but is not commonly used for the formation of definite adducts by colligation or coordination.

Aster The star-shaped structure seen in a cell during the prophase of mitosis; composed of a system of microtubules arranged in astral rays around the centrosome; may emanate from a centrosome or from a pole of a mitotic spindle.

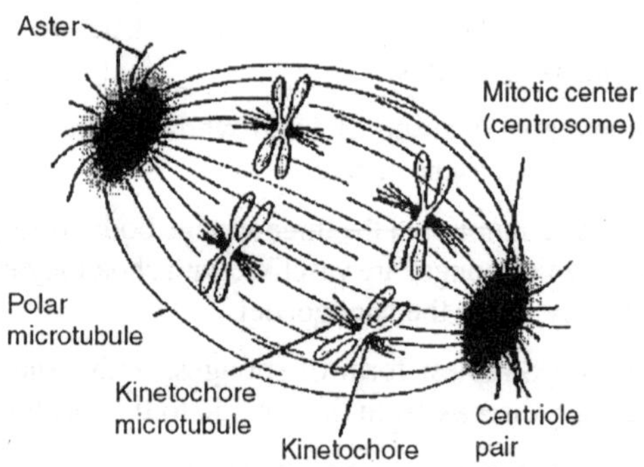

Fig. Aster

Asteroidea The class of echinoderms whose members have rays that are not sharply set off from the central disk; ambulacral grooves with tube feet; and suction disks on tube feet. Sea stars.

Asthma A distressing disease characterized by shortness of breath, wheezing, and bronchial muscle spasms.

Asymmetric induction The traditional term describing the preferential formation in a chemical reaction of one enantiomer or diastereoisomer over the other as a result of the influence of a chiral feature in the substrate, reagent, catalyst or environment. The term also refers to the formation of a new chiral feature preferentially in one configuration under such influence.

Asymmetry Without a balanced arrangement of similar parts on either side of a point or axis.

Atherosclerosis Disease characterized by fatty plaques forming in the inner lining of arteries.

Athlete's foot A disease caused by imperfect fungi that flourish under warm, wet conditions. Usually dissappears if the feet are kept dry.

Atmosphere A unit of pressure equal to 14.7 lbs/in^2 and equivalent to the pressure created by a 10 m column of water.

Atmospheric deposition Sedimentation of solids, liquids, or gaseous materials from the air. Snow, rain, and dust are the most familiar examples, but acids, metals, and toxic organic chemicals also can be transported by winds and deposited far from the source of origin.

Atmospheric pressure Ambient pressure created near the earth's surface by large air cells that circulate around the globe.

Atoke Anterior, nonreproductive part of a marine polychaete, as distinct from the posterior, reproductive part (epitoke) during the breeding season.

Atoll A ring-shaped chain of coral reefs from which a few low islands project above the sea surface.

Atom The smallest piece of matter which is still distinguishable as a particular chemical element. It consists of electrons orbiting a tiny nucleus which consists of densely packed protons and neutrons.

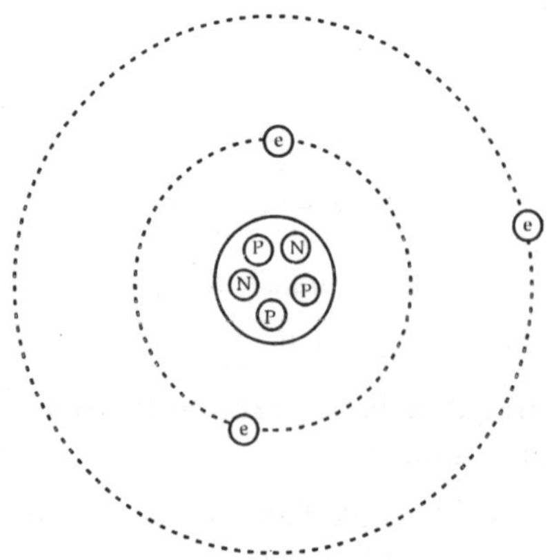

Fig. Basic model of the atom

Atom, models of After the discovery of the electron but before the discovery of the nucleus, the model of the atom was the plum pudding model in which the electrons were envisioned as plums in the pudding of protons. (Neutrons were not known to exist at the time.) After Rutherford's experiments, the model became known as the nuclear atom, the precursor to the picture of today. That the positively-charged nucleus is so small (in defiance of the electrical force which should tear the nucleus apart to separate the positive charges) was one of the clues that classical mechanics was incomplete, heralding the birth of quantum mechanics.

Atomic mass A mass unit determined by arbitrarily assigning the carbon-12 isotope a mass of 12 atomic mass units.

Atomic mass unit One-twelfth of the mass of carbon-12, or 1.6605655 x 10^{-27} kg.

Atomic nucleus Most of the mass of an atom; composed of protons and neutrons.

Atomic number Number of protons in the atomic nucleus; there is a characteristic number for each type of element.

Atomic orbital A one-electron wavefunction describing an electron in the effective field provided by a nucleus and the other electrons present.

Atomic weight The weight of an atom determined by adding the number of protons and neutrons (the mass of electrons is usually considered to be negligible).

ATP synthase A membrane-bound enzyme in mitochondria and chloroplasts that phosphorylates ADP to form ATP using energy from the passage of protons through the enzyme.

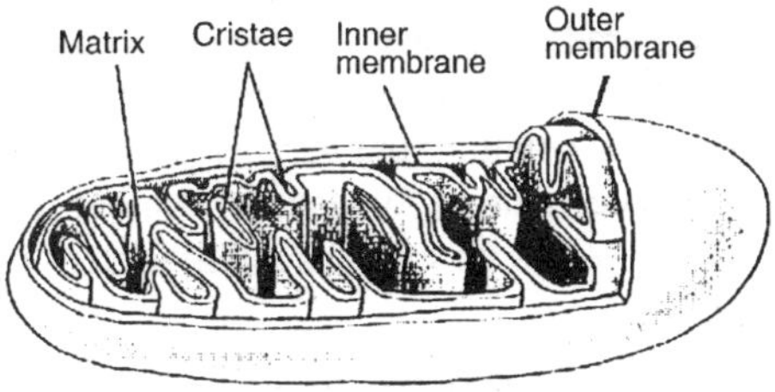

Fig. Cristae

Atrium One of the chambers of the heart; also, the tympanic cavity of the ear; also, the large cavity containing the pharynx in tunicates and cephalochordates.

Attachment A transformation by which one molecular entity (the substrate) is converted into another by the formation of one (and only one) two-centre bond between the substrate and another molecular entity and which involves no other changes in connectivity in the substrate. For example, the formation of an acyl cation by attachment of carbon monoxide to a carbenium ion (R^+):

$$R^+ + CO \longrightarrow (RCO)^+$$

The product of an attachment may also be the adduct of the two reactants, but not all adducts can be represented as the products of an attachment.

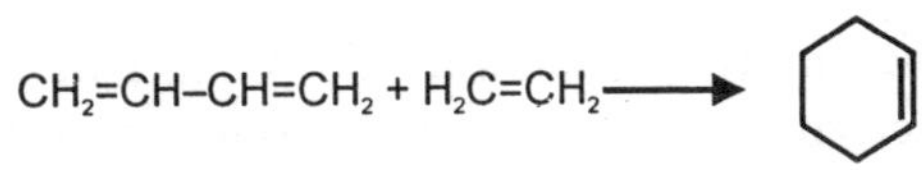

(For example, the Diels-Alder cycloaddition results in an adduct of buta-1,3-diene and ethene, but the reaction cannot be described as an attachment since bonds are formed between more than two centres.)

Atwood's machine A weight-and-pulley system devised to measure the acceleration due to gravity at earth's surface by measuring the net acceleration of a set of weights of known mass around a frictionless pulley.

Audio signal The alternating voltage proportional to the sound pressure produced in an electric circuit

Auricle The portion of the external ear not connected within the head. Also used to designate an atrium of a heart. In the class Turbellaria, the sensory lobes that project from the side of the head.

Auricularia A type of larva found in Holothuroidea.

Autocatalytic reaction A chemical reaction in which a product (or a reaction intermediate) also functions as catalyst. In such a reaction the observed rate of reaction is often found to increase with time from its initial value.

Autogamy Condition in which the gametic nuclei produced by meiosis fuse within the same organism that produced them to restore the diploid number.

Autoimmunity Immune response to ones' own proteins or other antigens.

Autoinfection Reinfection by a parasite juvenile without its leaving the host.

Automerization Synonymous with degenerate rearrangement.

Autoprotolysis A proton (hydron) transfer reaction between two identical molecules (usually a solvent), one acting as a Brønsted acid and the other as a Brønsted base. For example:

$$2\ H_2O \longrightarrow H_3O^+ + OH^-$$

Autoprotolysis constant The product of the activities (or, more approximately, concentrations) of the species produced as the result of autoprotolysis. For solvents in which no other ionization processes are significant the term is synonymous with "ionic product". The autoprotolysis constant for water, Kw, is equal to the product of activities

$$a(H_3O^+)a(OH^-) = 1.0 \times 10^{-14} \text{ at } 25^\circ C.$$

Autosome Any chromosome that is not a sex chromosome.

Autotomy The self amputation of an appendage. For example, the casting off of a section of a lizard's tail caught in the grasp of a predator. The autotomized appendage is usually regenerated.

Autotroph An organism that makes its organic nutrients from inorganic raw materials by using an external energy source, such as light energy.

Autotrophic Having the ability to synthesize food from inorganic compounds.

Autotrophic nutrition Nutrition characterized by the ability to use simple inorganic substances for the synthesis of more complex organic compounds, as in green plants and some bacteria.

Auxin A hormone that causes the bending response in Avena coleoptiles.

Auxospore The resistant stage of diatoms that restores the maximum size characteristic to the species.

Average velocity Total displacement divided by elapsed time.

Aves A class of vertebrates whose members are characterized by scales modified into feathers for flight, endothermy, and amniotic eggs. The birds.

Avicularium Modified zooid that is attached to the surface of the major zooid in Ectoprocta and resembles a bird's beak.

Avogadro's constant Avogadro's constant has the unit mole-1. It is not merely a number, and should not be called Avogadro's number. It is ok to say that the number of particles in a gram-mole is 6.02×10^{23}. Some older books call this value Avogadro's number, and when that is done, no units are attached to it. This can be confusing and misleading to students who are conscientiously trying to learn how to balance units in equations. One must specify whether the value of Avogadro's constant is expressed for a gram-mole or a kilogram-mole. A few books prefer a kilogram-mole. The unit name for a gram-mole is simply mol. The unit name for a kilogram-mole is kmol. When the kilogram-mole is used, Avogadro's constant should be written: 6.02252×10^{26} $kmol^{-1}$. The fact that Avogadro's constant has units further convinces us that it is not "merely a number." Though it seems inconsistent, the SI base unit is the gram-mole. Some textbooks still prefer to use the kilogram-mole, or use it and the gram-mole. This affects their quoted values for the universal gas constant and the Faraday Constant. Avogadro's constant, however, must be determined experimentally, for example by counting the number of atoms in a crystal. The value of Avogadro's number found in handbooks is an experimentally determined number.

The reciprocal of Avogadro's constant is numerically equal to the unified atomic mass unit, μ, that is, 1/12 the mass of the carbon 12 atom.

$1\mu = 1.66043 \times 10^{-27}$ kg $= 1/6.02252 \times 10^{23}$ mole^{-1}.

Axial cells Central cells of a dicyemid mesozoan.

Axial skeleton Portion of the skeleton that supports and protects the organs of the head, neck, and trunk.

Axil The upper angle between a stem.

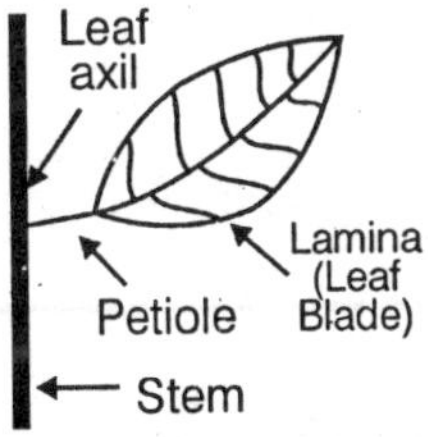

Fig. Axil

Axillary bud A bud, consisting of an apical meristem and leaf primordia, situated in the leaf axil.

Axis The second cervical vertebra; any axial part.

Axocoel The most anterior of three coelomic spaces that appear during larval echinoderm development.

Axolotl Larval stage of any of several species of the genus Ambystoma (such as Ambystoma tigrinum) exhibiting neotenic reproduction.

Axon Elongate extension of a neuron that conducts impulses away from the cell body and toward the synaptic terminals.

Axoneme The axial thread of the chromosome in which is located the axial combination of genes. The central core of a cilium or flagellum, consisting of a central pair of filaments surrounded by nine other pairs; also called axial filament.

Axopodium Fine, needlelike pseudopodium that contains a central bundle of microtubules. Also called axopod. Found in certain sarcodine protozoa.

Axostyle Tubelike organelle in some flagellate protozoa, extending from the area of the kinetosomes to the posterior end, where it often protrudes.

Azygous A blood vessel.

B

B cell A type of lymphocyte derived from bone marrow stem cells that matures into an immunologically competent cell under the influence of the bursa of Fabricius in the chicken, and the bone marrow in nonavian species; following interaction with antigen, it becomes a plasma cell, which synthesizes and secretes antibody molecules involved in humoral immunity; B lymphocyte.

Bacillary bands Lateral zones in the body wall of some nematodes, consisting of glandular and nonglandular cells of unknown function.

Bacillus A rod-shaped bacterium.

Back emf An induced emf in the armature of a motor that opposes the applied voltage.

Back reef The inner part of a barrier reefor atoll.

Backcross Crossing a hybrid offspring back to either parent.

Bacteria Prokaryotic and unicellular micro-organisms included in the kingdom Monera.

Bacteriochlorophyll A type of chlorophyll found only in bacterial systems.

Bacteriophage A virus that attacks bacterial cells.

Bacterioplankton The component of the plankton that consists of bacteria.

Baer's disc Large, ventral sucker of an aspido-gastrean trematode.

Balanced polymorphism Occurs when different phenotypic expressions are maintained at a relatively stable frequency in a population.

Baldwin's rules A set of empirical rules for certain formations of 3- to 7-membered rings. The predicted pathways are those in which the length and nature of the linking chain enables the terminal atoms to achieve the proper geometries for reaction. The disfavoured cases are subject to severe distortions of bond angles and bond distances.

Baleen plate A keratined growth in toothless whales.

Baleen The filtering plates that hang from the upper jaws of baleen whales.

Baleen whales The filter-feeding whales.

Ballonets Four inflated areas within the "head" of nematodes of the family Gnathostomatidae; each is connected to an internal cervical sac of unknown function.

Balmer series An equation which describes the emission spectrum of hydrogen when an electron is jumping to the second orbital; four of the lines are in the visible spectrum, and the remainder are in the ultraviolet.

Band spectrum An emission spectrum consisting of fluted bands of color The spectrum of a substance in the molecular state.

Band, energy Electrons orbit the nucleus of an atom in specific orbits determined by the quantization of their angular momentum. Each specific orbit has an associated energy. Since the orbits are discrete, the available energies are discrete. When atoms are brought together to form a material, the available energy levels adjust to accommodate the neighboring atoms. The energy levels tend to adjust into separated groups of energy levels called "energy bands." This is not inherently a spacial separation; rather it is a difference in the energy values which are available to the electrons. In fact, the electron can jump between bands and does so by gaining or losing energy. The jump is in energy-space, not in real-space. Although this jump does allow the electron to move through a different region of real-space, it is the change in energy which is the jump, not the motion in real-space. In this case, the electrons are said to be jumping across the energy gap.

Bar-built estuary An estuary that is formed when a barrier island or sand bar separates a section of the coast where fresh water enters.

Bark Those portions of a woody plant stem or trunk exterior to the vascular cambium.

Barnacles Crustaceans that live attached to surfaces and are typically enclosed by heavy calcareous plates.

Barometer A device used to measure the pressure of the atmosphere.

Baroreceptor A specialized nerve ending that is stimulated by changes in pressure.

Barr body A heterochromatic X chromosome usually found in the nucleus of female mammals.

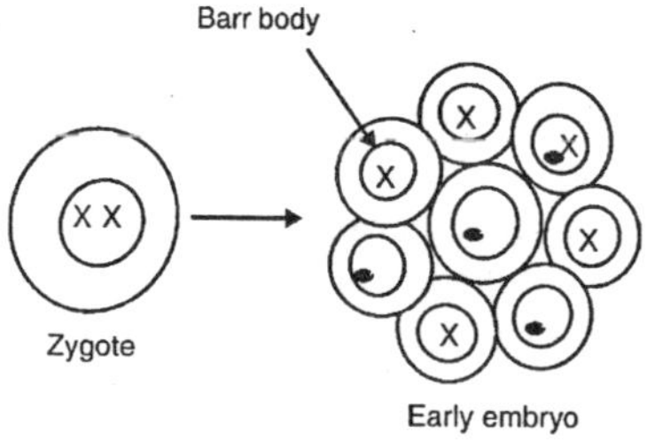

Fig. Barr body

Barrier island A long and narrow island that is built by waves along the coast.

Barrier reef A coral reef that runs approximately parallel to the shore and is separated from the shore by a lagoon.

Baryon These particles are not fundamental. These are made up of quarks. Baryons are a sub-category of hadrons; that is to say, all baryons are hadrons, but not all hadrons are baryons. A baryon is made up of three constituent quarks. These may or may not be the same flavor. The colors of the three quarks must combine to leave the hadron uncharged with respect to the strong interaction. It could still be electrically or otherwise charged.

Baryon decay The idea, predicted by several grand-unified theories, that a class of subatomic particles called baryons (of which the nucleons — protons and neutrons — are members) are not ultimately stable but indeed decay.

Basal body A centriole that has given rise to the microtubular system of a cilium or flagellum, and is located just beneath the plasma membrane. Serves as a nucleation site for the growth of the axoneme.

Basal rosette Leaves clustered at ground level; can be alternate or whorled, but with extremely short internodes.

Basalt A semiviscous layer of igneous rock underlying the granitic continental plates.

Base A chemical species or molecular entity having an available pair of electrons capable of forming a covalent bond with a hydron (proton).

Basic equation An equation that relates the unknown quantity with known quantities in a problem.

Basic law of electrostatics Similarly charged objects repel each other Oppositely charged objects attract each other.

Basic Possessing a large number of hydroxyl (OH^-) ions; a pH of more than 7.0.

Basic research Research conducted to elucidate basic concepts alone with no direct application in mind.

Basicity For Brønsted bases it means the tendency of a compound to act as hydron (proton) acceptor. The basicity of a chemical species is normally expressed by the acidity of the conjugate acid. For Lewis bases it relates to the association constants of Lewis adducts and -adducts.

Basidiocarp The densely packed hyphae of the fruiting body in the Basidiomycetes.

Basidiomycetes A group of fungi whose spores are borne in a basidium.

Basidiospores A haploid spore produced by a basidium in the Basidiomycetes.

Basidium Club-shaped reproductive cell in basidiomycetes undergoes meiosis to produce four haploid cells called basidiospores.

Basilar Foundation. Forming a foundation.

Basilar membrane A flat band of tissue that bears the auditory hair cells in the cochlea of the vertebrate ear; the membrane vibrates in response to pressure waves.

Basipetal Movement down the stem as with auxin moving from the shoot apex downward.

Basis (basipodite) Joint of a crustacean appendage from which the exopod and the endopod originate; that is, the joint between the coxa and the exopod and endopod.

Basophil White blood cell characterized by the presence of cytoplasmic granules that become stained by a basophilic dye.

Bast fiber Fibers located in phloem.

Bateman effect The large variation in the reproductive success of males compared to females in relation to the number of mates they have.

Batesian mimicry Evolution by one species to resemble the coloration, body shape, or behavior of a related species that is protected from predators by a venomous stinger, bad taste, or some other adaptation.

Bathochromic shift (effect) Shift of a spectral band to lower frequencies (longer wavelengths) owing to the influence of substitution or a change in environment. It is informally referred to as a red shift and is opposite to hypsochromic shift.

Bathyal zone The bottom between the shelf break and a depth of approximately 4,000 m (13,000 ft).

Bathypelagic zone The pelagic environment from a depth of 1,000 m.

B-carotene An important plant carotenoid and precursor of vitamin A.

BCS theory A theory put forth to explain both superconductivity and superfluidity. It suggests that in the superconducting (or superfluid) state electrons form cooper pairs, where two electrons act as a single unit. It takes a nonzero amount of energy to break such pairs, and the imperfections in the superconducting solid (which would normally lead to resistance) are incapable of breaking the pairs, so no dissipation occurs and there is no resistance.

Bdelloidea A class of rotifers containing members where there are no males; anterior end retractile and bearing two disks; mastax adapted for grinding; paired ovaries;, cylindrical body. Example: Rotaria.

Beam Several parallel rays of light considered collectively.

Beard worms Tube-dwelling invertebrates that lack a digestive system.

Beat The interference effect resulting from the superposition of two waves of slightly different frequencies propagating in the same direction The amplitude of the resultant wave varies with time.

Beauty criterion The idea that the more aesthetically pleasing a theory is, the better it is. Naturally this criterion does not stand up to the real test — whether or not predictions of a given theory agree with observational tests — but considering that it is a purely aesthetic quality that is being tested, many of the most successful theories (special relativity, general relativity, quantum electrodynamics, etc.) Match the criterion particularly well.

Becquerel; bq The derived SI unit of activity, defined as the activity of a radionuclide decaying at a rate, on the average, of one nuclear transition every 1 s; it thus has units of s^{-1}.

Bedrock Unbroken, solid rock overlain by sand, gravel, or soil.

Behavior genetics The study of the role that genes play in controlling behavior.

Behavioral ecology A subdiscipline within animal behavior that deals with the ways in which animals interact with their environment and the survival value of behavior as well as its contribution to reproductive success.

Behavioral isolating mechanisms Differences in behavior (usually courtship) that prevent genetic exchange between members of different populations or species.

Behaviorism A view of the actions of animals that postulates that behavior can be analyzed functionally in terms of stimulus and response combinations. Thus, most of behavior is a function of the experience of the organism in this view.

Bell's inequality A quantum mechanical theorem which demonstrates that if quantum mechanics were to rely on hidden variables, it must have nonlocal properties.

Bell-Evans-Polanyi principle The linear relation between energy of activation (e_a) and enthalpy of reaction (h_r) sometimes observed within a series of closely related reactions.

Benthic animals Refers to animals living in or on the bottom substrate of an ocean, lake, stream, or other body of water.

Benzyne 1,2-Didehydrobenzene (the aryne derived from benzene) and its derivatives formed by substitution. The terms m- and p-benzyne are occasionally used for 1,3- and 1,4-didehydro-benzene, respectively.

Bernoulli's equation In an irrotational fluid, the sum of the static pressure, the weight of the fluid per unit mass times the height, and half the density times the velocity squared is constant throughout the fluid.

Berry A fleshy, two- or multiple-carpeled ovary, each carpel having many seeds.

Best available, economically achievable technology (BAT) The best pollution control available; the Clean Water Act effectively negates this category by stipulating that equipment must be economically feasible.

Best practical control technology (BPT) The best technology for pollution control available at reasonable cost and operable under normal conditions.

Beta-carotene Yellow to red pigment in plants; one of the most important of the carotenoids; converts to vitamin A in the body.

Betatron A device that accelerates electrons by means of the transformer principle.

B-galactosidase The enzyme responsible for the splitting of lactose into glucose and galactose.

Biantitropical distribution A pattern of a species' geographical distribution that extends across comparable bands of latitude in both the Northern Hemisphere and the Southern Hemisphere.

Bicarbonate ion Carbonic acid ionizes to produce a bicarbonate ion, HCO_3 and a proton, H^+ in rainfall.

Biennial Plant that completes its life cycle within two growing seasons; example: carrot.

Bifunctional catalysis Catalysis (usually for hydron transfer) by a bifunctional chemical species involving a mechanism in which both functional groups are implicated in the rate-controlling step, so that the corresponding catalytic coefficient is larger than that expected for catalysis by chemical species containing only one of these functional groups.

Big bang theory The theory that a hypothetical cosmic explosion resulted in clouds of dust and gas from which the earth and solar system originated.

Bilateral body symmetry Animal body plan with mirror image left and right sides.

Bilateral symmetry A form of symmetry in which only the midsagittal plane will divide an organism into mirror images; bilateral symmetry is characteristic of actively moving organisms that have definite anterior (head) and posterior (tail) ends.

Bile A fluid secreted by the liver and poured into the small intestine via the bile duct; emulsifies fats.

Bilharziasis Disease caused by Schistosoma spp. Also called schistosomiasis.

Bilirubin A breakdown product of the heme group of hemoglobin, excreted in the bile.

Bimodal breathing The ability of an organism to exchange respiratory gases simultaneously with both air and water; usually using gills for water breathing and lungs for air breathing.

Binary fission Asexual reproduction in protists in which mitosis is followed by cytoplasmic division, producing two new organisms.

Binding energy Energy that must be applied to a nucleus to break it up

Binding site A specific region (or atom) in a molecular entity that is capable of entering into a stabilizing interaction with another molecular entity. An example of such an interaction is that of an active site in an enzyme with its substrate. Typical forms of interaction are by hydrogen bonding, coordination, and ion pair formation.

Two binding sites in different molecular entities are said to be complementary if their interaction is stabilizing.

Binomial nomenclature A system of naming species using two names, the first of which refers to the genus.

Binomial The two names, genus and species, comprising the scientific name.

Bioaccumulation Increasingly concentrated accumulation of substances, especially pollutants, at successively higher trophic levels in food chains.

Bioassay A quantitative assay of a particular substance using a portion of or an entire living organism.

Biocentric preservation A philosophy that emphasizes the fundamental right of living organisms to exist and to pursue their own goods.

Biocentrism The belief that all creatures have rights and values; being centered on nature rather than humans.

Biochemical taxonomy A biochemical approach to defining evolutionary relationships of organisms by determining molecular similarities and differences among organisms.

Biochemistry The chemistry of living organisms and of vital processes; also known as physiological or biological chemistry. The study of the molecular basis of life.

Biocide A broad-spectrum poison that kills a wide range of organisms.

Biodegradable A chemical that can be broken down by bacteria or other organisms.

Biodiversity The genetic, species, and ecological diversity of the organisms in a given area.

Biogenesis The doctrine that life originates only from preexisting life.

Biogenous sediment The type of sediment that is made up of the skeletons and shells of marine organisms.

Biogeochemical cycles Movement of matter within or between ecosystems; caused by living organisms, geological forces, or chemical reactions. The cycling of nitrogen, carbon, sulfur, oxygen, phosphorus, and water are examples.

Biogeographical area An entire self-contained natural ecosystem and its associated land, water, air, and wildlife resources.

Biogeography The study of the distribution of life on earth.

Biological clock An internal timing mechanism that involves both an internal self-sustaining pacemaker and cyclic environmental synchronizers.

Biological communication An action by one organism (or cell) that causes a reaction from another organism (or cell) in a fashion adaptive to either one or both of the participants.

Biological controls Use of natural predators, pathogens, or competitors to regulate pest populations.

Biological magnification The increased concentration of nonbiodegradable chemicals in the higher levels of the food chain.

Biological or biotic factors Organisms and products of organisms that are part of the environment and potentially affect the life of other organisms.

Biological oxygen demand (BOD) A standard test for measuring the amount of dissolved oxygen utilized by aquatic microorganisms over a five-day period.

Biological pests Organisms that reduce the availability, quality, or value of resources useful to humans.

Biological resources The earth's organisms.

Biological rhythm A cyclical pattern of behavior, occurring at some regular period (e.g., diurnal).

Biological species concept A reproductive community of populations (reproductively isolated from others) that occupies a specific niche in nature.

Biological vector Vector in which a disease organism lives or develops. Contrast with mechanical vector.

Biology The study of all living organisms.

Bioluminescence. Method of light production by living organisms in which usually certain proteins (luciferins), in the presence of oxygen and an enzyme (luciferase), are converted to oxyluciferins with the liberation of light.

Biomagnification Increase in concentration of certain stable chemicals (for example, heavy metals or fat-soluble pesticides) in successively higher trophic levels of a food chain or web.

Biomass fuel Organic material produced by plants, animals, or microorganisms that can be burned directly as a heat source or converted into a gaseous or liquid fuel.

Biomass pyramid A metaphor or diagram that explains the relationship between the amounts of biomass at different trophic levels.

Biomass The weight of total living organisms or of a species population per unit of area.

Biomes Distinctive associations of plant and animal populations; characterized by certain geographical boundaries and specific climatic and geographical features.

Bioregionalism Organization of human activities according to natural geographic or ecological boundaries and associations. This philosophy emphasizes a sense of place and living within the resources of one's local ecosystem.

Biosphere reserves World heritage sites identified by the IUCN as worthy for national park or wildlife refuge status because of high biological diversity or unique ecological features.

Biosphere The earth and all of its ecological interactions considered as a single system.

Biota All organisms in a given area.

Biotechnology The use of living organisms to provide products for humanity; using genetic engineering to create organisms with useful traits.

Biotic Of or relating to life.

Biotic factor Living component in the environment.

Biotic potential The maximum reproductive rate of an organism, given unlimited resources and ideal environmental conditions. Compare with environmental resistance.

Biot-Savart law A law which describes the contributions to a magnetic field by an electric current. It is analogous to coulomb's law. Mathematically, it is

$$Db = (\mu\ i)/(4\ \pi\ r^2)\ dl \times e$$

Where dl is the infinitesimal directed length of the electric current causing the magnetic field, i is the current running through that directed length, r is the distance from that directed length, and e is the unit vector directed from the test point to current-producing length.

Bipinnaria Free-swimming, ciliated, bilateral larva of the asteroid echinoderms; develops into the brachiolaria larva.

Biradical An even-electron molecular entity with two (possibly delocalized) radical centres which act nearly independently of each other, e.g.

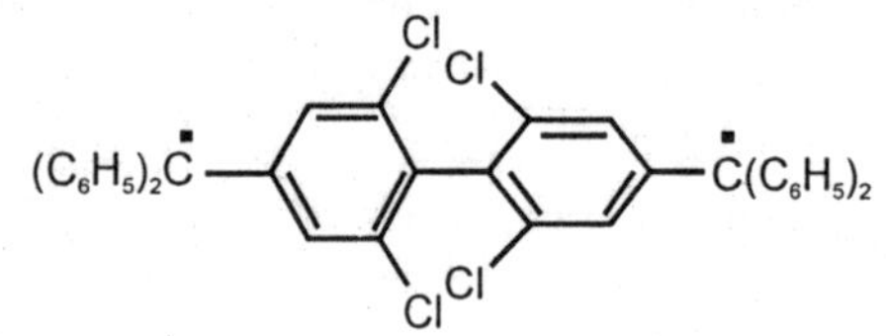

Species in which the two radical centres interact significantly are often referred to as "biradicaloids". If the two radical centres are located on the same atom, the species are more properly referred to by their generic names: carbenes, nitrenes, etc.

The lowest-energy triplet state of a biradical lies below or at most only a little above its lowest singlet state (usually judged relative to kBT, the product of the Boltzmann constant kB and the absolute temperature T). The states of those biradicals whose radical centres interact particularly weakly are most easily understood in terms of a pair of local doublets.

Theoretical descriptions of low-energy states of biradicals display the presence of two unsaturated valences (biradicals contain one fewer bond than permitted by the rules of valence): the dominant valence bond structures have two dots, the low energy molecular orbital configurations have only two electrons in two approximately nonbonding molecular orbitals, two of the natural orbitals have occupancies close to one, etc.

Birds Vertebrates that have feathers and lay eggs with calcified shells on land.

Birth control Any method used to reduce births, including celibacy, delayed marriage, contraception; devices or medication that prevent implantation of fertilized zygotes, and induced abortions.

Bisexual A flower having both stamens and pistil (both sexes); a perfect flower.

Bivalent The pairs of homologous chromosomes at synapsis in the first meiotic division, a tetrad.

Bivalves The clams, mussels, and other molluscs that possess a two-valved shell, filtering gills, and a shovel-like foot.

Bivalvia The class of molluscs whose members are enclosed in a shell consisting of two dorsally hinged valves, lack a radula, and possess a wedge-shaped foot. Clams, mussels, oysters.

Black body radiation: Objects which absorb all colors except green, which is reflected, look green. This is also true of blue (or whatever color) objects. Objects which can absorb all colors, thus appear black. Hence, a "black body" is one which absorbs all frequencies of light. Furthermore, objects which are heated will glow. The color of the glow will correspond to the temperature of the object - red is warm, yellow is hot, white is very hot, blue is extremely hot; the range goes well past the visible frequencies of light. In fact, every object has some ambiant temperature and therefore glows. (People glow in the infrared, hence the usefulness of IR goggles for night-vision.) Black Body radiation is this correspondance between the color of the glow and the temperature of the object. The use of a black body as described above idealizes the situation, removing complications which can arise if the object cannot absorb ambient light. For example, visible light does not heat a window but UV light can (because glass absorbs UV but not visible light). This effect was one of the clues that classical physics was incomplete, heralding the birth of quantum mechanics. It was first mathematically modeled by Planck with the claim that the energy in the absorbed light was quantized.

Black corals Colonial anthozoans that secrete a black protein skeleton.

Black fly fever Combination of symptoms resulting from sensitization to bites by black flies.

Black lung disease Inflammation and fibrosis caused by accumulation of coal dust in the lungs or airways.

Black smoker A chimney-like accumulation of mineral deposits that is found at hydrothermal vents.

Blackbody radiation The radiation — the radiance at particular frequencies all across the spectrum — produced by a blackbody — that is, a perfect radiator (and absorber) of heat. Physicists had difficulty explaining it until planck introduced his quantum of action.

Blackhead Disease of turkeys caused by the protozoan Histomonas meleagridis. Also called histomoniasis or infectious enterohepatitis.

Black-hole dynamic laws 1. First law of black hole dynamics For interactions between black holes and normal matter, the conservation laws of mass-energy, electric charge, linear momentum, and angular momentum, hold. This is analogous to the first law of thermodynamics.

2. Second law of black hole dynamics With black-hole interactions, or interactions between black holes and normal matter, the sum of the surface areas of all black holes involved can never decrease. This is analogous to the second law of thermodynamics, with the surface areas of the black holes being a measure of the entropy of the system.

Bladder worm The unilocular hydatid cyst of a tapeworm.

Blade Flat, green part of a leaf; expanded or flattened portion of a brown alga.

Blastocoel The internal cavity commonly formed by cell division early in embryonic development, prior to gastrulation.

Blastocyst An early stage of embryonic development consisting of a hollow ball of cells;a small ball of cells representing an early stage of embryonic development.

Blastoderm "Primary epithelium" formed in early embryonic development of many arthropods when the nuclei migrate to the periphery and undergo superficial cleavage; usually encloses the central yolk mass.

Blastomeres Any of the cells produced by cleavage of a zygote.

Blastopore External opening of the archenteron in the gastrula.

Bleaching The expulsion of zooxanthellae by reef corals in response to stress.

Blood A type of connective tissue with a fluid matrix called plasma in which blood cells are suspended. The fluid that circulates through the heart, arteries, capillaries, and veins.

Blood plasma The liquid, noncellular fraction of blood, including dissolved substances.

Blood pressure The force (energy) with which blood is pushed against the walls of blood vessels and circulated throughout the body when the heart contracts.

Blood type Characteristic of human blood given by the particular antigens on the membranes of the erythrocytes, genetically determined, causing agglutination when incompatible groups are mixed; the blood types are designated A, B, O, AB, Rh negative, Rh positive, and others.

Bloom A dense concentration of phytoplankton that occurs in response to optimum growth conditions.

Blowhole The nostrils, or nasal openings, of cetaceans.

Blubber The fat found between the skin and muscle of whales and other cetaceans, from which oil is made.

Blue revolution New techniques of fish farming that may contribute as much to human nutrition as miracle cereal grains but also may create social and environmental problems. Commercial fish farming already is a $5 billion per year industry, and domestic fish ponds supply as much as two-thirds of the protein consumed by subsistence farmers in many countries.

Blue-green algae Prokaryote organisms with photosynthetic phycobilin pigments.

Bluetongue Virus disease of ruminants transmitted by biting midges (Ceratopogonidae).

B-lymphocytes Cells that manufacture antibodies involved in immunity.

Bode's law, Titius-Tode law A mathematical formula which generates, with a fair amount of accuracy, the semimajor axes of the planets in order out from the sun. Write down the sequence 0, 3, 6, 12, 24,.. And add 4 to each term: 4, 7, 10, 16, 28,..

Then divide each term by 10. This leaves

0.4, 0.7, 1.0, 1.6, 2.8,..

Which is intended to give you the semimajor axes of the planets measured in astronomical units.

Bode's law had no theoretical justification when it was first introduced; it did, however, agree with the soon-to-be-discovered planet uranus' orbit. Similarly, it predicted a missing planet between mars and jupiter, and shortly thereafter the asteroids were found in very similar orbits (2.77 au actual for ceres; 2.8 au predicted). The series, however, seems to skip over Neptune's orbit. The form of Bode's law (that is, a roughly geometric series) is not surprising, considering our theories on the formation of solar systems, but its particular formulation is thought of as coincidental.

Bog An area of waterlogged soil that tends to be peaty; fed mainly by precipitation; low productivity; some bogs are acidic.

Bohr effect A characteristic of hemoglobin that causes it to dissociate from oxygen in greater degree at higher concentrations of carbon dioxide.

Bohr radius The distance corresponding the mean distance of an electron from the nucleus in the ground state of the hydrogen atom.

Boiling point The temperature at which the vapor pressure of a liquid equals the pressure of the atmosphere

Boltzmann constant; k A constant which describes the relationship between temperature and kinetic energy for molecules in an ideal gas. It is equal to 1.380622×10^{-23} j/k.

Bomb calorimeter An instrument used for measuring the caloric content of any organic material; measures energy release in combusting organic materials to CO_2 and H_2O.

Bond There is a chemical bond between two atoms or groups of atoms in case that the forces acting between them are such as to lead to the formation of an aggregate with sufficient stability to make it convenient for the chemist to consider it as an independent "molecular species".

Bond energy (mean bond energy) The average value of the gas-phase bond-dissociation energies (usually at a temperature of 298 K) for all bonds of the same type within the same chemical species. The mean bond energy for methane, for example, is one-fourth the enthalpy of reaction for

$$CH_4(g) \longrightarrow C(g) + 4H(g)$$

Tabulated bond energies are generally values of bond energies averaged over a number of selected typical chemical species containing that type of bond.

Bond number The number of electron-pair bonds between two nuclei in any given Lewis formula. For example in ethene the bond number between the carbon atoms is two, and between the carbon and hydrogen atoms is one.

Bond order A theoretical index of the degree of bonding between two atoms relative to that of a normal single bond, i.e. the bond provided by one localized electron pair. In the valence-bond theory it is a weighted average of the bond numbers between the respective atoms in the contributing structures. In molecular-orbital theory it is calculated from the weights of the atomic orbitals in each of the occupied molecular orbitals. For example, in valence-bond theory (neglecting other than Kekulé structures) the bond order between adjacent carbon atoms in benzene is 1.5; in Hückel molecular orbital theory it is 1.67. Other variants of molecular orbital theory provide other values for bond orders.

Bone cells The hard, rigid form of connective tissue constituting most of the skeleton of vertebrates; composed chiefly of calcium salts. Also called bone (osseous) tissue.

Bony fishes Fishes with a skeleton made mostly of bone; they also have gill covers.

Book gill Modifications of a horseshoe crab's exoskeleton into a series of leaflike plates that serve as a surface for gas exchange between the arthropod and the water (phylum Arthropoda, class Merostomata).

Borderline mechanism A mechanism intermediate between two extremes, for example a nucleophilic substitution intermediate between S_N1 and S_N2, or intermediate between electron transfer and S_N2.

Boreal Relating to a northern biotic area characterized by a predominance of coniferous forests and tundra.

Boreal forest A broad band of mixed coniferous and deciduous trees that stretches across northern North America (and also Europe and Asia); its northernmost edge, the taiga, intergrades with the arctic tundra.

Boring sponges Sponges that bore through calcareous skeletons and shells.

Boson Particles which have integral spin angular momentum are shown to behave differently than particles with half-integral spin angular momentum. Statistical mechanics can be used to describe the particles. The statistical rules which express how bosons interact are called Bose-Einstein statistics. The fundamental bosons which are found in nature are gauge bosons.

Botany The study of plants.

Bothria Dorsal or ventral grooves, which may be variously modified, on the scolex of a cestode.

Bothridium Muscular lappet on the dorsal or ventral side of the scolex of a tapeworm. Bothridia are often highly specialized, with many types of adaptations for adhesion.

Bottleneck effect Changes in gene frequency that result when numbers in a population are drastically reduced, and genetic variability is reduced as a result of the population being built up again from relatively few surviving individuals.

Boyle's law The product of the pressure and the volume of an ideal gas at constant temperature is a constant.

Brachiolaria This asteroid larva develops from the bipinnaria larva and has three preoral holdfast processes.

Brachiopoda A phylum of marine animals whose members possess a bivalved calcareous and/or chitinous shell that is secreted by a mantle and encloses nearly all of the body. Unlike the molluscs, the valves are dorsal and ventral. Possess a lophophore.

Brachium The arm from shoulder to elbow.

Brackett series (brackett) The series which describes the emission spectrum of hydrogen when the electron is jumping to the fourth orbital. All of the lines are in the infrared portion of the spectrum.

Bract A modified leaf or leaflike structure, usually much reduced in size.

Bradycardia Marked slowing of the heartbeat rate during a breath-hold dive.

Bradyzoite Small stage in various coccidia of the Isospora group that develops in a zoitocyst; similar to a merozoite.

Bragg's law When a beam of x-rays strikes a crystal surface in which the layers of atoms or ions are regularly separated, the maximum intensity of the reflected ray occurs when the complement of the angle of incidence, τηετα, the wavelength of the x-rays, lambda, and the distance betwen layers of atoms or ions, d, are related by the equation 2 d sin theta = n lambda, Where n is an integer.

Bran The husk of a cereal grain; including the pericarp and seed coat of a cereal grain. The bran is removed in the processing of refined grains.

Branch roots Lateral roots arising from the pericycle of a

Breaching Leaping into the air by whales.

Breakbone fever Another name for dengue, a virus disease transmitted by mosquitoes.

Bredt's rule A double bond cannot be placed with one terminus at the bridgehead of a bridged ring system unless the rings are large enough to accommodate the double bond without excessive strain. For example, while bicyclo[2.2.1]hept-1-ene is only capable of existence as a transient, its higher homologues having a double bond at the bridgehead position have been isolated.

Breeder reactor A nuclear reactor that produces fuel by bombarding isotopes of uranium and thorium with high-energy neutrons that convert inert atoms to fissionable ones. These reactors are inherently less stable and more dangerous than normal nuclear reactors.

Brewster's law The extent of the polarization of light reflected from a transparent surface is a maximum when the reflected ray is at right angles to the refracted ray.

Bridged carbocation A carbocation (real or hypothetical) in which there are two (or more) carbon atoms that could in alternative Lewis formulae be designated as carbenium centres but which is instead represented by a structure in which a group (a hydrogen atom or a hydrocarbon residue, possibly with substituents in non-involved positions) bridges these potential carbenium centres. One may distinguish "electron-sufficient bridged carbocations" and "electron-deficient bridged carbocations". Examples of the former are phenyl-bridged ions (for which the trivial name "phenonium ion" has been used), such as (A).

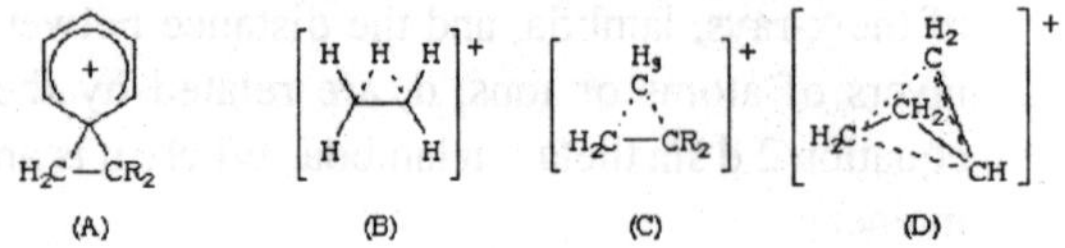

(A) (B) (C) (D)

These ions are straightforwardly classified as carbenium ions. The latter type of ion necessarily involves three-centre bonding. Structures (C) and (D) contain five-coordinate carbon atoms. The "hydrogen-bridged carbocation" (B) contains a two-co-ordinate hydrogen atom. Hypercoordination, which includes two-coordination for hydrogen and five- but also higher coordination for carbon is generally observed in bridged carbocations.

Bridging ligand A ligand attached to two or more, usually metallic, central atoms.

Brittle stars Echinoderms with five flexible arms that radiate from a conspicuous central disk and tube feet that are used in feeding.

Broadcast spawning The release of gametes into the water, allowing external fertilization to occur.

Bronchi The paired ventilatory tubes of a vertebrate that branch into each lung at the lower end of the trachea.

Bronchiole Small, thin-walled branch of the bronchus.

Bronchitis An inflammation of bronchial linings that causes persistent cough, copious production of sputum, and involuntary muscle spasms that constrict airways.

Bronchus Either of two primary divisions of the trachea that lead to the right and left lung.

Brønsted acid (Brönsted acid) A molecular entity capable of donating a hydron (proton) to a base, (i.e. a "hydron donor") or the corresponding chemical species. For example: H_2O, H_3O^+, CH_3CO_2H, H_2SO_4, HSO_4, HCl, CH_3OH, NH_3.

Brood patch The patch of feathers used by birds to incubate eggs; also known as incubation patch.

Brooding Parental care of developing young.

Brown algae (division Phaeophyta)Seaweeds with a predominance of yellow and brown pigments.

Brown fat Mitochondria-rich, heat-generating adipose tissue of endothermic vertebrates.

Brownian motion The continuous random motion of solid microscopic particles when suspended in a fluid medium due to the consequence of ongoing bombardment by atoms and molecules.

Brownian movement The irregular and random movement of small particles suspended in a fluid, known to be a consequence of the thermal motion of fluid molecules.

Bruce effect In mice, the effect of a strange male, or his odor, that causes females to abort and become receptive.

Bryophytes A group of nonvascular plants consisting of the hornworts, liverworts,

Bryozoans Colonial invertebrates in which the anus is located outside the edge of a lophophore.

Bubble chamber Instrument used for making the paths of ionizing particles visible as a trail of tiny bubbles in a liquid.

Bubo Swollen lymph node.

Buccal Referring to the mouth cavity.

Buccal cone Portion of the mouthparts of acarines composed of hypostome and labrum.

Buccal pump The mechanism by which lung ventilation occurs in amphibians; muscles of the mouth and pharynx create positive pressure to force air into the lungs.

Buccopharyngeal respiration The diffusion of gases across moist linings of the mouth and pharynx of amphibians.

Bud An apical meristem and the appendages which surround and enclose it. Also refers to the asexual reproductive outgrowth in yeasts and some bacteria.

Bud scale scar A scar or impression encircling the twig caused by the abscission of bud scales.

Bud scales Modified leaves surrounding and protecting a bud.

Budding A form of asexual reproduction in which new individuals develop from a portion of the parent, as in all bryozoans and in many protozoans, Cnidarians, and polychaetes.

Buffer Any substance or chemical compound that tends to keep ph levels constant when acids or bases are added.

Bulb An underground storage organ characterized by fleshy leaves attached to a stem base.

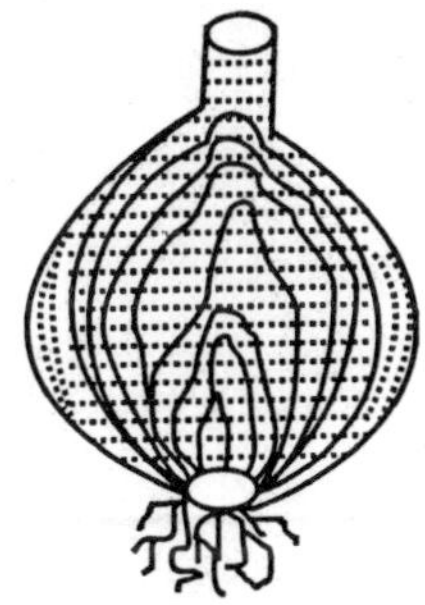

Fig. Bulb

Bulbourethral gland Gland that secretes a viscous fluid into the male urethra during sexual excitement.

Bulliform cell A large epidermal cell found on the upper surface of many grass leaves; turgor pressure in these cells controls the lateral rolling of the leaves during water stress.

Bunnett-Olsen equations The equations for the relation between $lg([sh^+]/[s]) + h_o$ and $h_o + lg[h^+]$ for base s in aqueous mineral acid solution, where h_o is Hammett's acidity function and $h_o + lg[h^+]$ represents the activity function $lg(\gamma_s\gamma_h+)/\gamma s_h{}^+$ for the nitroanilline reference bases to build h_o.

$$lg([sh^+]/[s]) - lg[h^+] = (\Phi - 1)(h_o + lg[h^+]) + pk_{sh}{}^+$$

$$lg([sh^+]/[s]) + h_o = \Phi(h_o + lg[h^+]) + pk_{sh}{}^+$$

Bursa A sac-like cavity. In ophiuroid echinoderms, pouches opening at bases of arms and functioning in respiration and reproduction (genitorespiratory bursae).

By-catch The less valuable catch that is captured while fishing for a more valuable catch.

Byssal thread a strong elastic fiber used by mussels to attach themselves to a solid substrate

C

C_3 pathway The Calvin cycle of carbon fixation in which CO_2 is incorporated into ribulose 1,5-bisphosphate.

C_4 pathway Also called the Hatch-Slack pathway of photosynthesis; CO_2 is initially fixed in mesophyll cells into malate and aspartate.

Cadre Sclerotized mouth lining of a pentastomid.

Cage An aggregate of molecules, generally in the condensed phase, that surrounds the fragments formed, for example, by thermal or photochemical dissociation. Because the cage hinders the separation of the fragments by diffusion, they may preferentially react with one another ("cage effect") but not necessarily to re-form the precursor species. For example,

R-N=N-R, heat $\longrightarrow$[R. + N2 + R.]cage $\longrightarrow$R-R + N_2

Cage compound A polycyclic compound having the shape of a cage. The term is also used for inclusion compounds.

Calabar swelling Transient subcutaneous nodule, provoked by the filarial nematode Loa loa.

Calcaneus Calc = heel bone.

Calcarea The class of sponges whose members are small and possess monaxon, triaxon, or tetraaxon calcium carbonate spicules.

Calcareous Made of calcium carbonate.

Calcareous ooze A type of biogenous sediment that is made of the calcium carbonate shells and skeletons of marine organisms.

Calciferous glands Glands in an earthworm that secrete calcium ions into the gut.

Calcitonin A thyroid hormone that lowers calcium and phosphate levels in the blood; also called thyrocalcitonin.

Calcium carbonate ($CaCO_3$) A mineral that is the major component of the shell, skeleton, and other parts of many organisms.

Callose A complex carbohydrate found in the sieve areas of sieve-tube elements; particularly abundant at the time of injury.

Callus An undifferentiated group of cells formed as a response to wounding (as at the base of a stem) or in tissue culture.

Calorie Unit of heat defined as the amount of heat required to heat 1 g of water from 14.5 to 15.5C; 1 cal = 4.184 joules in the International System of Units.

Calorimeter A heat-measuring device consisting of nested metal cups separated by an air space.

Calotte "Head" end of a dicyemid mesozoan.

Calvin Cycle Biochemical pathway in photosynthesis in which carbon from atmospheric CO_2 is fixed and reduced into carbohydrate.

Calypter Squama or lobe in the anal angle of a dipteran wing.

Calyx 1. A boat-shaped or cuplike central body of an entoproct or crinoid. The body and tentacles of an entoproct. 2. A cuplike set of ossicles that support the crown of a sea lily or feather star.

Cambrian A geological period of the Paleozoic beginning about 590 million years ago and lasting about 85 million years, during which time many divisions of protists occurred in the oceans.

Camerostome Ventral groove in the propodosoma of soft ticks wherein lies the capitulum.

Campestral Characteristic of rural locations, especially open country and grasslands.

Campodeiform Insect larva with well-defined head and thoracic appendages; typically predatory.

Canaliculus A small canal.

Cancellous Having a spongy or porous structure.

Cancer Invasive, out-of-control cell growth that results in malignant tumors.

Candela The fundamental SI unit of luminous intensity defined as the luminous intensity in a given direction of a source that emits monochromatic photons of frequency 540 x 10^{12} hz and has a radiant intensity in that direction of 1/683 w/sr.

Candle The unit of luminous intensity of a light source.

Canine Canis = a dog. A tooth.

Cannulation The introduction of a substance to a tissue (e.g., an area of the brain) via hollow electrode, fine tube (cannula), or needle.

Canopy The upper portion of a population of plants; the term is usually associated with forests and agricultural crops.

Capacitance The capacitance of a capacitor is measured by this procedure: Put equal and opposite charges on its plates and then measure the potential between the plates. Then C = |Q/V|, where Q is the charge on one of the plates. Capacitors for use in circuits consist of two conductors (plates). We speak of a capacitor as "charged" when it has charge Q on one plate, and -Q on the other. Of course the net charge of the entire object is zero; that is, the charged capacitor hasn't had net charge added to it, but has undergone an internal separation of charge. Unfortunately this process is usually called charging the capacitor, which is misleading because it suggests adding charge to the capacitor. In fact, this process usually consists of moving charge from one plate to the other. The capacitance of a single object, say an isolated sphere, is determined by considering the other plate to be an infinite sphere surrounding it. The object is given charge, by moving charge from the infinite sphere, which acts as an infinite charge reservoir ("ground"). The potential of the object is the potential between the object and the infinite sphere.

Capacitance depends only on the geometry of the capacitor's physical structure and the dielectric constant of the material medium in which the capacitor's electric field exists. The size of the capacitor's capacitance is the same whatever the charge and potential (assuming the dielectric constant doesn't change). This is true even if the charge on both plates is reduced to zero, and therefore the capacitor's potential is zero. If a capacitor with charge on its plates has a capacitance of, say, 2 microfarad, then its capacitance is also 2 microfarad when the plates have no charge. This should remind us that C = |Q/V| is not by itself the definition of capacitance, but merely a formula which allows us to relate the capacitance to the charge and potential when the capacitor plates have equal and opposite charge on them.

A common misunderstanding about electrical capacitance is to assume that capacitance represents the maximum amount of charge a capacitor can store. That is misleading because capacitors don't store charge (their total charge being zero) but their plates have equal and opposite charge. It is wrong because the maximum charge one may put on a capacitor plate is determined by the potential at which dielectric breakdown occurs. Compare: capacity.

We probably should avoid the phrase "charged capacitor" or "charging a capacitor". Some have suggested the alternative expression "energizing a capacitor" because the process is one of giving the capacitor electrical potential energy by rearranging charges in it.

Capacitor A combination of conducting plates separated by layers of a dielectric that is used to store an electric charge.

Capacity This word is properly used in names of quantities which express the relative amount of some quantity with respect to a another quantity upon which it depends. For example, heat capacity is dU/dT, where U is the internal energy and T is the temperature. Electrical capacity, usually called capacitance is another example: $C = |dQ/dV|$, where Q is the magnitude of charge on each capacitor plate and V is the potential difference between the plates.

Consistent use of the word "capacitance" for C avoids this conceptual error. But the same misconceptions can occur with the others, and we don't have other names for them which might help avoid this. Heat capacity isn't the maximum amount of heat something can have. That would also incorrectly suggest that heat is a "substance", which it isn't.

Capillarity The elevation or depression of liquids in small-diameter tubes

Capillary A small blood vessel that connects an arteriole and a venule; the functional unit of the circulatory system.

Capillary pores Small spaces in the soil that become filled with a fluid (such as water) because of the adhesion of particles to the matrix (solid substrate) and the cohesion of the water molecules to themselves.

Capitulum Anterior of two basic body regions of a mite or tick. Also called a gnathosoma.

Capsid The protein coat of a virus.

Capsule 1. A simple fruit that develops from a compound ovary with two or more carpels; capsules dehisce in many ways. Example: cotton.

2. The sporangium of a bryophyte.

Captacula Tentacles extending from head of scaphopod molluscs, used in feeding.

Captive breeding Raising plants or animals in zoos or other controlled conditions to produce stock for subsequent release into the wild.

Captodative effect Effect on the stability of a carbon centred radical determined by the combined action of a captor (electron withdrawing) and a dative (electron releasing) substituent, both attached to the radical centre. The term is also used for certain unsaturated compounds.

Carageenan A polysaccharide extracted from red algae and used for many industrial products.

Carapace Structure formed by posterior and lateral extension of dorsal sclerites of the head in many Crustacea, usually covering and/or fusing with one or more thoracic somites; considered as arising from a fold of head exoskeleton. Also a dorsal sclerotized plate often covering the idiosoma of acarines. The dorsal portion of the shell of a turtle. Formed from a fusion of vertebrae, ribs, and dermal bone.

Carbaminohemoglobin Compound formed by the union of carbon dioxide and hemoglobin.

Carbanion Generic name for anions containing an even number of electrons and having an unshared pair of electrons on a tervalent carbon atom (e.g. Cl_3C^- or $HC{\equiv}C^-$) or - if the ion is mesomeric - having a least one significant contributing structure with an unshared pair of electrons on a tervalent carbon atom, for example,

$CH_3C(\text{-}O\text{-}){=}CH\text{-}C({=}O)CH_3$ $CH_3C({=}O)\text{-}CH\text{—}C({=}O)CH_3$

Carbenium centre The three-coordinate carbon atom in a carbenium ion to which the excess positive charge of the ion (other than that located on heteroatoms) may be formally considered to be largely attributed, i.e., which has one vacant p-orbital. This formal attribution of charge often does not express the real charge distribution.

Carbenium ion A generic name for carbocations, real or hypothetical, that have at least one important contributing structure containing a tervalent carbon atom with a vacant p-orbital. (The name implies a protonated carbene or a substitution derivative thereof.)

The term was proposed (and rejected) as a replacement for the traditional usage of the name carbonium ion.

To avoid ambiguity, the name should not be used as the root for the systematic nomenclature of carbocations. The corresponding difficulty confused carbonium ion nomenclature for many years. For example, the term "ethylcarbonium ion" has at times been used to refer either to $CH_3CH_2^+$ (ethyl cation) or (correctly) to $CH_3CH_2CH_2^+$ (propyl cation).

Carbenoid A carbene like chemical species but with properties and reactivity differing from the free carbene itself, e.g.

$R_1R_2C(Cl)M$ (M = metal)

Carbocation A cation containing an even number of electrons with a significant portion of the excess positive charge located on one or more carbon atoms. This is a general term embracing carbenium ions, all types of carbonium ions, vinyl cations, etc. Carbocations may be named by adding the word "cation" to the name of the corresponding radical. Such names do not imply structure (e.g. whether three-coordinated or five-coordinated carbon atoms are present).

Carbohydrate Compounds of carbon, hydrogen, and oxygen having the generalized formula $(CH_2O)n$; aldehyde or ketone derivatives of polyhydric alcohols, with hydrogen and oxygen atoms attached in a 2:1 ratio.

Carbon (C) An element that is an essential constituent of all organic compounds.

Carbon cycle The circulation and reutilization of carbon atoms, especially via the processes of photosynthesis and respiration.

Carbon dating A procedure used to determine the age of fossils.

Carbon dioxide (CO_2) A colorless gas that is required in the process of photosynthesis.

Carbon fixation The conversion of inorganic carbon into energy-rich organic carbon, usually by photosynthesis.

Carbon monoxide (CO) Colorless, odorless, nonirritating but highly toxic gas produced by incomplete combustion of fuel, incineration of biomass or solid waste, or partially anaerobic decomposition of organic material.

Carbon sink Places of carbon accumulation, such as in large forests (organic compounds) or ocean sediments (calcium carbonate); carbon is thus removed from the carbon cycle for moderately long to very long periods of time.

Carbon skeleton The central portion of any organic molecule consisting only of the carbon molecule.

Carbon source Originating point of carbon that reenters the carbon cycle; cellular respiration and combustion.

Carbonic acid Atmospheric CO_2 dissolved in water to yield H_2CO_3, a weak acid.

Carboniferous A geological period of the Paleozoic beginning approximately 360 million years ago and lasting about 85 million years. During this time forests appeared and became dominant, composed of arborescent club mosses, horsetails, and ferns. Their extinction resulted in the formation of coal deposits.

Carbonium ion The term should be used with great care since several incompatible meanings are currently in use. It is not acceptable as the root for systematic nomenclature for carbocations. 1. In most of the existing literature the term is used in its traditional sense for what is here defined as carbenium ion.

2. A carbocation, real or hypothetical, that contains at least one five-coordinate carbon atom.

3. A carbocation, real or hypothetical, whose structure cannot adequately be described by two-electron two-centre bonds only. (The structure may involve carbon atoms with a coordination number greater than five.)

Carboxyl group An acid group attached to a molecule; -COOH.

Carbyne Generic name for the species HC and substitution derivatives thereof, such as EtO_2C-C containing an electrically neutral univalent carbon atom with three non-bonding electrons. Use of the alternative name "methylidyne" as a generic term is not recommended.

Carcinogens Substances that cause cancer.

Cardiac Belonging or relating to the heart.

Cardiac muscle Specialized type of muscle tissue found only in the heart.

Carnivore One of the flesh-eating mammals of the order Carnivora. Also, any organism that eats animals.

Carnot's theorem The theorem which states that no engine operating between two temperatures can be more efficient than a reversible engine.

Carotene A red, orange, or yellow pigment belonging to the group of carotenoids; precursor of vitamin A.

Carotenoids Class of plant pigments that includes carotenes and xanthophylls; most are yellow, orange, or red.

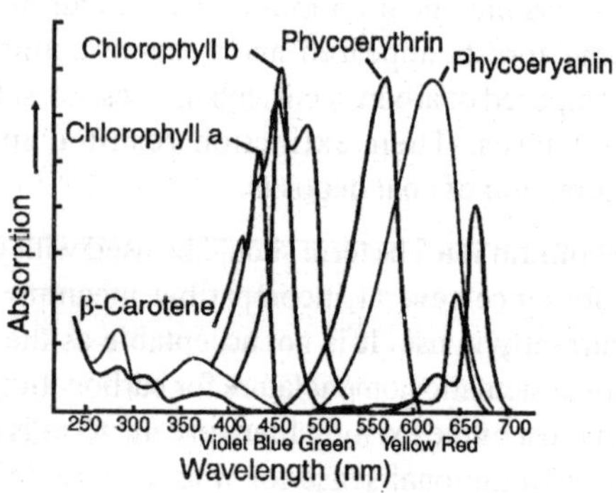

Fig. Carotenoids

Carpel The reproductive unit of angiosperms composed of a placental surface and ovules. A component of the gynoecium.

Fig. Carpel

Carrying capacity Maximum number of a population that can be supported by the environment over a given period of time.

Cartilage Type of connective tissue in which cells are located within lacunae and are separated by a semisolid matrix. Provides a site for muscle attachment, aids in movement of joints, and provides support.

Cartilaginous fishes Fishes with a skeleton made of cartilage: sharks, rays, skates, and ratfishes (or chimaeras).

Caryopsis Simple, dry, indehiscent fruit with a single seed that is fused to the ovary wall. Also called a grain. Example: wheat.

Cash crops Crops that are sold rather than consumed or bartered.

Casimir effect A quantum mechanical effect, where two very large plates placed close to each other will experience an attractive force, in the absence of other forces. The cause is virtual particle-antiparticle pair creation in the vicinity of the plates. Also, the speed of light will be increased in the region between the two plates, in the direction perpendicular to them.

Catalyst A substance that participates in a particular chemical reaction and thereby increases its rate but without a net change in the amount of that substance in the system. At the molecular level, the catalyst is used and regenerated during each set of microscopic chemical events leading from a molecular entity of reactant to a molecular entity of product.

Catalytic coefficient If the rate of reaction (v) is expressible in the form

$$v = (k_0 + \Sigma k_i[c_i]n_i) [a]^{\alpha} [b]^{\beta}..$$

where a, b,.. are reactants and ci represents one of a set of catalysts, then the proportionality factor k_i is the catalytic coefficient of the particular catalyst ci. Normally the partial order of reaction (n_i) with respect to a catalyst will be unity, so that ki is an ($\alpha+\beta+..+1$)th order rate coefficient. The proportionality factor k_0 is the ($\alpha+\beta+..$)th order rate coefficient of the uncatalysed component of the total reaction.

Catastrophic systems Dynamic systems that jump abruptly from one seemingly steady state to another without any intermediate stages.

Cathode When two metal plates are placed facing each other in an electrical circuit the negatively charged plate is called the cathode (the positively charged plate is called the anode).

Cathode rays When a cathode is heated with the application of a large voltage a beam flows from the cathode toward the anode. J.J. Thompson realized that the ray consisted of particles (rather than light as was presumed) and named them electrons. This effect was one of the clues that classical physics was incomplete, heralding the birth of quantum mechanics.

Cecum 1. Each arm of the blind-ending Y-shaped digestive tract of trematodes (phylum Platyhelminthes).

2. A region of the vertebrate digestive tract where fermentation can occur. It is located at the proximal end of the large intestine.

Cecum, caecum A blind pouch at the beginning of the large intestine; any similar pouch.

Cell enlargement The expansion in volume of a cell typically as a result of increasing turgor pressure.

Cell plate Double membrane across the equator of a dividing cell that develops from the phragmoplast; it marks where the new cell walls will form.

Cell wall The rigid outermost layer of the cells found in plants, some protists, and most bacteria. Found in plants composed principally of cellulose.

Cell-mediated immunity (CMI) Immunity in which antigen is bound to receptor sites on the surface of sensitized T lymphocytes that have been produced in response to prior immunizing experience with that antigen and in which manifestation is through macrophage response with no intervention of antibody.

Cellular immune response Binding of antigen with receptor sites on sensitized T lymphocytes to cause release of lymphokines that affect macrophages, a direct response with no intervention of antibody. Also, the entire process by which the body responds to an antigen, resulting in a condition of cell-mediated immunity.

Cellular respiration The process in which a cell breaks down sugar or other organic compounds to release energy used for cellular work; may be anaerobic or aerobic, depending on the availability of oxygen.

Celsius scale The temperature scale using the ice point as 0° and the steam point as 100°, with 100 equal divisions, or degrees, between; formerly the centigrade scale.

Cement glands Glands in a male acanthocephalan that produce secretions sealing the female reproductive tract after copulation.

Cementation Compacting and natural "gluing together" of sedimentary deposits to form sedimentary rocks.

Cenozoic A geological era beginning 65 million years ago and extending to the present, characterized by major adaptive radiation of specialized life forms and formation of biomes.

Center of curvature The center of the sphere of which the mirror or lens surface forms a part.

Center of gravity The point at which all of the weight of a body can be considered to be concentrated.

Central dogma The relationship among the steps from DNA to the production of a protein. The synthesis of messenger RNA from DNA, and the movement mRNA out of the nucleus (transcription). Carries the genetic code into the cytoplasm. Within the cytoplasm, ribosomes and transfer RNA convert this genetic code in messenger RNA into a protein (translation).

Central nervous system (CNS) That part of the nervous system that is condensed and centrally located; for example, the brain and spinal cord of vertebrates and the brain and ganglia of insects.

Central place foragers Animals that collect more than one food item at a time and bring the items to a central location for storage or feeding to offspring.

Centrifugal force When a non-inertial rotating coordinate system is used to analyze motion, Newton's law F = ma is not correct unless one adds to the real forces a fictitious force called the centrifugal force. The centrifugal force required in the non-inertial system is equal and opposite to the centripetal force calculated in the inertial system. Since the centrifugal and centripetal forces are concepts used in two different formulations of the problem, they can not in any sense be considered a pair of reaction forces. Also, they act on the same body, not different bodies.

Centrifugal pseudoforce A pseudoforce that occurs when one is moving in uniform circular motion. One feels a "force" directed outward from the center of motion.

Centriole A minute cytoplasmic organelle usually found in the centrosome and considered to be the active division center of the animal cell; organizes spindle fibers during mitosis and meiosis. Same structure as basal body or kinetosome.

Centrosome Minute body in cytoplasm of many plant and animal cells that contains one or two centrioles and is the center of dynamic activity in mitosis.

Cephalaspidomorphi The class of vertebrates characterized by the absence of paired appendages and the presence of sucking mouthparts with teeth and a rasping tongue. Lampreys.

Cephalic Having to do with, or toward, the head of an animal.

Cgs The system of units based upon the fundamental metric units: centimeter, gram and second.

Chaetotaxy Taxonomic study of the location and arrangement of bristles on an insect. Especially important in the order Diptera.

Chaff Bracts surrounding a cereal grain; removed during threshing.

Chagas' disease Disease of humans and other mammals caused by Trypanosoma cruzi.

Chagoma Reddish nodule that forms at the site of entrance of Trypanosoma cruzi into the skin.

Chain reaction A reaction in which one or more reactive reaction intermediates (frequently radicals) are continuously regenerated, usually through a repetitive cycle of elementary steps (the "propagation step"). For example, in the chlorination of methane by a radical mechanism, Cl. is continuously regenerated in the chain propagation steps:

$Cl^{\cdot} + CH_4 \longrightarrow HCl + H_3C^{\cdot}$

$H_3C^{\cdot} + Cl_2 \longrightarrow CH_3Cl + Cl^{\cdot}$

In chain polymerization reactions, reactive intermediates of the same types, generated in successive steps or cycles of steps, differ in relative molecular mass, as in

$RCH_2C.HPh + H_2C{=}CHPh \longrightarrow RCH_2CHPhCH_2C.HPh$

Chain transfer The abstraction, by the radical end of a growing chain-polymer, of an atom from another molecule. The growth of the polymer chain is thereby terminated but a new radical, capable of chain propagation and polymerization, is simultaneously created. For the example of alkene polymerization cited for a chain reaction, the reaction

$RCH_2C.HPh + CCl_4 \longrightarrow RCH_2CHClPh + Cl_3C.$

Represents a chain transfer, the radical $Cl_3C.$ inducing further polymerization:

$H_2C{=}CHPH + Cl_3C. \longrightarrow Cl_3CCH_2C.HPh$

$Cl_3CCH_2C.HPh + H_2C{=}CHPH \longrightarrow Cl_3CCH_2CHPhCH_2C.HPh$

The phenomenon occurs also in other chain reactions such as cationic polymerization.

Chalazal Ovule end opposite the micropyle.

Chalimus Specialized, parasitic copepodid, found in the copepod order Siphonostomatoida; attached to its host by an anterior "frontal filament" that is secreted by the frontal gland.

Chandrasekhar limit A limit which mandates that no white dwarf (a collapsed, degenerate star) can be more massive than about 1.4 masses solar. Any degenerate mass more massive must inevitably collapse into a neutron star.

Chaotic systems Systems that exhibit variability, which may not be necessarily random, yet whose complex patterns are not discernible over a normal human timescale.

Chaparral A relatively dry biome that occurs in the southwestern United States. Plants are low and shrubby and have tough, waxy leaves. Animals include insects, rodents, rabbits, lizards, snakes, and mule deer.

Charge That property of matter which allows the matter to interact via one of the fundamental forces. The word "charge" is generally used to denote electrical charge, but it may also refer to gravitational charge (usually called mass), to weak charge, or to color charge. As a verb, this also denotes the building up of charge.

Charge carrier, negative A negative charge carrier is an electron which is in the conduction band, the flow of which explicitly creates the electrical current.

Charge carrier, positive A positive charge carrier is a hole which "gets filled" by electrons. As an electron flows to the right to fill a hole, it leaves a hole behind on the left; in effect the hole has "moved" to the left because the electron has moved to the right. The flow of these holes are explicitly the cause of the electrical current.

Charge population The net electric charge on a specified atom in a molecular entity, as determined by some prescribed definition, e.g. that of Mulliken (1955).

Charge, electrical That property of matter which allows the matter to interact via the electrical interaction. (Electrical charge is expressed in units of coulmb which were named for Charles A. Coulomb.) The electrical charge can be either positive or negative such that like charges repel and opposite charges attract. The electrical charge of some common particles are:

Electron ($q_e = -1.60 \times 10^{-19}$ Coulombs)

Proton ($q_p = +1.60 \times 10^{-19}$ Coulombs)

Neutron ($q_n = 0.00$ Coulombs)

Quarks (depending on the type, either 1/3(qe) or 2/3(-qe))

Charge-transfer complex A ground state adduct which exhibits an observable charge transfer absorption band.

Charging. In protein synthesis, a reaction catalyzed by tRNA synthetase, in which an amino acid is attached to its particular tRNA molecule.

Charles' law The volume of an ideal gas at constant pressure is proportional to the thermodynamic temperature of that gas.

Chelation The formation or presence of bonds (or other attractive interactions) between two or more separate binding sites within the same ligand and a single central atom. A molecular entity in which there is chelation (and the corresponding chemical species) is called a "chelate". The terms bidentate (or didentate), tridentate, tetradentate.. multidentate are used to indicate the number of potential binding sites of the ligand, at least two of which must be used by the ligand in forming a "chelate". For example, the bidentate ethylenediamine forms a chelate with CuI in which both nitrogen atoms of ethylenediamine are bonded to copper. (The use of the term is often restricted to metallic central atoms.)

The phrase "separate binding sites" is intended to exclude cases such as $[PtCl_3(CH_2=CH_2)]$, ferrocene, and (benzene)tricarbonylchromium in which ethene, the cyclopentadienyl group, and benzene, respectively, are considered to present single binding sites to the respective metal atom, and which are not normally thought of as chelates.

Chemical bond The force that holds molecules together.

Chemical change A change in which new substances with new properties are formed.

Chemical element A substance that cannot be separated into different substances by ordinary chemical methods.

Chemical equivalent The quantity of an element, expressed in grams, equal to the ratio of its atomic weight to its valence.

Chemical flux A concept related to rate of reaction, particularly applicable to the progress in one direction only of component reaction steps in a complex system or to the progress in one direction of reactions in a system at dynamic equilibrium (in which there are no observable concentration changes with time). Chemical flux (Φ) is a derivative with respect to time, and has the dimensions of amount of substance per unit volume transformed per unit time.

Chemical relaxation If the equilibrium mixture of a chemical reaction is disturbed by a sudden change, especially of some external parameter (such as temperature, pressure, or electrical field strength), the system will readjust itself to a new position of the chemical equilibrium or return to the original position, if the perturbation is temporary. The readjustment is known as chemical relaxation.

In many cases, and in particular when the displacement from equilibrium is slight, the progress of the system towards equilibrium can be expressed as a first-order law

$$[C_t - (C_{eq})_2] = [(C_{eq})_1 - (C_{eq})_2]\exp(-t/)$$

where $(C_{eq})_1$ and $(Ceq)_2$ are the equilibrium concentrations of one of the chemical species involved in the reaction before and after the change in the external parameter, and Ct is its concentration at time t. The time parameter t, named relaxation time, is related to the rate constants of the chemical reaction involved.

Measurements of the relaxation times by relaxation methods [involving a temperature jump (T-jump), pressure jump, electric field jump or a periodic disturbance of an external parameter, as in ultrasonic techniques] are commonly used to follow the kinetics of very fast reactions.

Chemical shift The variation of the resonance frequency of a nucleus in nuclear magnetic resonance (nmr) spectroscopy in consequence of its magnetic environment. The chemical shift of a nucleus, , is expressed in ppm by its frequency, $_{cpd}$, relative to a standard, ref, and defined as

$$= 10^6(_{cpd} - _{ref})/_o$$

Where $_o$ is the operating frequency of the spectrometer. For ^1H and ^{13}C nmr the reference signal is usually that of tetramethylsilane. Other references are used in the older literature and in other solvents, such as D_2O.

If a resonance signal occurs at lower frequency or higher applied field than an arbitrarily selected reference signal, it is said to be upfield, and if resonance occurs at higher frequency or lower applied field, the signal is downfield. Resonance lines upfield from $sime_4$ have positive, and resonance lines downfield from $sime_4$ have negative -values.

Chemical species An ensemble of chemically identical molecular entities that can explore the same set of molecular energy levels on the time scale of the experiment. The term is applied equally to a set of chemically identical atomic or molecular structural units in a solid array. For example, two conformational isomers may be interconverted sufficiently slowly to be detectable by separate NMR spectra and hence to be considered to be separate chemical species on a time scale governed by the radiofrequency of the spectrometer used. On the other hand, in a slow chemical reaction the same mixture of conformers may behave as a single chemical species, i.e. there is virtually complete equilibrium population of the total set of molecular energy levels belonging to the two conformers.

Except where the context requires otherwise, the term is taken to refer to a set of molecular entities containing isotopes in their natural abundance. The wording of the definition given in the first paragraph is intended to embrace both cases such as graphite, sodium chloride, or a surface oxide, where the basic structural units may not be capable of isolated existence, as well as those cases where they are. In common chemical usage, generic and specific chemical names (such as radical or hydroxide ion) or chemical formulae refer either to a chemical species or to a molecular entity.

Chemistry The science dealing with the elements and atomic relations of matter, and various elemental compounds. The study of the properties of substances and how substances react with one another.

Chemoselective Chemoselectivity is the preferential reaction of a chemical reagent with one of two or more different functional groups. A reagent has a high chemoselectivity if reaction occurs with only a limited number of different functional groups. For example, sodium tetrahydroborate is a more chemoselective reducing agent than is lithium tetrahydroaluminate. The concept has not been defined in more quantitative terms. The term is also applied to reacting molecules or intermediates which exhibit selectivity towards chemically different reagents.

Some authors use the term chemospecificity for 100% chemoselectivity. However, this usage is discouraged.

Cherenkov radiation Radiation emitted by a massive particle which is moving faster than light in the medium through which it is travelling. No particle can travel faster than light in vacuum, but the speed of light in other media, such as water, glass, etc., are considerably lower. Cherenkov radiation is the electromagnetic analogue of the sonic boom, though cherenkov radiation is a shockwave set up in the electromagnetic field.

Chiasma A decussation or X-shaped crossing; the places where pairs of homologous chromatids remain in contact during late prophase to anaphase of the first meiotic division. The chiasma indicates where an exchange of homologous segments has taken place between nonsister chromatids by crossing over.

Chitin 1.The polysaccharide found in the exoskeleton of arthropods.

2. High molecular weight polymer of N-acetyl glucosamine linked by 1,4-b-glycosidic bonds.

Chitons (class Polyplacophora) Molluscs that have a shell divided into eight overlapping plates.

Chlorophyll Green pigment found in plants and in some animals; necessary for photosynthesis.

Chordamesoderm Tissue in the amphibian gastrula that forms between ectoderm and endoderm in the dorsal lip region of the blastopore; develops into the mesoderm and notochord.

Chromatid A replicated chromosome joined to its sister chromatid by the centromere; separates and becomes daughter chromosome at anaphase of mitosis or anaphase of the second meiotic division.

Chromoplast Membrane-bound organelle containing pigments other than chlorophyll; the pigments may be yellow, orange, or red.

Chromosome Rodlike structure that appears in the nucleus of a cell during mitosis; contains the genes responsible for heredity. Structure composed of a very long DNA molecule and associated proteins that carries part (or all) of the hereditary information of an organism.

Chronic effects Long-lasting results of exposure to a toxin; can be a permanent change caused by a single, acute exposure or a continuous, low-level exposure.

Chronic food shortages Long- term undernutrition and malnutrition; usually caused by people's lack of money to buy food or lack of opportunity to grow it themselves.

Chronology protection conjecture The concept that the formation of any closed timelike curve will automatically be destroyed by quantum fluctuations as soon as it is formed. In other words, quantum fluctuations prevent time machines from being created.

Cilia Microscopic, hairlike processes on the exposed surfaces of certain eukaryotic cells. Cilia contain a core bundle of microtubules and are capable of performing repeated beating movements. They are also responsible for the swimming of many single-celled organisms.

Ciliary comb One of eight bands of cilia fused at the base that is found in comb jellies.

Ciliary creeping The principal means of nemertine locomotion.

Ciliary organelles Organelles of specialized function formed by the fusion of cilia.

Ciliates A group of protozoans that have cilia.

Ciliophora The protozoan phylum characterized by members with simple or compound cilia at some stage in their life history; heterotrophs with a well-developed cytostome and feeding organelles; at least one macronucleus and micronucleus present. Examples: Paramecium, Stentor, Vorticella, Balantidium.

Cilium A short, hair-like flagellum that is found in large numbers and used in movement, for pushing food particles, and in other functions.

Cinclides Small pores in the external body wall of sea anemones for extrusion of acontia.

Cine-substitution A substitution reaction (generally aromatic) in which the entering group takes up a position adjacent to that occupied by the leaving group.

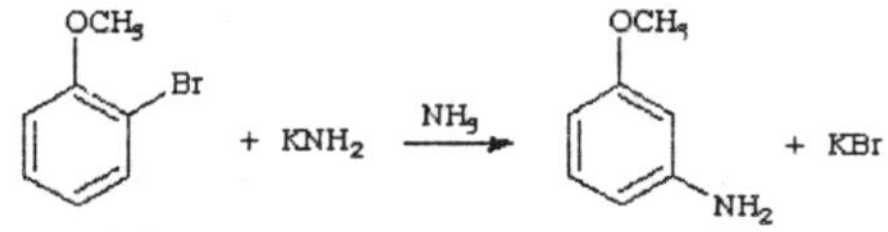

Circadian rhythms Daily cycles of activity. Circadian rhythms are usually based upon photoperiods.

Circuit The route laid out with wires which connect circuit components along which the electrical current flows. As one follows the electrons around the circuit, the electrical potential increases, but the current does not diminish. This is equivolent to saying that the charges are not used up, but the electrical potential energy is. However, as with water flowing through a pipe, it is possible to split the current amoung a variety of paths. Circuit analysis is usually concerned with the values of the current through the components and voltages between points along the circuit, such as across the components.

Circular motion Motion of a body along a circular path.

Circulation cells Air movement in a circular pattern from the earth's surface up into the outer atmosphere and then back down again; produce high and low pressure systems and dictate precipitation patterns at different latitudes.

Circulatory system Pertaining to the circulation. Also cardiovascular system.

Circumcision The removal of all or part of the prepuce or foreskin.

Cirri Any of various slender or filamentous, usually flexible appendages, such as one of the compound organelles composed of groups of fused cilia seen in certain peritrichious ciliate protozoa that are used for locomotion; an eversible penis in flatworms; a fingerlike projection of a polychete parapodium.

Citric acid cycle A series of chemical reactions in the mitochondrion by which various molecules are oxidized and energy is released from them; Krebs cycle, TCA or tricarboxylic acid cycle.

Classical (Pavlovian) conditioning A type of learning whereby an unconditioned stimulus (US) that elicits a specific response (unconditioned response [UCR]) is paired with a neutral stimulus (becomes the conditioned stimulus [CS]) so that the response becomes conditioned (CR).

Cloaca Posterior chamber of digestive tract in many vertebrates, receiving feces and urogenital products. In certain invertebrates, a terminal portion of digestive tract that serves also as respiratory, excretory, or reproductive tract.

Clone A series of identical cells or individuals that have developed from a single cell or individual.

Closed canopy A forest where tree crowns spread over 20 percent of the ground; has the potential for commercial timber harvests.

Closed circulatory system A circulatory system in an animal in which blood is confined to vessels throughout its circuit.

Closed system A physical system on which no outside influences act; closed so that nothing gets in or out of the system and nothing from outside can influence the system's observable behavior or properties.

Obviously we could never make measurements on a closed system unless we were in it, for no information about it could get out of it! In practice we loosen up the condition a bit, and only insist that there be no interactions with the outside world which would affect those properties of the system which are being studied.

Besides, when the experimenter is a part of the system, all sorts of other problems arise. This is a dilemma physicists must deal with: the fact that if we take measurements, we are a part of the system, and must be very certain that we carry out experiments so that fact doesn't distort or prejudice the results.

Coastal shelf The shallow region of the ocean surrounding a large landmass.

Coccyx Anthropoid caudal vertebrae.

Cochlea A tubular cavity of the inner ear containing the essential organs of hearing; occurs in crocodiles, birds, and mammals; spirally coiled in mammals.

Co-composting Microbial decomposition of organic materials in solid waste into useful soil additives and fertilizer; often, extra organic material in the form of sewer sludge, animal manure, leaves, and grass clippings are added to solid waste to speed the process and make the product more useful.

Cocoon Protective covering of a resting or developmental stage, sometimes used to refer to both the covering and its contents; for example, the cocoon of a moth or the protective covering for the developing embryos in some annelids.

Codominance A condition in which both alleles of a heterozygous pair are expressed independently.

Codon A sequence of three bases on messenger RNA that specifies the position of an amino acid in a protein.

Coelom A fluid-filled body cavity, lined by mesodermlying between the gut and the outer body wall musculature that is lined with derivatives of the embryonic mesoderm.

Coelomic fluid Thc fluid that fills thc coclom of echinoderms and other invertebrates.

Coelomocyte Another name for amebocyte; primitive or undifferentiated cell of the coelom and the water-vascular system.

Coelomoduct A duct that carries gametes or excretory products (or both) from the coelom to the exterior.

Coelozoic Living in the lumen of a hollow organ, such as the intestine.

Coenecium, coenoecium The common secreted investment of an ectoproct colony; may be chitinous, gelatinous, or calcareous.

Coenenchyme Extensive mesogleal tissue between the polyps of an alcyonarian (phylum Cnidaria) colony.

Coenocyte Organisms possessing a large mass of cytoplasm with many nuclei.

Coenocytic A tissue in which the nuclei are not separated by cell membranes; syncytial.

Coenosarc The inner, living part of hydrocauli in hydroids.

Coenurus Tapeworm metacestode in the family Taeniidae, in which several scolices bud from an internal germinative membrane; not enclosed in an internal secondary cyst.

Coenzyme An organic nonprotein molecule, frequently a phosphorylated derivative of a water-soluble vitamin, that binds with the protein molecule (apoenzyme) to form the active enzyme (holoenzyme). Examples include biotin, NAD^+, and coenzyme A.

Coevolution The change in gene frequencies resulting from two species, acting as strong selective forces on one another.

Cofactor A metal ion or inorganic ion with which an enzyme must unite in order to function.

Cogeneration The simultaneous production of electricity and steam or hot water in the same plant.

Colonial hypothesis A hypothesis formulated to explain the origin of multicellularity from protist ancestors; animals may have been derived when protists associated together and cells became specialized and interdependent.

Commensalism A relationship in which one individual lives close to or on another and benefits, and the host is unaffected; often symbiotic.

Common misuse We often hear non-scientists say such things as "The car was going at a high rate of speed." This is redundant at best, since it merely means "The car was moving at high speed." It is the sort of mistake made by people who don't think while they talk.

Common name A regional name for well-known plants; in the language of the region, rather than in Latin, and not necessarily paralleling any scientific name.

Commonality In learning, the inference of a phyletic, evolutionary relationship among the species exhibiting a particular type of learning.

Common-ion effect (on rates) A REDUCTION IN THE rate of certain reactions of a substrate RX in solution caused by the addition to the reaction mixture of an electrolyte solute containing the "common ion" X^- (or X^+). For example, the rate of solvolysis of diphenylmethyl chloride in acetone-water is reduced by the addition of salts of the common ion Cl^- which causes a decrease in the quasi-equilibrium concentration of the diphenylmethyl cation. This phenomenon is a direct consequence of the mass-law effect on ionization equilibria in electrolytic solution. More generally, the common-ion effect is the influence of the "common ion" on the reactivity due to the shift of the dissociation equilibrium. It may also lead to an enhancement of the rate of reaction.

Comparative anatomy The study of animal structure in an attempt to deduce evolutionary pathways in particular animal groups.

Comparative embryology The study of animal development in an attempt to deduce evolutionary pathways in particular animal groups.

Comparative method A comparison of the behavior of two or more species for the purpose of either elucidating some common aspects of the ecology and evolution of behavior or exploring the mechanisms underlying behavior.

Complementary DNA (cDNA) DNA prepared by transcribing the base sequence from mRNA into DNA by reverse transcriptase; also called copy DNA.

Complementary strand Two polynucleotide chains in which the pairing of adenine is always with thymine (in DNA) or uracil (in RNA), and guanine is always paired with cytosine.

Composite reaction A chemical reaction for which the expression for the rate of disappearance of a reactant (or rate of appearance of a product) involves rate constants of more than a single elementary reaction. Examples are "opposing reactions" (where rate constants of two opposed chemical reactions are involved), "parallel reactions" (for which the rate of disappearance of any reactant is governed by the rate constants relating to several simultaneous reactions to form different respective products from a single set of reactants), and stepwise reactions.

Compression A fossil formed when carbonized plant material is still present in the original shape but is greatly compressed and reduced in size by pressure.

Compression wood The reaction wood produced along the lower side of leaning trees, straightening the trunk by expanding and pushing the tree upright.

Comproportionation The reverse of disproportionation. The term "symproportionation" is also used.

Compton effect An effect that demonstrates that photons (the quantum of electromagnetic radiation) have momentum. A photon fired at a stationary particle, such as an electron, will impart momentum to the electron and, since its energy has been decreased, will experience a corresponding decrease in frequency.

Concave Surface with center of curvature on the same side as the observer.

Concave lens A lens that diverges parallel light rays (assuming the outside refractive index to be smaller).

Concave mirror A mirror that converges parallel light rays incident on its surface.

Concentration gradient The difference in concentration in two parts of a system.

Concentricycloidea The class of echinoderms whose members are characterized by two concentric water-vascular rings encircling a disklike body; no digestive system; and internal brood pouches. Sea daisies.

Concerted process Two or more primitive changes are said to be concerted (or to constitute a concerted process) if they occur within the same elementary reaction. Such changes will normally (though perhaps not inevitably) be "energetically coupled". (In the present context the term "energetically coupled" means that the simultaneous progress of the primitive changes involves a transition state of lower energy than that for their successive occurrence.) In a concerted process the primitive changes may be synchronous or asynchronous.

Conchae Thin bones in the nasal chamber.

Concomitant immunity Premunition.

Concurrent forces Forces with lines of action that pass through the same point.

Condensation The aggregation of water molecules from vapor to liquid or solid when the saturation concentration is exceeded.

Condensation nuclei Tiny particles that float in the air and facilitate the condensation process.

Condensation reaction A (usually stepwise) reaction in which two or more reactants (or remote reactive sites within the same molecular entity) yield a single main product with accompanying formation of water or of some other small molecule, e.g. ammonia, ethanol, acetic acid, hydrogen sulfide.

The mechanism of many condensation reactions has been shown to comprise consecutive addition and elimination reactions, as in the base-catalysed formation of (E)-but-2-enal (crotonaldehyde) from acetaldehyde, via 3-hydroxybutanal (aldol). The overall reaction in this example is known as the aldol condensation.

The term is sometimes also applied to cases where the formation of water or another simple molecule does not occur, as in "benzoin condensation".

Conditioned response (CR) The behavior pattern that becomes conditioned during classical conditioning.

Conditioned stimulus (CS) The previously neutral stimulus that, through classical conditioning, now elicits the conditioned response.

Conditioning A form of learning in which a behavior is associated with a reward such as food.

Conduct The conduction of electricity refers to the process of electrical charge flowing through a material. The conduction of the charge carriers is the electrical current.

Conduction The conveyance of energy, such as heat, sound, or electricity. The direct transfer of thermal motion (heat) between molecules of the environment and those on the body surface of an animal.

Conductor A material which has a small energy gap is a conducting material. A conductor conducts electricity very well; it conducts better than either an insulator or a semi-conductor.

Condyles Bearing surfaces between arthropod joints that provide the fulcra on which the joints move.

Cone A vertebrate photoreceptor cell with a tapered outer segment; unlike rod cells, cone cells are usually sensitive

Cone cell A color-sensitive photoreceptor cell concentrated in the retina.

Configuration (electronic) A distribution of the electrons of an atom or a molecular entity over a set of one-electron wavefunctions called orbitals, according to the Pauli principle. From one configuration several states with different multiplicities may result.

Conformation The spatial arrangements of atoms affording distinction between stereoisomers which can be interconverted by rotation about formally single bonds. Some authorities extend the term to include inversion at trigonal bipyramidal centres and other "polytopal rearrangements".

Confusion effect The reduced capture efficiency experienced by a predator caused by high densities or swarms of prey.

Conidiophore Hyphae on which one or many conidia are produced.

Conidiosporangium A structure that produces conidiospores.

Conidiospore An asexual spore produced singly

Conidium Asexual reproductive spores of some ascomycetes and imperfect fungi.

Conifer Any of a group of plants that produce a strobilus or cone as a reproductive structure.

Coniferous forest A community type dominated by cone-bearing gymnosperms mostly in the northern hemisphere.

Conifers Needle-bearing trees that produce seeds in cones.

Conjugate acid-base pair The Brønsted acid BH^+ formed on protonation of a base B is called the conjugate acid of B, and B is the conjugate base of BH^+. (The conjugate acid always carries one unit of positive charge more than the base, but the absolute charges of the species are immaterial to the definition.) For example: the Brønsted acid HCl and its conjugate base Cl^- constitute a conjugate acid-base pair.

Conjugated system, conjugation In the original meaning a conjugated system is a molecular entity whose structure may be represented as a system of alternating single and multiple bonds: e.g.

$CH_2{=}CH{=}CH{=}CH_2$ $CH_2{=}CH{-}C{\equiv}N$

In such systems, conjugation is the interaction of one p-orbital with another across an intervening sigma bond in such structures. (in appropriate molecular entities d-orbitals may be involved.) The term is also extended to the analogous interaction involving a p-orbital containing an unshared electron pair, e.g.

$:Cl{-}CH{=}CH_2$

Conjugation Temporary union of two ciliate protozoa while they are exchanging chromatin material and undergoing nuclear phenomena resulting in binary fission. Also, formation of cytoplasmic bridges between bacteria for transfer of plasmids.

Connective tissue A basic type of tissue that includes bone, cartilage, and various fibrous tissues. Connective tissue serves to support and bind tissues together.

Connectivity In a chemical context, the information content of a line formula, but omitting any indication of bond multiplicity.

Conoid Truncated cone of spiral fibrils located within the polar rings of the suborder Eimeriina.

Conservation To protect, conserve the natural world and all its components.

Conservation laws A law which states that, in a closed system, the total quantity of something will not increase or decrease, but remain exactly the same; that is, its rate of change is zero. For physical quantities, it states that something can neither be created nor destroyed.

Conservation of angular momentum The total angular momentum of a closed system remains constant.

There are several other laws that deal with particle physics, such as conservation of baryon number, of strangeness, etc., which are conserved in some fundamental interactions (such as the electromagnetic interaction) but not others (such as the weak interaction).

Conservation of electric charge The total electric charge of a closed system remains constant.

Conservation of linear momentum The total linear momentum of a closed system remains constant.

Conservation of mass-energy The total mass-energy of a closed system remains constant.

Conservation of matter In any chemical reaction, matter changes form; it is neither created nor destroyed.

Conservative forces Forces for which the law of conservation of mechanical energy holds true; gravitational forces and electrostatic forces.

Conserved A quantity is said to be conserved if under specified conditions it's value does not change with time.

Example: In a closed system, the charge, mass, total energy, linear momentum and angular momentum of the system are conserved. (Relativity theory allows that mass can be converted to energy and vice-versa, so we modify this to say that the mass-energy is conserved.)

Conspecific A member of the same species.

Constancy principle One of the postulates of a. Einstein's special theory of relativity, which puts forth that the speed of light in vacuum is measured as the same speed to all observers, regardless of their relative motion. One of the results of this postulate (one of the predictions of special relativity) is that no massive particle can be accelerated to (or beyond) lightspeed, and thus the speed of light also represents the ultimate cosmic speed limit. Only massless particles (collectively called luxons, including photons, gravitons, and possibly neutrinos, should they prove to indeed be massless) travel at lightspeed, and all other particles must travel at slower speeds.

Constant (sigma constant) Specifically the substituent constant for meta- and for para-substituents in benzene derivatives as defined by Hammett on the basis of the ionization constant of a substituted benzoic acid in water at 25°C, i.e. lg(Ka/KaO), where Ka is the ionization constant of a m- or p-substituted benzoic acid and KaO that of benzoic acid itself.

The term is also used as a collective description for related electronic substituent constants based on other standard reaction series, of which, +, - and o are typical; also constants which represent dissected electronic effects such as I and R. For this purpose it might be better always to spell out the term in full, i.e. as "Hammett sigma constant", and restrict -constants to the scale of substituent constants which is based on benzoic acid.

Constant proportions, rule of A principle that states that the relative amounts of ions in seawater are always the same.

Constitution The description of the identity and connectivity (and corresponding bond multiplicities) of the atoms in a molecular entity (omitting any distinction from their spatial arrangement).

Constraints In optimality models, the limits placed on an animal because of its inherent abilities.

Consumer A heterotroph. A primary consumer feeds directly on a primary producer, whereas a secondary consumer feeds on a primary consumer.

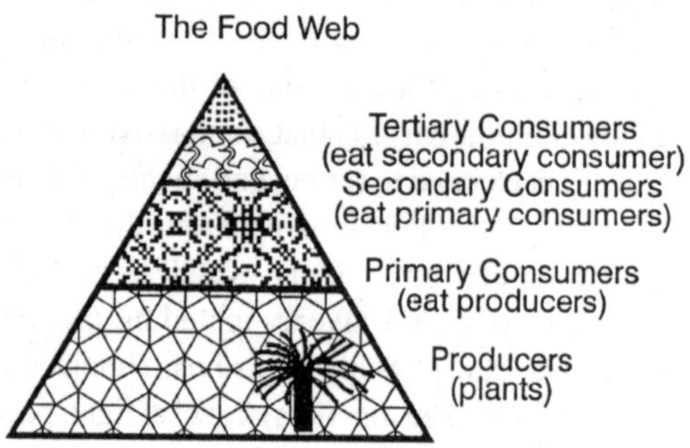

Fig. Consumer

Consummatory behavior Actions of an animal completing a behavior sequence, as in consuming food, mating, etc.

Consumption The fraction of withdrawn water that is lost in transmission or that is evaporated, absorbed, chemically transformed, or otherwise made unavailable for other purposes as a result of human use.

Contagious Capable of being transmitted through direct contact. Also used to describe population distributions that are aggregated, such as in an area.

Contaminative antigen Antigen borne by the parasite that is common to both the host and the parasite but that genetically is of host origin.

Context In communication, stimuli other than the signal that are impinging on the receiver that might alter the meaning of a signal.

Contiguity The association of events in time with particular reference here to classical conditioning.

Continental boundary current A surface ocean current flowing generally north or south along a continental edge.

Continental drift The breakup and movement of land masses of the earth. The earth had a single landmass about 250 million years ago. This mass broke apart into continents, which have moved slowly to their present positions.

Continental margin The edge of a continent; the zone between a continent and the deep-sea floor.

Continental rise The gently sloping area at the base of the continental slope.

Continental shelf The shallow, gently sloping section of the continental margin that extends from the shore to the point where the slope gets steeper.

Continental slope The relatively steep portion of the sea bottom between the outer edge of the continental shelf and the deep ocean basin.

Continuous feeder Usually slow-moving sessile animals that feed all of the time.

Continuous spectrum A spectrum without dark lines or bands or in which there is an uninterrupted change from one color to another.

Continuous variable One that can assume any value whatsoever between certain limits, versus a discrete variable that can assume only specific values (usually integers).

Contour feathers Feathers that cover the body, wings, and tail of a bird. Contour feathers provide flight surfaces and are responsible for plumage colors.

Contour plowing Plowing along hill contours; reduces erosion.

Contractile root A specialized root, common in bulbs, that contracts, pulling the stem deeper into the ground.

Contractile vacuole A clear fluid-filled cell vacuole in protozoa and a few lower metazoa; takes up water and releases it to the outside in a cyclical manner, for osmoregulation and some excretion.

Contributing structure The definition is based on the valence-bond formulation of the quantum mechanical idea of the wavefunction of a molecule as composed of a linear combination of wavefunctions, each representative of a formula containing bonds that are only single, double or triple with a particular pairing of electron spins. Each such formula represents a contributing structure, also called "resonance structure" to the total wavefunction, and the degree to which each contributes is indicated by the square of its coefficient in the linear combination. The contributing structures, also called "canonical forms", themselves thus have a purely formal significance: they are the components from which wavefunctions can be built. Structures may be covalent (or non-polar) or ionic (or polar). The representation is frequently kept qualitative so that we speak of important or major contributing structures and minor contributing structures.

Control That part of a scientific experiment to which the experimental variable is not applied but which is similar to the experimental group in all other respects.

Control group An observed, unmanipulated set of test subjects. In an experimental manipulation, we must also perform the actions involved in that manipulation on a second test group of organisms in order to ascertain whether any effects of the experimental treatment are due to that treatment and not to the actions involved in providing the treatment.

Control rods Neutron-absorbing material inserted into spaces between fuel assemblies in nuclear reactors to regulate fission reaction.

Conurbation An area with more than one million people that is formed by the merger of two or more population centers and that meets other specified requirements.

Convection The act of conveying or transmission. The movement of air (or a liquid) over the surface of a body and contributes to heat loss (if the air is cooler than the body) or heat gain (if air is warmer than the body).

Convection currents Rising or sinking air currents that stir the atmosphere and transport heat from one area to another. Convection currents also occur in water.

Convective mixing The vertical mixing of water masses driven by wind stresses or density changes at the sea surface.

Conventional pollutants The seven substances (sulfur dioxide, carbon monoxide, particulates, hydrocarbons, nitrogen oxides, photochemical oxidants, and lead) that make up the largest volume of air quality degradation; identified by the Clean Air Act as the most serious threat of all pollutants to human health and welfare; also called criteria pollutants.

Convergent evolution Groups of unrelated organisms becoming similar in appearance because of evolutionary change in response to similar environments.

Converging lens A lens that is thicker in the middle than it is at the edge and bends incident parallel rays toward a common point.

Convex Surface with center of curvature on the opposite side from the observer.

Convex lens A lens that converges parallel light rays (assuming the outside refractive index to be smaller) Convex mirror A mirror that diverges parallel light rays incident on its surface.

Cooperation Behavior patterns that are mutually beneficial to the cooperators.

Cooperative breeding A breeding system in which nonparents share in rearing the young.

Coordinate system In order to describe where things are, one needs to lay out the basic directions (equivolent to North, South, East, and West on a map). This is a coordinate system. There are two main differences between the directions on a map and in a coordinate system: First, a coordinate system has an origin — a place from which every other place is referenced; the map-equivolent would be if everything were measured from city hall - "The bank is three blocks south of city hall." "The grocery store is two block east and one block north of city hall." Etc. Second, a mathematical coordinate system would treat the north-south pair as a single dimension. One would be the "positive direction" and the other would be the "negative direction." (It is arbitrary which is which, but there is usually a convention.) i.e. If we call north the "positive North-South direction," then south is the negative North-South direction." Similarly, if east is the "positive East-West direction," then west is the "negative East-West direction." Finally, it is important in mathematics that the dimensions of a coordinate system be perpendicular. (North-South is perpendicular to East-West.)

Coordination The formation of a covalent bond, the two shared electrons of which have come from only one of the two parts of the molecular entity linked by it, as in the reaction of a Lewis acid and a Lewis base to form a Lewis adduct; alternatively, the bonding formed in this way. In the former sense, it is the reverse of unimolecular heterolysis. "Coordinate covalence" and "coordinate link" are synonymous (obsolescent) terms. The synonym "dative bond" is obsolete.

(The origin of the bonding electrons has by itself no bearing on the character of the bond formed. Thus, the formation of methyl chloride from a methyl cation and a chloride ion involves coordination; the resultant bond obviously differs in no way from the C-Cl bond in methyl chloride formed by any other path, e.g. by colligation of a methyl radical and a chlorine atom.)

The term is also used to describe the number of ligands around a central atom without necessarily implying two-electron bonds.

Coordination number The coordination number of a specified atom in a chemical species is the number of other atoms directly linked to that specified atom. For example, the coordination number of carbon in methane is four, and it is five in protonated methane, CH_5^+. (The term is used in a different sense in the crystallographic description of ionic crystals.)

Coordinatively saturated A transition metal complex that has formally 18 outer shell electrons at the central metal atom.

Coordinatively unsaturated A transition metal complex that possesses fewer ligands than exist in the coordinatively saturated complex. These complexes usually have fewer than 18 outer shell electrons at the central metal atom.

Copepodid Juvenile stage that succeeds the naupliar stages in copepods, often quite similar in body form to the adult.

Copepods Small, mostly planktonic crustaceans.

Copernican principle The idea, suggested by copernicus, that the sun, not the earth, is at the center of the universe. We now know that neither idea is correct (the sun is not even located at the center of our galaxy, much less the universe), but it set into effect a long chain of demotions of earth's and our place in the universe, to where it is now: on an unimpressive planet orbiting a mediocre star in a corner of a typical galaxy, lost in the universe.

Coprolite Fossilized fecal material.

Coprophagy Feeding on dung or excrement as a normal behavior among animals; reinjestion of feces.

Copulation Sexual union to facilitate the reception of sperm by the female.

Coracidium Larva with a ciliated epithelium hatching from the egg of certain cestodes; a ciliated free-swimming oncosphere.

Coral knoll (pinnacle)A column of coral within the lagoon of an atoll.

Coral reefs Prominent oceanic features composed of hard, limy skeletons produced by coral animals; usually formed along edges of shallow, submerged ocean banks or along shelves in warm, shallow, tropical seas.

Coral rubble Coral fragments.

Coralline algae Any red alga that is impregnated with calcium carbonate. Coralline algae often contribute to coral reefs.

Corallite The calcareous skeletal cup in which a coral polyp sits.

Cord grasses Salt-tolerant grasses, species of Zostera, that inhabit salt marshes.

Core The dense, intensely hot mass of molten metal, mostly iron and nickel, thousands of kilometers in diameter at the earth's center.

Core area The area of heaviest use within the home range.

Core region The primary industrial region of a country; usually located around the capital or largest port; has both the greatest population density and the greatest economic activity of the country.

Corepressor A substance that inhibits production of a particular enzyme.

Coriolis effect The apparent change in direction of a moving object (to the left in the Southern Hemisphere and to the right in the Northern Hemisphere) due to the rotation of the earth.

Coriolis pseudoforce A pseudoforce which arises because of motion relative to a frame which is itself rotating relative to second, inertial frame. The magnitude of the coriolis "force" is dependent on the speed of the object relative to the noninertial frame, and the direction of the "force" is orthogonal to the object's velocity.

Corium The deep layer of the skin; dermis.

Cork Suberized cells on the outer surface of woody stems and roots; produced by the cork cambium.

Cork cambium Meristematic tissue that produces cork cells on its outer surface and phelloderm on its inner surface.

Corm Underground, enlarged, food-storing stem covered by papery leaves.

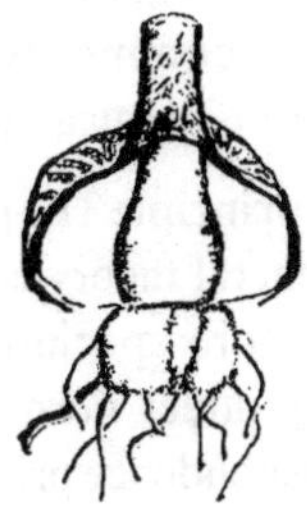

Fig. Corm

Cornea The outer transparent coat of the eye.

Corneum Epithelial layer of dead, keratinized cells.

Cornucopian fallacy The belief that nature is limitless in its abundance and that perpetual growth is not only possible but essential.

Corolla Collective term for the petals of a flower.

Corona A crown; an encircling structure. The ciliated organ at the anterior end of rotifers used for swimming or feeding.

Coronary Encircling as a crown. 1. Vessels encircling the heart.

2. A condition caused by a decreased blood flow to the heart muscle.

3. Specifically, a blocked coronary artery or a branch of a coronary artery.

Corpora allata Paired invertebrate endocrine glands, located in the thoracic region, which secrete juvenile hormone that prevents maturation.

Corpora cardiaca Paired invertebrate endocrine glands, located near the heart, which are involved in regulation of metabolism. They receive neurosecretions from the brain and release endocrine products into the hemocoel.

Correlation analysis The use of empirical correlations relating one body of experimental data to another, with the objective of finding quantitative estimates of the factors underlying the phenomena involved. Correlation analysis in organic chemistry often uses linear free-energy relations for rates or equilibria of reactions, but the term also embraces similar analysis of physical (most commonly spectroscopic) properties and of biological activity.

Correspondence principle The principle that when a new, more general theory is put forth, it must reduce to the more specialized (and usually simpler) theory under normal circumstances. There are correspondence principles for general relativity to special relativity and special relativity to newtonian mechanics, but the most widely known correspondence principle (and generally what is meant when one says "correspondence principle") is that of quantum mechanics to classical mechanics.

Corridor A strip of natural habitat that connects two adjacent nature preserves to allow migration of organisms from one place to another.

Cortex The outer layer of a structure.Ground tissue located between the vascular bundles and epidermis of stems and roots.

Cortisol A glucocorticoid secreted by the adrenal cortex.

Cortisone An animal hormone not synthesized by plants.

Cosmic background radiation; primal glow The background of radiation mostly in the frequency range 3×10^{11} to 3×10^{8} hz discovered in space in 1965. It is believed to be the cosmologically redshifted radiation released by the big bang itself. Presently it has an energy density in empty space of about 4×10^{-14} J/m^3.

Cosmic censorship conjecture The conjecture, so far totally undemonstrated within the context of general relativity, that all singularities (with the possible exception of the big bang singularity) are accompanied by event horizons which completely surround them at all points in time. That is, problematic issues with the singularity are rendered irrelevant, since no information can ever escape from a black hole's event horizon.

Cosmic rays High-energy nuclear particles apparently originating from outer space.

Cosmological constant; lambda The constant introduced to the Einstein field equation, intended to admit static cosmological solutions. At the time the current philosophical view was the steady-state model of the universe, where the universe has been around for infinite time. Early analysis of the field equation indicated that general relativity allowed dynamic cosmological models only (ones that are either contracting or expanding), but no static models. Einstein introduced the most natural abberation to the field equation that he could think of: the addition of a term proportional to the spacetime metric tensor, g, with the constant of proportionality being the cosmological constant:

$$G + \lambda g = 8 \pi t.$$

Hubble's later discovery of the expansion of the universe indicated that the introduction of the cosmological constant was unnecessary; had Einstein believed what his field equation was telling him, he could have claimed the expansion of the universe as perhaps the greatest and most convincing prediction of general relativity; he called this the "greatest blunder of my life."

Cosmological redshift An effect where light emitted from a distant source appears redshifted because of the expansion of spacetime itself.

Cosmopolitan Worldwide in distribution.

Cosmotron A high-energy synchrotron.

Cost of meiosis The reduction in numbers of alleles passed on to subsequent generations as a result of reproducing sexually compared to asexually.

Cost/benefit analysis An evaluation of large-scale public projects by comparing the costs and benefits that accrue from them.

Costa Prominent striated rod in some flagellate protozoa that courses from one of the kinetosomes along the cell surface beneath the recurrent flagellum and undulating membrane.

Costal Pertaining to a rib.

Cotyledon A seed leaf; the first leaf formed in a seed.

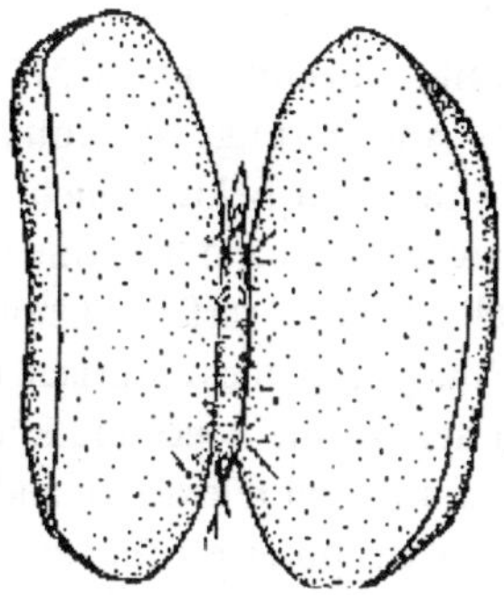

Two cotyledons (seed leaves) of a dicotyledon

Fig. Cotyledon

Cotylocidia Larva of Aspidobothria.

Coulomb's law The primary law for electrostatics, analogous to newton's law of universal gravitation. It states that the force between two point charges is proportional to the algebraic product of their respective charges as well as proportional to the inverse square of the distance between them; mathematically,

Coulomb's law of electrostatics The force between two point charges is directly proportional to the product of their magnitudes and inversely proportional ta the square of the distance between them.

Coulomb's law of magnetism The force between two magnetic poles is directly proportional to the strengths of the poles and inversely proportional to the square of their distance apart

Counteracting osmolyte strategy An osmolyte (ion) that counteracts another ion.

Countercurrent An ocean current that flows directly back into another current; also, in some animals, paired blood vessels containing blood flowing in opposite directions.

Countercurrent exchange mechanism The passive exchange of something between fluids moving in opposite directions past each other.

Counterillumination The emission of light by midwater animals to match the background light.

Countershading Contrasting coloration that helps conceal the animal (e.g., the darkly pigmented top and lightly pigmented bottom of frog embryos).

Couple Two forces of equal magnitude acting in opposite directions in the same plane, but not along the same line.

Coupling constant (spin-spin coupling constant), J (SI unit: Hz (NMR)) A quantitative measure for nuclear spin-spin, nuclear-electron (hyperfine coupling) and electron-electron (fine coupling in EPR) coupling in magnetic resonance spectroscopy. The "indirect" or scalar NMR coupling constants are in a first approximation independent of the external magnetic field and are expressed in Hz.

Coursers Predators that chase their prey over long distances.

Courtship behavior Behavior related to the attraction of the opposite sex and mating.

Covalent bond A region of relatively high electron density between nuclei which arises at least partly from sharing of electrons and gives rise to an attractive force and characteristic internuclear distance.

Cover crops Plants, such as rye, alfalfa, or clover, that can be planted immediately after harvest to hold and protect the soil.

Coxa, coxopodite The proximal joint of an insect or arachnid leg; in crustaceans, the proximal joint of the protopod.

Coxae The hip.

Coxal glands Excretory organs of arachnids, consisting of a sac, tubule, and opening on the coxa.

Crack Form of purified cocaine with widespread street use.

Crasulacean acid metabolism (CAM Pathway)A variation of the C4 pathway that functions in a number of cacti and succulents; allows for the fixation of carbon dioxide during the night, then in the daytime the carbon dioxide is transferred to the Calvin Cycle.

Creatine phosphate. High-energy phosphate compound found in the muscle of vertebrates and some invertebrates, used to regenerate stores of ATP.

Creeping eruption Skin condition caused by hookworm larvae not able to mature in a given host.

Cremaster Suspender.

Crepuscular Daily cycles with peak activity around dusk and/or dawn.

Crest A region of upward displacement in a transverse wave.

Cretin A human with severe mental, somatic, and sexual retardation resulting from hypothyroidism during early stages of development.

Cricoid A ring, circle.

Crinoidea The class of echinoderms whose members are attached by a stalk of ossicles or are free living. Possess a reduced central disk. Sea lilies and feather stars.

Crinoids The echinoderms with a small, cup-shaped body and feathery arms: the sea lilies and feather stars.

Crista A crest or ridge on a body organ or organelle; a platelike projection formed by the inner membrane of mitochondrion.

Cristae The folded-membrane inner structure of mitochondria.

Critical angle That limiting angle of incidence in the optically denser medium that results in an angle of refraction of 90o

Critical depth the depth at which photosynthesis equals cell respiration

Critical factor The single environmental factor closest to a tolerance limit for a given species at a given time.

Critical mass The amount of a particular fissionable material required to make a fission reaction self-sustaining

Critical micelle concentration (CMC) There is a relatively small range of concentrations separating the limit below which virtually no micelles are detected and the limit above which virtually all additional surfactant molecules form micelles. Many properties of surfactant solutions, if plotted against the concentration, appear to change at a different rate above and below this range. By extrapolating the loci of such a property above and below this range until they intersect, a value may be obtained known as the critical micellization concentration (critical micelle concentration), symbol CM, abbreviation CMC. As values obtained using different properties are not quite identical, the method by which the CMC is determined should be clearly stated.

Critical period A discrete portion of a sensitive period during development, during which the probability for forming and reinforcing the behavior is greatest.

Critical point The upper limit of the temperature-pressure curve of a substance.

Critical pressure The pressure needed to liquefy a gas at its critical temperature.

Critical temperature The temperature to which a gas must be cooled before it can be liquefied by pressure.

Critical velocity Velocity below which an object moving in a vertical circle will not describe a circular path

Croplands Lands used to grow crops.

Cropper a deep-sea animal in which the roles of predator and deposit feeder have merged

Crossing-over The exchange of material between homologous chromosomes, during the first meiotic division, resulting in a new combination of genes.

Cross-pollinated Pertaining to a flower having pollen deposited on it from a different flower of another plant.

Crown A molecular entity comprising a monocyclic ligand assembly that contains three or more binding sites held together by covalent bonds and capable of binding a guest in a central (or nearly central) position. The adducts formed are sometimes known as "coronates". The best known members of this group are macrocyclic polyethers, such as "18-crown-6", containing several repeating units $-CR_2-CR_2O-$ (where R is most commonly H), and known as crown ethers.

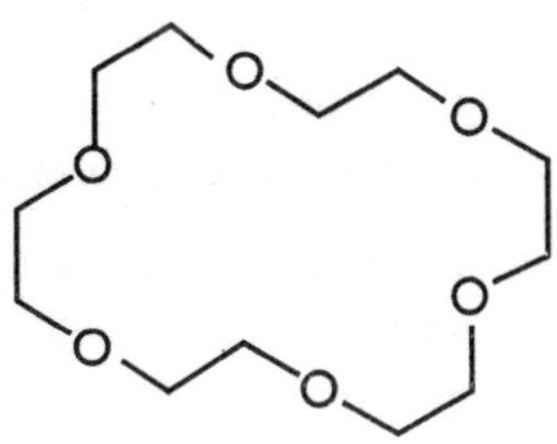

"18-crown-6"

Crude birth rate The number of births in a year divided by the midyear population.

Crude death rate The number of deaths per thousand persons in a given year; also called crude mortality rate.

Crura Branches of the intestine of a flatworm.

Crust The cool, lightweight, outermost layer of the earth's surface that floats on the soft, pliable underlying layers; similar to the "skin" on a bowl of warm pudding.

Crustacea The subphylum of mandibulate arthropods whose members are characterized by having two pairs of antennae, one pair of mandibles, two pairs of maxillae, and biramous appendages. Crabs, crayfish, lobsters.

Crustaceans (subphylum Crustacea)Arthropods that have two pairs of

Cryptand A molecular entity comprising a cyclic or polycyclic assembly of binding sites that contains three or more binding sites held together by covalent bonds, and which defines a molecular cavity in such a way as to bind (and thus "hide" in the cavity) another molecular entity, the guest (a cation, an anion or a neutral species), more strongly than do the separate parts of the assembly (at the same total concentration of binding sites). The adduct thus formed is called a "cryptate". The term is usually restricted to bicyclic or oligocyclic molecular entities.

Example

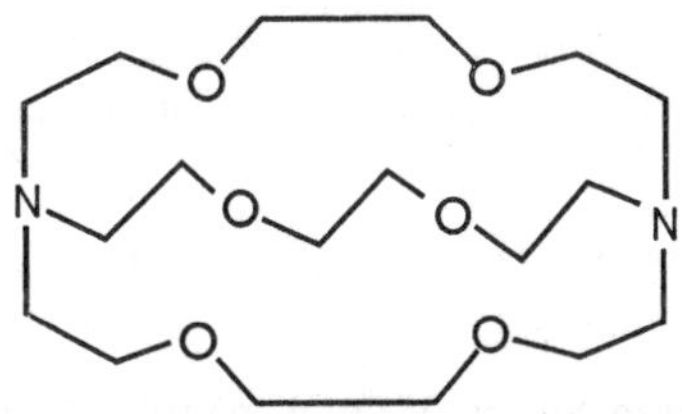

Corresponding monocyclic ligand assemblies (crowns) are sometimes included in this group, if they can be considered to define a cavity in which a guest can hide. The terms "podand" and "spherand" are used for certain specific ligand assemblies. Coplanar cyclic polydentate ligands, such as porphyrins, are not normally regarded as cryptands.

Cryptic coloration A color pattern that allows an organism to blend with the surroundings.

Cryptobiotic Living in concealment; refers to insects and other animals that live in secluded situations, such as underground or in wood; also tardigrades and some nematodes, rotifers, and others that survive harsh environmental conditions by assuming for a time a state of very low metabolism.

Cryptogonochorism Separate sexes joined or associated to form the appearance of hermaphroditism.

Cryptomonads(phylum Cryptophyta)Unicellular, eukaryotic members of the phytoplankton that have two flagella and no skeleton.

Cryptoniscus Intermediate, free-swimming larval stage of the isopod suborder Epicaridea, developing after microniscus; attaches to definitive host.

Cryptozoite Preerythrocytic schizont of Plasmodium spp.

Crystal A solid that consists of a regular pattern of molecules.

Crystalline style A proteinaceous, rodlike structure in the digestive tract of a bivalve (Mollusca) that rotates against a gastric shield and releases digestive enzymes.

Crystallized song The final stage in song development, when the song variation (repertoire) has decreased and become more fixed.

Ctene Bands of cilia found on the body surfaces of ctenophores.

Ctenidium Series of stout, peglike spines on the head (genal ctenidium) and first thoracic tergite (pronotal ctenidium) of many fleas.

Ctenoid scales Thin, overlapping dermal scales of the more advanced fishes; exposed posterior margins have fine, toothlike spines.

Ctenophora The phylum of animals whose members are characterized by biradial symmetry, diploblastic organization, colloblasts, and meridionally arranged comb rows. Comb jellies.

Cuboid Cubo = like a cube. Square.

Cuboidal Cubo = a cube.

Cubozoa The class of Cnidarians whose members have prominent cuboidal medusae with tentacles that hang from the corner of the medusa. Small polyp, gametes gastrodermal in origin. Chironex.

Culm Hollow, jointed stem of grasses.

Cultivar Abbreviation for cultivated variety of plant.

Cultural eutrophication An increase in biological productivity and ecosystem succession caused by human activities.

Cupula Small inverted cup-like structure housing another structure; gelatinous matrix covering hair cells in lateral line and equilibrium organs.

Curare A variety of poisonous plant extracts from the bark, roots, stems, and tendrils of several woody lianas, including Chondodendron tomentosum.

Curie The quantity of any radioactive nuclide that has a disintegration rate of 3.7×10^{10} becquerels.

Curie constant A characteristic constant, dependent on the material in question, which indicates the proportionality between its susceptibility and its thermodynamic temperature.

Curie-Weiss law A more general form of curie's law, which states that the susceptibility, khi, of an paramagnetic substance is related to its thermodynamic temperature t by the equation.

Currency The resource that is being maximized in an optimality model; usually assumed to be energy in an optimal foraging model.

Current The time rate at which charge passes through a circuit element or through a fixed place in a conducting wire, $I = dq/dt$.

Current sensitivity Current per unit scale division of an electric meter. Cut-off bias The smallest negative grid voltage, for a given plate voltage, that causes a vacuum tube to cease to conduct.

Current, electrical The flow (usually, the motion through a wire) of electrical charge (referred to as charge carriers). While current is simply the flow of charge, in general, people also refer to AC and DC. Any moving charge is a current; there are a variety of convenient ways to move a charge. AC (alternating current) is what comes from the outlets in your house. This is where the electrical charges oscillate back and forth rather than flow throughout the wires. DC (direct current) is what comes from batteries; this is where the electrical charges flow through the wire from one battery terminal to the other battery terminal. (Notice that the notion "AC current" is redundant, because "AC current" means "Alternating Current current"!) Although in a wire, it is a flow of electrons which causes the current, technically, the current is the flow of positive charge. It can be shown that positive charge moving to the right has the same effect as negative charge moving to the left. Both of these describe a current to the right. Electrical current is expressed in units of ampere which were named for Andre Marie Ampere.

Curtin-Hammett principle In a chemical reaction that yields one product (X) from one conformational isomer (A') and a different product (Y) from another conformational isomer (A") (and provided these two isomers are rapidly interconvertible relative to the rate of product formation, whereas the products do not undergo interconversion) the product composition is not in direct proportion to the relative concentrations of the conformational isomers in the substrate; it is controlled only by the difference in standard free energies (dG) of the respective transition states.

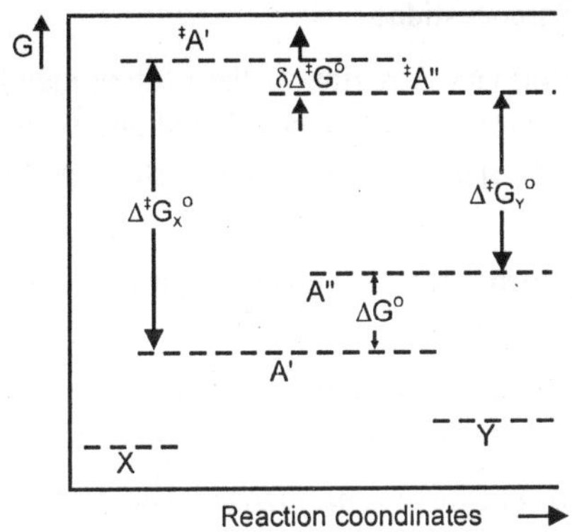

It is also truc that thc product composition is formally related to the relative concentrations of the conformational isomers A' and A" (i.e. the conformational equilibrium constant) and the respective rate constants of their reactions; these parameters are generally - though not invariably - unknown.

Cutaneous Pertaining to the skin (e.g., a skin infection).

Cutaneous respiration Exchange of gases across thin, moist surfaces of the skin. Also cutaneous exchange or integumentary exchange.

Cuticle Anoncellular, protective, organic layer secreted by the external epithelium (hypodermis) of many invertebrates; refers to the epidermis or skin in higher animals. Waterproof layer of cutin on leaves and nonwoody stems.

Cuticulin Protein component of arthropod exoskeletons.

Cutin Waxy material secreted by epidermal cells.

Cut-off frequency A characteristic threshold frequency of incident light below which, for a given material, the photoelectric emission of electrons ceases

Cut-off potential A negative potential on the collector of a photoelectric cell that reduces the photoelectric current to zero.

Cuttings Portions of stems, usually consisting of two or three nodes, used for propagation by the production adventitious roots.

Cyanobacteria (blue-green algae) The prokaryotic and generally photosynthetic organisms included in the kingdom Monera.

Cyanogenic glycoside Glycoside that releases cyanide.

Cyanophyte A cyanobacterium, blue-green alga.

Cybotactic region That part of a solution in the vicinity of a solute molecule in which the ordering of the solvent molecules is modified by the presence of the solute molecule. The term solvent "cosphere" of the solute has also been used.

Cycads Members of the Cycadophyta, which have unbranched stems and a terminal grouping of leathery, compound leaves.

Cycle A series of changes produced in sequence that recur periodically

Cyclic photophosphorylation The formation of ATP by activation of PS I, but failure to activate PS II. Consequently ATP is formed, but NADP is not reduced.

Cyclin. A protein important in the control of the cell division cycle and mitosis.

Cyclization Formation of a ring compound from a chain by formation of a new bond.

Cycloelimination The reverse of cycloaddition. The term is preferred to the synonyms "cycloreversion", "retro-addition", and "retrocycloaddition".

Cycloid scales Thin, overlapping dermal scales of the more primitive fishes; posterior margins are smooth.

Cyclosis The spontaneous movement of cytoplasm and some organelles within the cell.

Cyclotron A device for accelerating charged atomic particles by means of D-shaped electrodes.

Cydippid larva Free-swimming larva of most ctenophores; superficially similar to the adult.

Cynodonts A group of mammal-like carnivorous synapsids of the Upper Permian and Triassic.

Cypris Postnaupliar larva of barnacles (crustacean subclass Cirripedia) in which the carapace largely envelops the body; so called because of its resemblance to the ostracod genus Cypris.

Cyrtocyte A protonephridial cell with a single flagellum enclosed in a cylinder of cytoplasmic rods.

Cyst A secreted covering that protects many small invertebrates, including some protozoans, rotifers, and nematodes, from environmental stresses, such as desiccation and overcrowding.In diatoms, a resistant and photosynthetically inactive cell.

Cystacanth Juvenile stage of an acanthocephalan that is infective to the definitive host.

Cysticercoid A type of juvenile tapeworm composed of a solid-bodied cyst containing an invaginated scolex; contrast with cysticercus.

Cysticercosis Infection with the larval forms (Cysticercus cellulosae) of Taenia solium.

Cysticercus Metacestode developing from the oncosphere in most Cyclophyllidea; usually has a tail and a well-formed scolex and characterized by a fluid-filled oval body with an invaginated scolex; cysticercoid.

Cystid In an ectoproct, the dead secreted outer parts plus the adherent underlying living layers.

Cystogenic cells Secretory cells in a cercaria that produce a metacercarial cyst.

Cytochrome Several iron-containing pigments that serve as electron carriers in aerobic respiration.

Cytokine A molecule secreted by an activated or stimulated cell, for example, macrophages, that causes physiological changes in certain other cells.

Cytokinesis The division of the cytoplasm of a cell into two parts, as distinct from the division of the nucleus (which is mitosis).

Cyton Cell body; contains nucleus and some other organelles but excludes processes extending from cell. For example, the neurocyton is the nerve cell body, excluding the axon and dendrites.

Cytophaneres Fibers radiating out from a zoitocyst into surrounding muscle; found in some species of Sarcocystidae.

Cytopharynx A region of the plasma membrane and cytoplasm of some ciliated and flagellated protists specialized for endocytosis. A permanent oral canal.

Cytoplasm The living matter of the cell, excluding the nucleus.

Cytoplasmic inclusion The basic food material or stored product of a cell's metabolic activities.

Cytoproct Site on a protozoan where undigestible matter is expelled.

Cytopyge A region of the plasma membrane and cytoplasm of some ciliated protists specialized for exocytosis of undigested wastes.

Cytosine A nitrogen base found in both DNA and RNA.

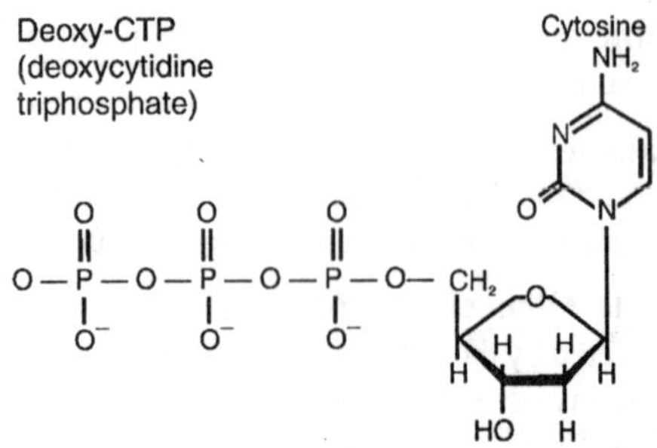

Fig. Cytosine

Cytoskeleton In the cytoplasm of eukaryotic cells, an internal framework of microtubules, microfilaments, and intermediate filaments by which organelles and other structures are anchored, organized, and moved about.

Cytosol Contents of the main compartment of the cytoplasm, excluding membrane-bound organelles.

Cytosome The cell body inside the plasma membrane.

Cytotoxic T cells A special T cell activated during cell-mediated immune responses that recognizes and destroys virus-infected cells.

D

Dactylozooid A polyp of a colonial hydroid specialized for defense or killing food.

Daily torpor Daily sluggishness that some animals experience. A period of inactivity that is normally induced by cold; observed in numerous ectotherms and also endotherms; often used to describe the specific physiological state of endotherms with a circadian cycle of lowered body temperature and depressed metabolic rate.

Dalton Atomic Weight Unit The mass of one proton or one neutron.

Dalton's law of partial pressures The total pressure of a mixture of ideal gases is equal to the sum of the partial pressures of its components; that is, the sum of the pressures that each component would exert if it were present alone and occuped the same volume as the mixture.

Damping The reduction in amplitude of a wave due to the dissipation of wave energy.

Dark reaction that part of the photosynthetic process that, in the absence of light, utilizes the preformed high-energy molecules ATP and $NADPH_2$ to synthesize complex organic molecules.

Darwinism Theory of evolution emphasizing common descent of all living organisms, gradual change, multiplication of species and natural selection.

Dauer juvenile Nematode juvenile in which development is arrested during unsuitable conditions and resumed when conditions improve.

Daughter sporocysts In digenetic trematodes the embryonic cells that develop from sporocysts and give rise to rediae.

Davisson-Germer experiment An experiment that conclusively confirmed the wave nature of electrons; diffraction patterns were observed by an electron beam penetrating into a nickel target.

Day length The number of hours that sunlight illuminates a given area on earth; dependent on the angle of the earth relative to the rays of the sun.

Day-neutral plant A plant whose flowering is not controlled by the length of day.

De Broglie wavelength The prediction that particles also have wave characteristics, where the effective wavelength of a particle would be inversely proportional to its momentum, where the constant of proportionality is the planck constant.

Deamination reaction A reaction in which an amino group,-NH_2, is enzymatically removed from a compound.

Debt-for-nature swap Forgiveness of international debt in exchange for nature protection in developing countries. Environmental groups and nongovernmental organizations often pay banks to write off uncollectable debts of developing countries at a steep discount in exchange for a promise by the debtor country to establish nature preserves.

Decacanth Ten-hooked larva that hatches from the egg of a cestodarian tapeworm. Also called a lycophora.

Decapods Crustaceans with five pairs of walking legs and a well-developed carapace. The group includes the shrimps, lobsters, hermit crabs, and crabs.

Decarboxylation The removal of a single carbon atom as CO_2 from an organic molecule.

Decay constant The ratio between the number of nuclei decaying per second and the total number of nuclei

Decibel A unit of sound intensity level The smallest change of sound intensity that the normal human ear can detect.

Deciduous Referring to plants that lose all their leaves during the cool season; as opposed to evergreen plants.

Deciduous forest Forest biome in which the dominant trees are deciduous.

Declination The angle between magnetic north and the true north from any surface location; also called variation

Decline spiral A catastrophic deterioration of a species, community, or whole ecosystem; accelerates as functions are disrupted or lost in a downward cascade.

Decomposer Mostly heterotrophic bacteria and fungi that obtain organic nutrients by breaking down the remains or products of other organic compounds; their activities help cycle the simple compounds back to the autotrophs.

Decomposition, chemical The breakdown of a single entity (normal molecule, reaction intermediate, etc.) into two or more fragments.

Decongestant An agent that relieves nasal or respiratory congestion.

Deduction Reasoning from the general to the particular, that is, from given premises to their necessary conclusion.

Deep (bottom) layer The deepest and coldest of the three layers of the ocean.

Deep ecology A philosophy that calls for a profound shift in our attitudes and behavior based on voluntary simplicity; rejection of anthropocentric attitudes; intimate contact with nature; decentralization of power; support for cultural and biological diversity; a belief in the sacredness of nature; and direct personal action to protect nature, improve the environment, and bring about fundamental societal change.

Deep scattering layer (DSL) A sound-reflecting layer of many types of organisms that migrates daily from the mesopelagic zone to the epipelagic zone.

Deep sea The dark waters below the mesopelagic zone.

Deep-sea fan A fan-like accumulation of sediment at the base of a submarine canyon.

Dees The electrodes of a cyclotron

Deficit irrigation The concept of low levels of irrigation to achieve a moderate level of productivity; even though yields are not maximized, limited water supplies are conserved.

Definitive host Host in which a parasite achieves sexual maturity. If there is no sexual reproduction in the life of the parasite, the host most important to humans is the definitive host.

Degeneracy The genetic code is said to be degenerate because more than one three-base sequence in DNA can code for one amino acid.

Degenerate rearrangement A molecular rearrangement in which the principal product is indistinguishable (in the absence of isotopic labelling) from the principal reactant. The term includes both "degenerate intramolecular rearrangements" and reactions that involve intermolecular transfer of atoms or groups ("degenerate intermolecular rearrangements"): both are degenerate isomerizations. The occurrence of degenerate rearrangements may be detectable by isotopic labelling or by dynamic NMR techniques.

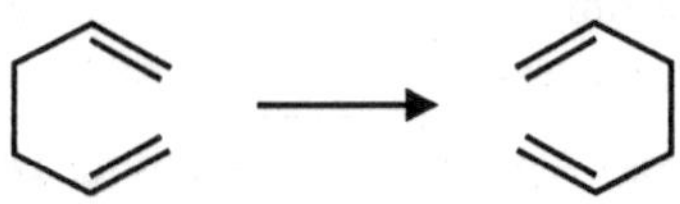

Degradation Deterioration in water quality due to contamination or pollution; makes water unsuitable for other desirable purposes.

Dehiscent fruit Fruit that splits open at maturity, facilitating seed dispersal.

Dehiscent Mature fruit that splits open to release the seed.

Deirid Sensory papilla on each side near the anterior end of some nematodes.

Delaney Clause A controversial amendment to the Federal Food, Drug, and Cosmetic Act, added in 1958, prohibiting the addition of any known cancer-causing agent to processed foods, drugs, or cosmetics.

Delayed fertilization Occurs when fertilization of an egg does not occur immediately following coitus, but may be delayed for weeks or months.

Delayed implantation A pattern in the reproductive cycle of some mammals causing the blastocyst to remain dormant in the female's uterus for some time before implantation on the uterine wall.

Delayed response problem An operant conditioning procedure in which the test animal sees the experimenter set up the test situation, including what will be the correct response, and then must wait until the experimenter provides the animal with an opportunity to respond.

Delayed type hypersensitivity (DTH) Manifestation of cell-mediated immunity, distinguished from immediate hypersensitivity in that maximal response is reached about 24 hours or more after intradermal injection of the antigen. The lesion site is infiltrated primarily by monocytes and macrophages.

Deletion A mutation resulting from the loss of a small segment of DNA.

Deletion mutant A mutation in which a base pair is deleted

Delocalization A quantum mechanical concept most usually applied in organic chemistry to describe the pi bonding in a conjugated system. This bonding is not localized between two atoms: instead, each link has a "fractional double bond character" or bond order. There is a corresponding "delocalization energy", identifiable with the stabilization of the system compared with a hypothetical alternative in which formal (localized) single and double bonds are present. Some degree of delocalization is always present and can be estimated by quantum mechanical calculations. The effects are particularly evident in aromatic systems and in symmetrical molecular entities in which a lone pair of electrons or a vacant p-orbital is conjugated with a double bond (e.g. carboxylate ions, nitro compounds, enamines, the allyl cation). Delocalization in such species may be represented by partial bonds or as resonance (here symbolized by a two-headed arrow) between contributing structures.

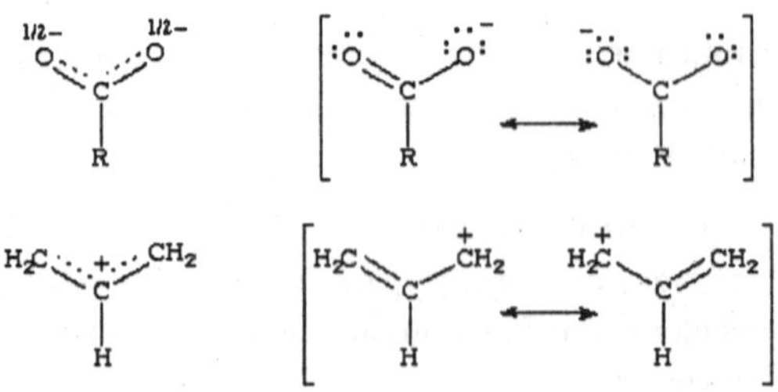

These examples also illustrate the concomitant delocalization of charge in ionic conjugated systems. Analogously, delocalization of the spin

of an unpaired electron occurs in conjugated radicals.

Delta Fan-shaped sediment deposit found at the mouth of a river.

Demand The amount of a product that consumers are willing and able to buy at various possible prices, assuming they are free to express their preferences.

Deme A small, local subpopulation. Isolated subpopulations sometimes display genetic changes that may contribute to evolutionary change in the subpopulation.

Demersal fish A bottom-dwelling fish.

Demographic transition A pattern of falling death and birth rates in response to improved living conditions; could be reversed in deteriorating conditions.

Demography Vital statistics about people: births, marriages, deaths, etc.; the statistical study of human populations relating to growth rate, age structure, geographic distribution, etc., and their effects on social, economic, and environmental conditions.

Demospongiae The class of poriferans whose members have monaxon or tetraaxon siliceous spicules or spongin. Leuconoid body forms are present and vary in size from a few centimeters to 1 m in height.

Denaturation Disruption of bonds holding a protein in its three-dimensional form, such that its polypeptide chain(s) unfolds partially or completely. Denaturation can be caused by changes in ph, salt concentration, or environmental temperature.

Denature To change the configuration of a protein molecule such that it loses specificity and no longer functions as an enzyme.

Dendrite Any of nerve cell processes that conduct impulses toward the cell body.

Dendrochronology The science of studying growth rings of trees to determine past conditions.

Dendroclimatology Study of the annual rings of trees in order to interpret climatic changes in the past.

Dengue Virus disease transmitted by mosquitoes.

Denitrifying bacteria Free-living soil bacteria that convert nitrates to gaseous nitrogen and nitrous oxide.

Density The weight (or more correctly the mass) of a given volume of a substance.

Dental formula A notation that indicates the number of incisors, canines, premolars, and molars in the upper and lower jaw of a mammal.

Denticles (denticulate) Small, toothlike projections.

Deoxyribonucleic acid A polymer of deoxyribonucleotides that is in the form of a double helix; DNA is the genetic molecule of life in that it codes for the sequence of amino acids in proteins; contains nucleotide monomers with deoxyribose sugar and nitrogenous bases adenine (A), cytosine (C), guanine (G), and thymine (T).

Deoxyribose A 5-carbon sugar having 1 oxygen atom less than ribose; a component of deoxyribonucleic acid (DNA).

Dependency ratio The number of nonworking members compared to working members for a given population.

Dependent variable What may change (e.g., aggression) in response to manipulation of an independent variable (e.g., food supply) by the experimenter.

Deplasmolysis Water flowing into a plasmolized cell to correct the process that would cause cellular death from dehydration.

Depolarized The reduction in the difference in charge (potential) between the outside and inside of a membrane.

Deposit feeding The type of feeding whereby an animal obtains its nutrients from the sediments of soft-bottom habitats (mud or sands) or terrestrial soils.

Depressant Psychoactive drug that has a sedative effect on the central nervous system; actions include dulling mental awareness and inducing sleep.

Deprivation experiments Experimental manipulations involving removal of particular types of stimuli (e.g., social, sensory, motor) to ascertain the effects later in development.

Derive To derive a result or conclusion is to show, using logic and mathematics, how a conclusion follows logically from certain given facts and principles.

Derived unit A unit of measure that consists of combinations of fundamental units.

Dermal branchiae Thin folds of the body wall of a sea star that extend between ossicles and function in gas exchange and other exchange processes.

Dermal Pertaining to the skin; cutaneous.

Dermal tissue Tissue that covers surfaces in plants.

Dermatitis Infection or inflammation of the skin.

Dermatophytes Certain imperfect fungi that cause skin diseases in humans; includes the organisms causing ringworm and athlete's foot.

Dermis The layer of the skin deep to the epidermis, consisting of a dense bed of vascular connective tissue.

Desalinization Removal of salt from water by distillation, freezing, or ultrafiltration.

Desert A type of biome characterized by low moisture levels (below 25 cm per year), and infrequent and unpredictable precipitation. Daily and seasonal temperatures fluctuate widely. Deserts are characterized by cacti and desert shrubs or sagebrush. Animals include birds, rodents, reptiles, and numerous species of arthropods.

Desertification A process by which fragile, semiarid ecosystems lose productivity because of loss of plant cover, soil erosion, salinization, or waterlogging. Usually associated with human misuse.

Desiccation Dehydration. The process of drying.

Desmosome Buttonlike plaque serving as an intercellular connection.

Despot A dominant individual that controls resources.

Detachment The reverse of an attachment.

Detailed balancing, principle of When equilibrium is reached in a reaction system (containing an arbitrary number of components and reaction paths), as many atoms, in their respective molecular entities, will pass forward, as well as backwards, along each individual path in a given finite time interval. Accordingly, the reaction path in the reverse direction must in every detail be the reverse of the reaction path in the forward direction (provided always that the system is at equilibrium). The principle of detailed balancing is a consequence for macroscopic systems of the principle of microscopic reversibility.

Determinate cleavage The type of cleavage, usually spiral, in which the fate of the blastomeres is determined very early in development; mosaic cleavage.

Determinate growth Pertaining to a leaf or stem that stops growing after differentiating into a terminal reproductive unit.

Determinism principle The principle that if one knows the state to an infinite accuracy of a system at one point in time, one would be able to predict the state of that system with infinite accuracy at any other time, past or future. For example, if one were to know all of the positions and velocities of all the particles in a closed system, then determinism would imply that one could then predict the positions and velocities of those particles at any other time. This principle has been disfavored due to the advent of quantum mechanics, where probabilities take an important part in the actions of the subatomic world, and the uncertainty principle implies that one cannot know both the position and velocity of a particle to arbitrary precision.

Detritivore Organisms that consume organic litter, debris, and dung.

Detritus Dead organic matter and the decomposers that live on it.

Deuteromycetes A miscellaneous assemblage of fungi also termed the imperfect fungi because the sexual reproductive features are either not known, not used, or have been reduced

Deuterostomes Animals in which the anus forms from, or in the region of, the blastopore; often characterized by enterocoelous coelom formation, radial cleavage, and the presence of a dipleurulalike larval stage.

Deuterostomia A group of higher phyla in which cleavage is indeterminate (regulative) and primitively radial. The endomesoderm is enterocoelous, and the mouth is derived away from the blastopore. Includes Echinodermata, Chordata, and a number of minor phyla. Compare with Protostomia.

Deuterotoky Type of parthenogenesis in which all individuals are uniparental but in which both males and females occur.

Deutomerite Posterior half of a cephaline gregarine protozoan.

Deutonymph In the life cycle of some mesostigmatid mites, a nonfeeding stage that molts into the adult.

Deutovum Incompletely developed larva that hatches from the egg of a chigger mite.

Development The summation of all activities leading to changes in cells, tissues, organs, or organisms; a genetically controlled sequence.

Device, electrical A collection of electrical circuits which perform a specific task depending on the specific device. Some common tasks are: amplification, generation, modulation, and detection of electrical signals. A microphone is an example of a device called a transducer which converts energy from one form to another. Multiple devices are connected with circuits to form an electrical system.

Devonian A geological period of the Paleozoic era beginning about 400 million years ago. During this period major diversification of the early land plants occurred.

Dew point The temperature at which a given amount of water vapor will exert equilibrium vapor pressure.

Dextral Pertaining to the right; in gastropods, shell is dextral if opening is to right of columella when held with spire up and facing observer.

Diabetes Condition characterized by a high blood glucose level and the appearance of glucose in the urine due to a deficiency of insulin or the inability of body cells to respond to insulin; diabetes mellitus.

Diamagnetic Substances having a negative magnetic susceptibility are diamagnetic. They are repelled out of a magnetic field.

Diapause A period of arrested development in the life cycle of insects and certain other animals in which physiological activity is very low and the animal is highly resistant to unfavorable external conditions.

Diapause phase A period of dormancy, common in insect species, which occurs during the more rigorous portions of the annual climatic cycle.

Diaphragm The domed respiratory muscle between thoracic and abdominal compartments of mammals.

Diapolar cells Ciliated somatodermal cells located between the parapolar and uropolar cells of a mesozoan.

Diapsids Amniotes in which the skull bears two pairs of temporal openings; includes reptiles (except turtles) and birds.

Diastereoisomerism Stereoisomerism other than enantiomerism. Diastereoisomers (or diastereomers) are stereoisomers not related as mirror images. Diastereoisomers are characterized by differences in physical properties, and by some differences in chemical behaviour toward achiral as well as chiral reagents.

Diastole Phase of the cardiac cycle during which a heart chamber wall is relaxed; also diastolic pressure.

Diastolic pressure The blood pressure measurement during the interval between heartbeats; it is the second number shown in a blood pressure reading.

Diatomaceous earth A powdery, soil-like material formed by the glass cell walls of dead diatoms deposited on the marine floor.

Diatomaceous ooze A biogenous sediment that consists mostly of the siliceous frustules of diatoms. It is known as diatomaceous earth when found inland.

Diatoms Unicellular and eukaryotic autotrophs with a siliceous frustule; mostly planktonic.

Dichotomous Pertaining to the division or forking of a single axis into two branches.

Dichroism A property of certain crystalline substances in which one polarized component of incident light is absorbed and the other is transmitted.

Dicotyledon A class of angiosperms in which the seedlings typically possess two cotyledons; commonly abbreviated to dicot.

Dieback A sudden population decline; also called a population crash.

Diecdysis Condition in which ecdysis processes are going on continuously and one ecdysis cycle grades rapidly into another.

Dielectric An electric insulator A nonconducting medium.

Dielectric constant A measure for the effect of a medium on the potential energy of interaction between two charges. It is measured by comparing the capacity of a capacitor with and without the sample present.

Differentially permeable membrane A membrane that permits the passage of certain types of particles and inhibits the passage of others; also termed selectively permeable membrane.

Differentiation The chemical and physical changes associated with the developmental process of an organism or cell.

Diffraction The spreading of a wave disturbance into a region behind an obstruction.

Diffraction angle The angle that a diffracted wavefront forms with the grating plane.

Diffraction grating An optical surface, either transmitting or reflecting, with several thousand equally spaced and parallel grooves ruled in it.

Diffuse-porous wood Secondary xylem characterized by the same-sized vessels and tracheids throughout the growing season so that growth rings are difficult or impossible to detect.

Diffusion The random movement of molecules from one location to another because of random thermal molecular motion; net diffusion always occurs from a region of higher concentration to a region of lower concentration.

Digestion The process by which larger molecules of food substances are broken down into smaller molecules that can be taken up by the digestive system; hydrolysis.

Digestive gland An enzyme-producing gland in several groups of invertebrates where digestion and absorption take place.

Digitigrade Walking on the digits with the posterior part of the foot raised; compare plantigrade.

Dihybrid A hybrid whose parents differ in two distinct characters; an offspring having two different alleles at two different loci, for example, A/a B/b.

Dihybrid cross A genetic cross between parents that differ for two traits.

Dikaryon Mycelium of some fungi that have two separate haploid nuclei in each cell.

Dimension 1. The dimension of a variable expresses the type of quantity being measured. This is occaisionally confused with the units of a variable. If one is discussing a distance or a length, then the measurement can be taken in any of several units, but there is a specific dimension. Feet, yards, miles, meters, and kilometers are all examples of units which measure the dimension of length. Gallons and liters are units which measure the dimension of volume (length cubed). Dimension

2. The dimension defines the directions through which something can move within a coordinate system. The dimension encorporates two directions: up-down, right-left, north-south, east-west, forward-backward, etc. One direction is "positive" and the other is "negative." It is generally arbitrary which is referred to as "positive," but there is usually a guiding convention for consistency.

Dimensional analysis The performance of indicated mathematical operations in a problem with the measurement units alone.

Diminishing returns A condition in which unrestrained population growth causes the standard of living to decrease to a subsistence level where poverty, misery, vice, and starvation makes life permanently drab and miserable. This dreary prophecy has led economics to be called "the dismal science."

Dimorphic Having more than one form, size, or appearance; usually referring to the difference between males and females of a species.

Dimorphism Existence within a species of two distinct forms according to color, sex, size, organ structure, and so on. Occurrence of two kinds of zooids in a colonial organism.

Dinoflagellates Unicellular, eukaryotic, mostly autotrophic organisms with two unequal flagella.

Diode A two-terminal device that will conduct electric current more easily in one direction than in the other

Dioecious Characterized by having separate sexes; that is, an individual is either male or female, but never both. "Gonochoristic" means the same thing.

Diphycercal A tail that tapers to a point, as in lungfishes; vertebral column extends to tip without upturning.

Diphyodont Having deciduous and permanent sets of teeth successively.

Diploblastic Animals whose body parts are organized into layers that are derived embryologically from two tissue layers: ectoderm and endoderm. Animals in the phyla Cnidaria and Ctenophora are diploblastic.

Diploid (2n) cell A cell, such as a body cell, that contains two similar sets of chromosomes, one from each parent. Compare haploid.

Diplopoda The class of arthropods whose members are characterized by having two pairs of legs per apparent segment and a body that is round in cross section. Millipedes.

Diplostomulum Strigeoid metacercaria in the family Diplostomatidae.

Dipolar aprotic solvent A solvent with a comparatively high relative permittivity (or dielectric constant), greater than ca. 15, and a sizable permanent dipole moment, that cannot donate suitably labile hydrogen atoms to form strong hydrogen bonds, e.g. dimethyl sulfoxide. The term (and its alternative "polar aprotic solvent") is a misnomer and is therefore discouraged. Such solvents are usually not aprotic, but protophilic (and at most weakly protogenic). In describing a solvent it is better to be explicit about its essential properties, e.g. dipolar and non-protogenic.

Dipolar bond A bond formed (actually or conceptually) by coordination of two neutral moieties, the combination of which results in charge-separated structures.

Dipole-dipole interaction Intermolecular or intramolecular interaction between molecules or groups having a permanent electric dipole moment. The strength of the interaction depends on the distance and relative orientation of the dipoles. The term applies also to intramolecular interactions between bonds having permanent dipole moments.

Diporpa Larval stage in the life cycle of the monogenean Diplozoon.

Dirac constant A sometimes more convenient form of the planck constant, defined as

Hbar = h/(2 pi).

Direct current An essentially constant value current in which the movement of charge is in only one direction

Direct development In arthropods refers to development in which a juvenile hatches from the egg, and the juvenile is not distinctly different from adult except in size and maturity.

Direct fitness A measure of an individual's potential to contribute genes to future generations via personal reproduction.

Direct flight Insect flight that is accomplished by flight muscles acting on wing bases and in which a single nerve impulse results in a single wing cycle; also called synchronous flight.

Direct proportion The relation between two quantities whose graph is a straight line.

Directional selection Natural selection that occurs when individuals at one phenotypic extreme have an advantage over individuals with more common phenotypes.

Dirt Displaced soil.

Disaccharide Sugar consisting of two monosaccharides; example: sucrose is composed of glucose and fructose.

Discharge The amount of water that passes a fixed point in a given amount of time; usually expressed as liters or cubic feet of water per second.

Discontinuous feeder An animal that does not feed all the time; instead it generally eats large meals sporadically and does not spend time in the continuous pursuit of prey.

Discrimination The ability to detect differences between two or more stimuli and make choices based on those differences in a learning situation.

Disease A deleterious change in the body's condition in response to destabilizing factors, such as nutrition, chemicals, or biological agents.

Dispersal A more or less permanent movement of an individual from an area, such as movement of a juvenile away from its place of birth.

Dispersion The process of separating polychromatic light into its component wavelengths.

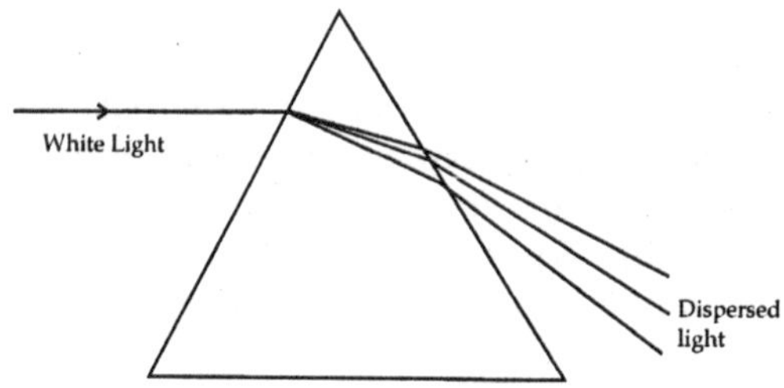

Fig. Dispersion of light beam

Displacement 1. A change of position in a particular direction

2. Distance of a vibrating particle from the midpoint of its vibration.

Displacement activity An innate or stereotypic response to a stimulus that seems inappropriate or irrelevant to the situation.

Display Any behavior pattern especially adapted in physical form or frequency to function as a social signal in communication.

Disruptive coloration A color pattern that helps break the outline of an organism.

Disruptive selection Natural selection that occurs when individuals of the most common phenotypes are at a disadvantage; produces contrasting subpopulations.

Dissipative forces Forces for which the law of conservation of mechanical energy does not hold true; frictional forces.

Dissociation 1. The separation of a molecular entity into two or more molecular entities (or any similar separation within a polyatomic molecular entity). Examples include unimolecular heterolysis and homolysis, and the separation of the constituents of an ion pair into free ions.

2. The separation of the constituents of any aggregate of molecular entities.

In both senses dissociation is the reverse of association.

Dissolved organic matter (DOM) Organic matter that is dissolved in water rather than being in particles.

Dissolved oxygen (DO) contentamount of oxygen dissolved in a given volume of water at a given temperature and atmospheric pressure; usually expressed in parts per million (ppm).

Distal Away from the point of attachment of a structure on the body (e.g., the toes are distal to the knee).

Distal cytoplasm Distal cytoplasmic layer in the tegument of Monogenea, Digenea, and Cestoidea.

Distillation Boiling of a liquid to evaporation and subsequent condensation of the vapors for the purposes of purification and concentration.

Distilled spirit Alcoholic beverage with an alcoholic content between 80-100 proof, obtained by distillation of a beer or wine.

Distome Fluke with two suckers, oral and ventral.

Distonic radical cation A radical cation in which charge and radical sites are separated.

Disulfide bond A linkage between the sulfur atoms of two different amino acids in a protein.

Diurnal Pertaining to daily cycles or events.An animal with an activity period during the light portion of the daily cycle.

Diurnal tide A tidal pattern with a high and a low tide each day.

Dive reflex the suite of internal responses, including bradycardia and peripheral circulation shutdown, that occurs during dives by an air-breathing vertebrate

Diverging lens A lens that is thicker at the edge than it is in the middle and bends incident parallel rays so that they appear to come from a common point.

Diversity (species diversity, biological diversity)The number of species present in a community (species richness), as well as the relative abundance of each species.

Diving reflex The reflex certain animals have to stay underwater for prolonged periods of time.

Division Taxonomic rank that includes related classes; synonymous to phylum used in animal systematics.

Dizygotic Twins that arise from two different zygotes, hence, individuals no more closely related than two different-aged siblings.

DNA Deoxyribonucleic acid; the long, double-helix molecule in the nucleus of cells that contains the genetic code and directs the development and functioning of all cells.

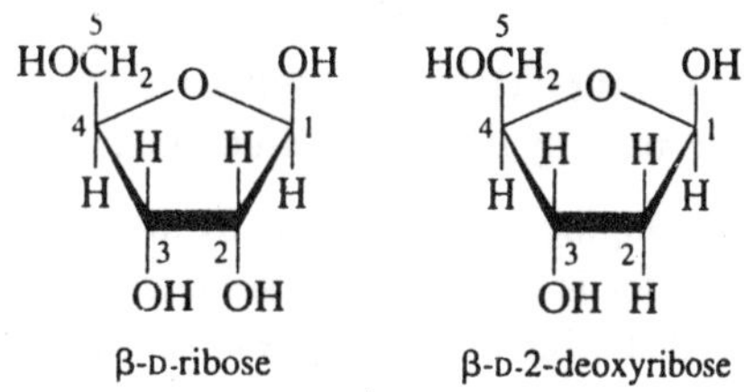

Fig. DNA

Doctrine of signatures Early concept that the Creator placed certain items on earth for humans and identified their intended use by their shape.

Domain A microscopic magnetic region composed of a group of atoms whose magnetic fields are aligned in a common direction.

Domestic sewage Wastewater from homes and non-industrial buildings. Compare industrial sewage.

Domesticated plant A plant that has been genetically changed from the wild type due to artificial selection.

Dominance hierarchy. A social ranking, formed through agonistic behavior, in which individuals are associated with each other so that some have greater access to resources than do others.

Dominant 1. The most prevalent species in a plant community.

2. The allele that has its trait expressed.

3. A gene that has its phenotypic expression appear in the offspring, regardless of the nature of its allelic partner.

Dominant plants Those plant species in a community that provide a food base for most of the community; they usually take up the most space and have the largest biomass.

Dominant species A species that exerts an overriding influence in determining the characteristics of a community.

Dominant trait A trait whose phenotype is determined by a single allele at a particular locus.

Donor A substance with five valence electrons per atom which when added to a semiconductor crystal provides free electrons in the lattice structure of the crystal.

Donor number (DN) A quantitative measure of Lewis basicity devised by Gutmann.

Doping Doping is the process by which a semi-conductor is suffused with some element other than that which comprises the semi-conductor itself. This is the introduction of impurities; generally, the different elements are vastly outnumbered by the original element.

Doping, acceptor This is the process of doping a semi-conductor with an impurity which has fewer protons per atom (and thus fewer electrons per atom). Generally, the atoms in a semicondutor will form bonds with neighboring atoms in such a manner that each atom sees a "closed shell". Since the acceptor atom introduces fewer electrons, the shell is not closed; all of the electrons are tightly bound. If an additional electron can be found, it will tend to fill the shell. Occaisionally, it is possible for an electron in a neighboring closed shell to fill this shell leaving a new unclosed shell behind. The "motion" of this closed shell is referred to as the motion of the positive charge carriers. An acceptor-doped semiconductor is called a p-type semiconductor because it has positive charge carriers.

Doping, donor This is the process of doping a semi-conductor with an impurity which has more protons per atom (and thus more electrons per atom). Generally, the atoms in a semicondutor will form bonds with neighboring atoms in such a manner that each atom sees a "closed shell". Since the donor atom introduces an additional electron, this electron is loosely bound; it is more free to move than the electrons which form the closed shell. This additional electron is easily put into the conduction band and can therefore conduct electricity easily. A donor-doped semiconductor is called an n-type semiconductor because it has negative charge carriers.

Doppler effect Waves emitted by a moving object as received by an observer will be blueshifted (compressed) if approaching, redshifted (elongated) if receding. It occurs both in sound as well as electromagnetic phenomena, although it takes on different forms in each.

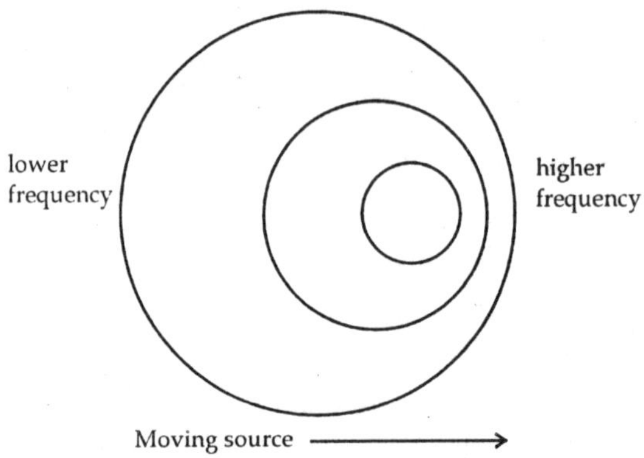

Fig. Doppler effect

Dormant Having reduced metabolic and respiratory activity.

Dorsal plate Dorsal plate on the body of a mesostigmatid mite.

Dorsal The back of an animal; usually the upper surface; synonymous with posterior for animals that walk upright.

Double bond A covalent bond sharing two pairs of electrons.

Double fertilization The fusion of egg and sperm resulting in a zygote, and the simultaneous fusion of sperm with two polar nuclei resulting in the formation of endosperm that characterizes all angiosperms.

Double helix A helix composed of two molecules winding around each other, as in DNA.

Double refraction The separation of a beam of unpolarized light into two refracted plane-polarized beams by certain crystals such as quartz and calcite.

Dourine Disease of horses and other equids caused by Trypanosoma equiperdum.

Down feathers Feathers that provide insulation for adult and immature birds.

Down syndrome. A congenital syndrome including mental retardation, caused by the cells in a person's body having an extra chromosome 21; also called trisomy 21.

Drag Resistance to movement through water or any other medium.

Drift net A very long fishing net that is allowed to drift for a long time before it is pulled on board.

Drift tubes Charged cylinders used to accelerate charged subatomic particles in a linear accelerator.

Drip irrigation Uses pipe or tubing perforated with very small holes to deliver water one drop at a time directly to the soil around each plant. This conserves water and reduces soil waterlogging and salinization.

Drives Manifestations of specific motivations corresponding to specific needs.

Driving force (also called the affinity) of a reaction, A (SI unit: kj mol^{-1}) The decrease in Gibbs energy on going from the reactants to the products of a chemical reaction.

Dropsy An old term for congestive heart failure which produces edema and resulting decrease in manual dexterity, causing one to drop things.

Drought An environmental condition in which precipitation is not sufficient to maximize biological productivity.

Drowned river valley (or coastal plain) estuaryan estuary that is formed when sea level rose at the end of the last glacial age.

Drupe A fleshy fruit with a one-carpeled ovary and only one seed. The endocarp is hard and stony, tightly enclosing the seed; the mesocarp is fleshy, and the exocarp is soft and thin; example: cherry.

Dry alkali injection Spraying dry sodium bicarbonate into flue gas to absorb and neutralize acidic sulfur compounds.

Dry fruit Fruit in which the cells of the pericarp are dry (dead) at maturity.

Dual substituent-parameter equation In a general sense, this is any equation which expresses substituent effects in terms of two parameters. However, in practice the term is used more specifically for an equation for summarizing the effects of meta- or para-substituents (i = m or p) x on chemical reactivity, spectroscopic properties, etc.

Dual-gland adhesive organ Organs in the epidermis of most turbellarians, with three cell types; viscid and releasing gland cells and anchor cells.

Ductility The property of a metal that enables it to be drawn through a die to form a wire.

Dulong-Petit law The molar heat capacity is approximately equal to the three times the ideal gas constant:

$$C = 3 R.$$

Duo-adhesive gland Platyhelminth tegument gland with two types of cells, one producing an adhesive substance and the other a releasing substance.

Duodenum The first and shortest portion of the small intestine lying between the pyloric end of the stomach and the jejunum.

Duplications The presence of two copies of one or more loci in a chromosome.

Dura mater The outermost meninx.

Dyad One of the groups of two chromosomes formed by the division of a tetrad during the first meiotic division.

Dyotropic rearrangement An uncatalyzed process in which two sigma bonds simultaneously migrate intramolecularly, e.g.

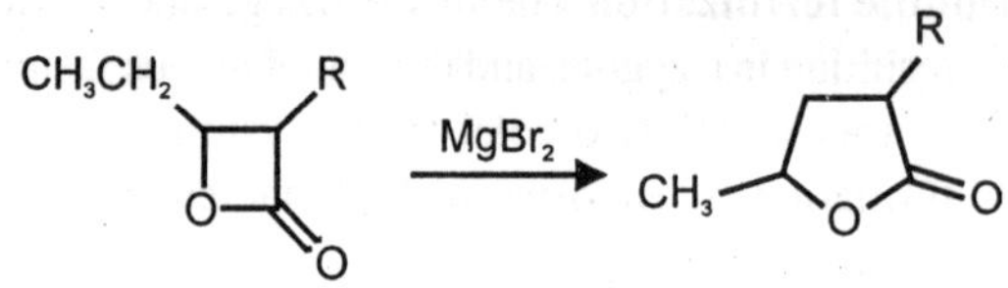

Dyspnea Difficult or labored breathing.

E

Ear stone A calcareous body in the inner ear of fishes and other vertebrates that is used in the maintenance of equilibrium.

Earlywood The large, thin-walled xylem cells produced early in the growing season, which appear less dense than latewood.

Eccrine Applies to a type of mammalian sweat gland that produces a watery secretion.

Ecdysiotropin Hormone secreted in brain of insects that stimulates prothoracic gland to secrete molting hormone. Prothoracicotropic hormone; brain hormone.

Ecdysis 1. The shedding of the arthropod exoskeleton to accommodate increased body size or a change in morphology (as may occur in molting from immature to adult); to molt; may also refer to the shedding of the outer epidermis of the skin of reptiles. 2. The shedding of the cuticle in aschelminths in order to grow. Also called molting.

Ecdysone Molting hormone of arthropods, stimulates growth and ecdysis, produced by prothoracic glands in insects and Y organs in crustaceans.

Echinodermata The phylum of coelomate animals whose members are pentaradially symmetrical as adults and possess an endoskeleton covered by epithelium and a water-vascular system. Sea stars, sea urchins, sea cucumbers, sea lilies.

Echinoderms Invertebrates with radial symmetry and a water vascular system.

Echinoidea The class of echinoderms whose members are globular or disk shaped, possess moveable spines, and a skeleton of closely fitting plates. Sea urchins and sand dollars.

Echiura A phylum of protostomate, marine animals whose members burrow in mud or sand or live in rock crevices. They possess a spatula-shaped proboscis and are 15 to 50 cm in length. Spoon worms.

Echiurans Burrowing invertebrates that have an unsegmented body and a non-retractable proboscis.

Echolocation A method of locating objects by determining the time it takes for an echo to return and the direction from which it returns. As in bat echolocation.

Eclipsed antigen Antigen borne by the parasite that is common to both the host and the parasite but that genetically is of parasite origin.

Eclosion The process whereby the adult form of an insect emerges from the pupa.

Ecocentric A philosophy that claims moral values and rights for both organisms and ecological systems and processes.

Ecocline The gradient between adjacent biomes; a gradient of environmental conditions.

Ecofeminism A pluralistic, nonhierarchical, relationship-oriented philosophy that suggests how humans could reconceive themselves and their relationships to nature in nondominating ways as an alternative to patriarchal systems of domination.

Ecojustice Justice in the social order and integrity in the natural order.

Ecological adaptation Short-term changing responses expressed by an individual organism to its environment.

Ecological development A gradual process of environmental modification by organisms.

Ecological equivalents Different species that occupy similar ecological niches in similar ecosystems in different parts of the world.

Ecological niche The functional role and position of a species (population) within a community or ecosystem, including what resources it uses, how and when it uses the resources, and how it interacts with other populations.

Ecological pyramid of biomass Total biomass of all organisms at each trophic level in a food chain; typically biomass declines with successively higher trophic levels.

Ecological pyramid of energy Total energy content of all organisms at each trophic level in a food chain; since only 5%-20% of energy is passed between trophic levels; energy content declines at successively higher trophic levels.

Ecological pyramid of numbers Number of organisms supported at each trophic level in a food chain; typically, fewer organisms are supported at successively higher trophic levels.

Ecological succession The sequential replacement of one vegetative community by another through a series of stages; succession ends when the climax community is established.

Ecology The scientific study of relationships between organisms and their environment. It is concerned with the life histories, distribution, and behavior of individual species as well as the structure and function of natural systems at the level of populations, communities, and ecosystems.

Economic defendability Defense of a resource that yields benefits that outweigh the costs of defending it.

Economic development A rise in real income per person; usually associated with new technology that increases productivity or resources.

Economic growth An increase in the total wealth of a nation; if population grows faster than the economy, there may be real economic growth, but the share per person may decline.

Economic sanctions Using the threat or action of withholding food, medical care, or other critical societal needs to force compliance with one's policies.

Ecosystem An ecological unit consisting of both the biotic communities and the nonliving (abiotic) environment, which interact to produce a stable system.

Ecosystem management An integration of ecological, economic, and social goals in a unified systems approach to resource management.

Ecosystem restoration To reinstate an entire community of organisms to as near its natural condition as possible.

Ecotage Direct action (guerrilla warfare) or sabotage in defense of nature; also called monkey wrenching.

Ecotone The transition zone between two adjacent communities.

Ecotourism A combination of adventure travel, cultural exploration, and nature appreciation in wild settings.

Ectocommensal Commensal symbiont that lives on the outer surface of its host.

Ectoderm Outer layer of cells of an early embryo (gastrula stage); one of the germ layers, also sometimes used to include tissues derived from ectoderm.

Ectognathous Derived character of most insects; mandibles and maxillae not in pouches.

Ectolecitha l Yolk for nutrition of the embryo contributed by cells that are separate from the egg cell and are combined with the zygote by envelopment within the eggshell.

Ectomycorrhizae Association of a fungus with a vascular plant root or rhizome system in which the fungus surrounds, but does not penetrate the living cells (protoplast) of the system.

Ectoneural Oral (chief) nervous system in echinoderms.

Ectoparasite Parasite that lives on the outer surface of its host.

Ectopic Infection in a location other than normal or expected.

Ectoplasm The cortex of a cell or that part of cytoplasm just under the cell surface; contrasts with endoplasm.

Ectoprocta A phylum of animals whose members are colonial and fresh water or marine. Anus ends outside a ring of tentacles. Lophophore used in feeding. Moss animals or bryozoans.

Ectotherm An organism whose internal temperature varies with that of the environment. Compare endotherm.

Ectothermic Having a variable body temperature derived from heat acquired from the environment; contrasts with endothermic.

Ectotympanic ecto = outer; tympani = a drum.

Eddington limit The theoretical limit at which the photon pressure would exceed the gravitational attraction of a light-emitting body. That is, a body emitting radiation at greater than the eddington limit would break up from its own photon pressure.

Eddy currents Closed loops of induced current set up in a piece of metal when there is relative motion between the metal and a magnetic field The eddy currents are in such direction that the resulting magnetic forces oppose the relative motion.

Edema Escape of fluid from blood into interstitial space, causing swelling.

Edge effects A change in species composition, physical conditions, or other ecological factors at the boundary between two ecosystems. Some organisms flourish at this boundary and benefit from processes such as habitat fragmentation that increase edge area. Other organisms are harmed by increasing edge effects.

Educt Term used mainly in the German literature for starting material (reactant). It should be avoided in English, because there it means "something that comes out" and not "something that goes in". The German use of the term is in fact also incorrect.

Edwards-Casimir quantum vacuum drive A hypothetical drive exploiting the peculiarities of quantum mechanics by restricting allowed wavelengths of virtual photons on one side of the drive (the bow of the ship); the pressure generated from the unrestricted virtual photons toward the aft generates a net force and propels the drive.

Effect A positive deviation of an -nucleophile (a nucleophile bearing an unshared pair of electrons on an atom adjacent to the nucleophilic site) from a Brønsted-type plot of lg knuc vs. pKa constructed for a series of related normal nucleophiles. More generally, it is the influence of the atom bearing a lone pair of electrons on the reactivity at the adjacent site. The use of the term has been extended to include the effect of any substituent on an adjacent reactive centre, for example in the case of the "-silicon effect".

Effective charge Change in effective charge is a quantity obtained by comparison of the polar effect of substituents on the free energies of rate or equilibrium processes with that on a standard ionization equilibrium. Provided the effective charge on the states in the standard equilibrium are defined, then it is possible to measure effective charges for states in the reaction or equilibrium under consideration.

Effective molarity (or effective concentration) The ratio of the first-order rate constant of an intramolecular reaction involving two functional groups within the same molecular entity to the second-order rate constant of an analogous intermolecular elementary reaction. This ratio has the dimension of concentration. The term can also apply to an equilibrium constant.

Effector An organ, tissue, or cell that becomes active in response to stimulation.

Efferent Leading or conveying away from some organ, for example, nerve impulses conducted away from the brain, or blood conveyed away from an organ; contrasts with afferent.

Effluent sewerage A low-cost alternative sewage treatment for cities in poor countries that combines some features of septic systems and centralized municipal treatment systems. A septic tank near each residence collects and processes solid wastes_and must be pumped periodically just like a septic tank_while liquid effluent is collected and piped to a central treatment plant. This avoids drainfields in urban areas and yet allows sewer pipes and treatment plants to be downsized, saving money.

Egestion Act of casting out indigestible or waste matter from the body by any normal route.

Egestion vacuole A membrane-bound vesicle within the cytoplasm of a protist that functions in expelling wastes.

Egg apparatus Egg cell and adjacent synergids in the embryo sac (female gametophyte) of angiosperms.

Egg The middle of three haploid cells at the micropylar end of the embryo sac; when fertilized, it will form the zygote.

Ehrenfest paradox The special relativistic "paradox" involving a rapidly rotating disc. Since any radial segment of the disc is perpendicular to the direction of motion, there should be no length contraction of the radius; however, since the circumference of the disc is parallel to the direction of motion, it should contract.

Eighteen-electron rule An electron-counting rule to which an overwhelming majority of stable diamagnetic transition metal complexes adhere. The number of nonbonded electrons at the metal plus the number of electrons in the metal-ligand bonds should be 18. The 18 electron rule in transition metal chemistry is a full analogue of the "Lewis octet rule".

Einstein field equation The cornerstone of Einstein's general theory of relativity, relating the gravitational tensor **g** to the stress-energy tensor **t** by the simple equation

$$\mathbf{G} = 8\pi\mathbf{t}.$$

Einstein-Podolsky-Rosen effect Consider the following quantum mechanical thought-experiment: take a particle which is at rest and has spin zero. It spontaneously decays into two fermions (spin 1/2 particles), which stream away in opposite directions at high speed. Due to the law of conservation of spin, we know that one is a spin +1/2 and the other is spin -1/2. Which one is which? According to quantum mechanics, neither takes on a definite state until it is observed (the wavefunction is collapsed). The epr effect demonstrates that if one of the particles is detected, and its spin is then measured, then the other particle — no matter where it is in the universe — instantaneously is forced to choose as well and take on the role of the other particle. This illustrates that certain kinds of quantum information travel instantaneously; not everything is limited by the speed of light.

However, it can be easily demonstrated that this effect does not make faster-than-light communication or travel possible.

Ekman spiral The spiral change in the movement of water in the water column when the water is pushed by wind. The Ekman layer is that part of the water column affected by wind; Ekman transport is the net water movement 90from the wind direction.

El Nino A climatic change marked by shifting of a large warm water pool from the western Pacific Ocean towards the east. Nutrient-rich upwelling currents along the coast of South America are blocked by this sea change and fisheries fail catastrophically. An El Niño event normally is accompanied by droughts in Australia and Southeast Asia together with heavy rain and snow in western North America.

El Nino Southern Oscillation (ENSO) Large-scale changes in the normal weather patterns of the Pacific basin and adjacent regions. El Niño is a warming of the surface currents in the eastern Pacific, only one of the many consequences of ENSO.

Elastic collision A collision in which objects rebound from each other without a loss of kinetic energy.

Elastic limit The condition in which a substance is on the verge of becoming permanently deformed.

Elastic potential energy The potential energy in a stretched or compressed elastic object.

Elasticity The ability of an object to return to its original size or shape when the external forces producing distortion are removed.

Elaters The spiral walls that aid in the release and dispersal of spores from the sporangia in horsetails (Equisetum) or the elongate, hygroscopic cells with spiral wall thickenings that aid in the release of spores in liverworts.

Electric current The rate of flow of charge past a given point in an electric circuit.

Electric field intensity The force per unit positive charge at a given point in an electric field.

Electric ground 1. A conductor connected with the earth to establish zero (ground) potential

2. A common return to an arbitrary zero potential

Electrical synapse A synapse at which local currents resulting from electrical activity flow between the two neurons through gap junctions joining them.

Electricity This word names a branch or subdivision of physics, just as other subdivisions are named 'mechanics', 'thermodynamics', 'optics', etc.

Electrification The process of charging a body by adding or removing electrons

Electrochemical cell A cell in which chemical energy is converted to electric energy by a spontaneous electron transfer reaction.

Electrochemical equivalent The mass of an element, in grams, deposited by one coulomb of electric charge.

Electrocyclic reaction A molecular rearrangement that involves the formation of a sigma bond between the termini of a fully conjugated linear pi-electron system (or a linear fragment of a pi-electron system) and a decrease by one in the number of pi bonds, or the reverse of that process. For example:

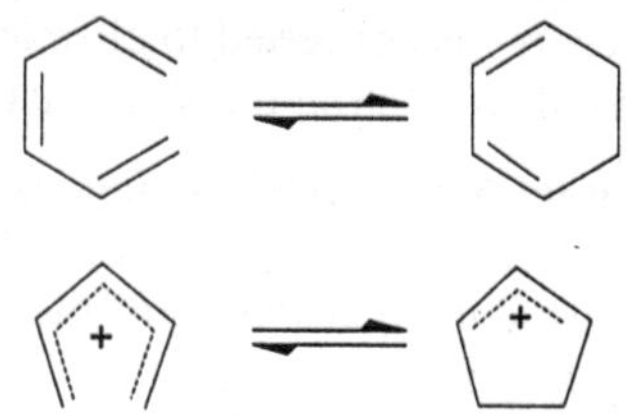

The stereochemistry of such a process is termed "conrotatory" or antarafacial if the substituents at the interacting termini of the conjugated system both rotate in the same sense, e.g.

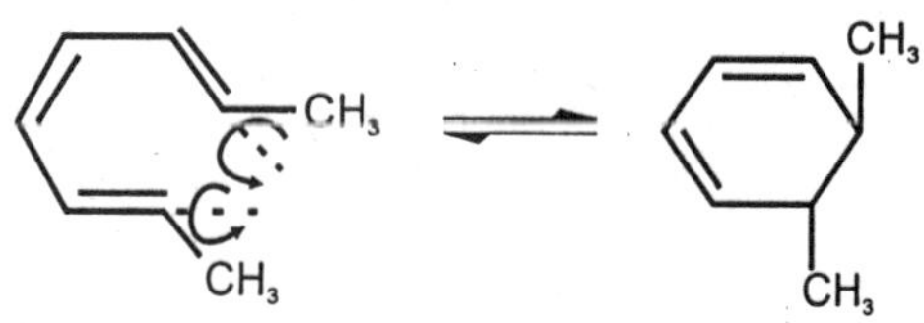

or "disrotatory" (or suprafacial) if one terminus rotates in a clockwise and the other in a counter-clockwise sense.

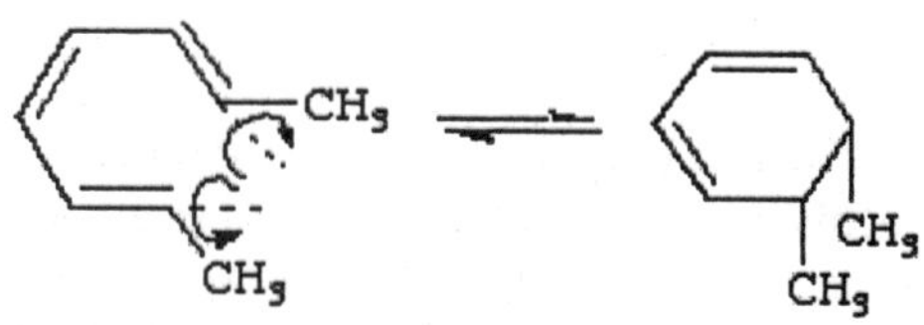

Electrode A conducting element in an electric cell, electronic tube, or semiconductor device.

Electrode stimulation The stimulation of a neuron by artificially depolarizing the membrane with an electrode.

Electrodynamic In contrast to electrostatics, this is the study of the interactions between electrical charges which are allowed to move.

Electrofuge A leaving group that does not carry away the bonding electron pair. For example, in the nitration of benzene by NO_2^+, H^+ is the electrofuge.

Electrolysis The conduction of electricity through a solution of an electrolyte or through a fused ionic compound, together with the resulting chemical changes.

Electrolyte A substance whose solution conducts an electric current Electrolytic cell A cell in which electric energy is converted to chemical energy by means of an electron-transfer reaction.

Electromagnet A magnet whose field is produced by an electric current, and which is largely demagnetized upon cessation of the current, is an electromagnet.

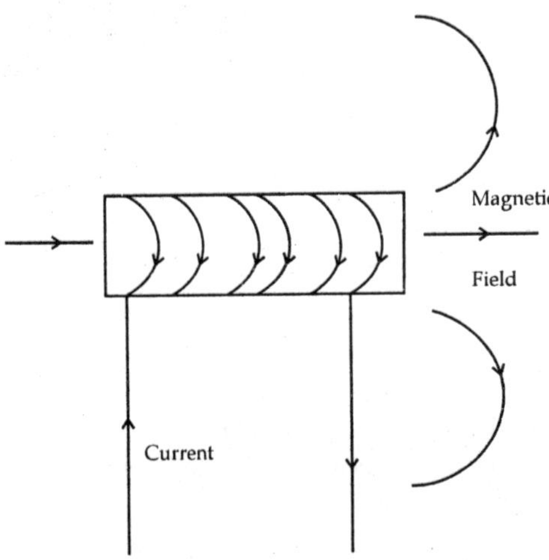

Fig. Electromagnet

In order to obtain the strongest field possible, highly permeable soft iron or steel is employed. In an electromagnet the current flows through a solenoid, which is a conductor wound in the form of a helix, and which produces a strong magnetic field coaxial with the helix. The core is placed inside the helix in order to give a magnetic path of the least reluctance. Electromagnets are found in a number of different forms, such as the plain solenoid with cylindrical core, or the horseshoe electromagnet, much used in electric bells, telegraph instruments, and telephones. Very powerful electromagnets are often used to move masses of iron, such as scrap iron, and have the advantage that the loading or unloading of the crane to which the magnet is attached is simply a matter of applying or disconnecting the electric current. For practical approximate calculations, we ignore the fringing of the field at the air gap, and compute the flux by the Bosanquet law (analogous to the Ohm law).

Electromagnetic induction The process by which an emf is set up in a conducting circuit by a changing magnetic flux linked by the circuit.

Electromagnetic interaction The interaction that keeps electrons in orbit and forms bonds between atoms and molecules.

Electromagnetic spectrum The entire radiation spectrum from the high-energy levels of cosmic rays to the low-energy levels of radiowaves; a small portion in the center provides wavelengths visible to humans.

Electromagnetic waves Transverse waves having an electric component and a magnetic component, each being perpendicular to the other and both perpendicular to the direction of propagation.

Electromeric effect A molecular polarizability effect occurring by an intramolecular electron displacement (sometimes called the "conjugative mechanism" and, previously, the "tautomeric mechanism") characterized by the substitution of one electron pair for another within the same atomic octet of electrons. It can be indicated by curved arrows symbolizing the displacement of electron pairs, as in

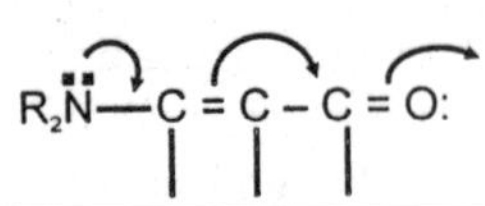

which represents the hypothetical electron shift

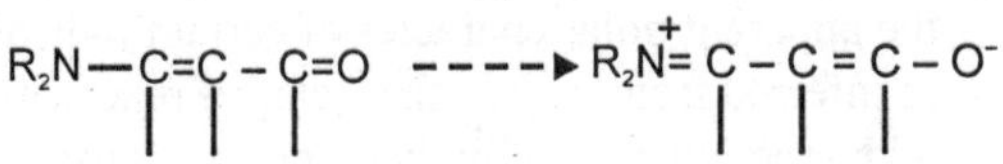

The term has been deemed obsolescent or even obsolete. It has long been custom to use phrases such as "enhanced substituent resonance effect" which imply the operation of the electromeric effect, without using the term, and various modern theoretical treatments parametrize the response of substituents to "electronic demand", which amounts to considering the electromeric effect together with the inductomeric effect.

Electron A negatively charged subatomic particle having a rest mass of 9.109534×10^{-31} kg and a charge of $1.6021892 \times 10^{-19}$ c.

Electron acceptor 1. A substance to which an electron may be transferred; for example 1,4-dinitrobenzene or the dication 1,1'-dimethyl-4,4'-bipyridyldiium.

Electron affinity The energy released when an additional electron (without excess energy) attaches itself to a molecular entity (usually an electrically neutral molecular entity). (The direct measurement of this quantity involves molecular entities in the gas phase.)

Electron attachment The transfer of an electron to a molecular entity, resulting in a molecular entity of (algebraically) increased negative charge.

Electron density If p(x,y,z) dx dy dz is the probability of finding an electron in the volume element dx dy dz at the point of a molecular entity with coordinates x,y,z, then p(x,y,z) is the electron density at this point. For many purposes (e.g., x-ray scattering, forces on atoms) the system behaves exactly as if the electrons were spread out into a continuously distributed charge. The term has frequently been wrongly applied to negative charge population.

Electron detachment The reverse of an electron attachment.

Electron donor 1. A molecular entity that can transfer an electron to another molecular entity, or to the corresponding chemical species.

Electron shell A region about the nucleus of an atom in which electrons move and which is made up of electron orbitals.

Electron transfer The transfer of an electron from one molecular entity to another, or between two localized sites in the same molecular entity.

Electron transport chain In a cell membrane, electron carriers and enzymes positioned in an organized array that enhance oxidation-reduction reactions; such systems function in the release of energy that is used in ATP formation and other reactions.

Electron Transport System The final stage of respiration; a series of enzymes and coenzymes on the inner membrane of mitochondria functioning in the transfer of electrons and the resulting synthesis of ATP.

Electron volt The energy required to move an electron between two points that have a difference af potential of one volt.

Electron-deficient bond A single bond between adjacent atoms that is formed by less than two electrons, as in B_2H_6:

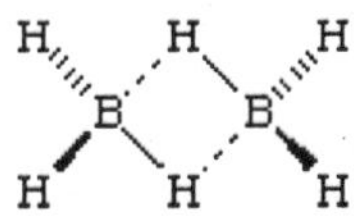

The B-H-B bonds are called a "two-electron three-centre bonds".

Electron-donor-acceptor complex A term sometimes employed instead of charge-transfer complex or Lewis adduct.

Electronegativity A measure of the power of an atom or a group of atoms to attract electrons from other parts of the same molecular entity. The concept has been quantified by a number of authors, including especially Pauling and Mulliken.

Electroneutrality principle The principle expresses the fact that all pure substances carry a net charge of zero.

Electronic effect of substituents: symbols and signs The inductive effect has universally been represented by the symbol i. This is now commonly taken to include both through-bonds and through-space transmission, but i is also used specifically for through-bonds transmission; through-space transmission is then symbolized as f (for field effect). The symbols for the influence of substituents exerted through electron delocalization have variously been m (mesomeric), e (electromeric), t (tautomeric), c (conjugative), k (konjugativ), and r (resonance). Since the present fashion is to use the term resonance effect, r is the most commonly used symbol, although m is still seen quite often.

Both the possible sign conventions are in use. The ingold sign convention associates electronegativity (relative to hydrogen atom) with a negative sign, electropositivity with a positive sign. Thus the nitro group is described as electron- withdrawing by virtue of its - i and - m effects; chloro is described as a - i, +m substituent, etc. For correlation analysis and linear free-energy relationships this convention has been found inconvenient, for it is in contradiction to the sign convention for polar substituent constants (σ-constants).

Electrophile, electrophilic An electrophile (or electrophilic reagent) is a reagent that forms a bond to its reaction partner (the nucleophile) by accepting both bonding electrons from that reaction partner.

An "electrophilic substitution reaction" is a heterolytic reaction in which the reagent supplying the entering group acts as an electrophile. For example

$C_6H_6 + NO_2^+$ (electrophile) $\rightarrow C_6H_5NO_2 + H^+$ (electrofuge)

Electrophilic reagents are Lewis acids. "Electrophilic catalysis" is catalysis by Lewis acids.

The term "electrophilic" is also used to designate the apparent polar character of certain radicals, as inferred from their higher relative reactivities with reaction sites of higher electron density.

Electrophilicity 1. The property of being electrophilic.

2. The relative reactivity of an electrophilic reagent. (It is also sometimes referred to as "electrophilic power".) Qualitatively, the concept is related to Lewis acidity. However, whereas Lewis acidity is measured by relative equilibrium constants, electrophilicity is measured by relative rate constants for reactions of different electrophilic reagents towards a common substrate (usually involving attack at a carbon atom).

Electrophoresis The separation of different proteins or nucleic acids within a gel matrix that is subjected to an electric field; separation is based on size and/or charge of the molecule(s).

Electroreceptor A receptor that senses changes in an electrical current usually in the surrounding water; also called an ampullary organ.

Electroscope A device used to observe the presence of an electrostatic charge

Electrostatic In contrast to electrodynamics, this is the study of the interactions between stationary electrical charges.

Electrostatic precipitators The most common particulate controls in power plants; fly ash particles pick up an electrostatic surface charge as they pass between large electrodes in the effluent stream, causing particles to migrate to the oppositely charged plate.

Element Building blocks of matter that cannot be broken down into a simpler substance. Each element is composed of just one type of atom.

Element effect The ratio of the rate constants of two reactions that differ only in the identity of the element of the atom in the leaving group, e.g., kBr/kCl. As for isotope effects, a ratio of unity is regarded as a "null effect".

Elementary colors The six regions of color in the solar spectrum observed by the dispersion of sunlight: red, orange, yellow, green, blue, indigo, and violet.

Elementary reaction A reaction for which no reaction intermediates have been detected or need to be postulated in order to describe the chemical reaction on a molecular scale. An elementary reaction is assumed to occur in a single step and to pass through a single transition state.

Eleocyte Fat-containing cells in annelids that originate from the chlorogogen tissue.

Elephantiasis Disfiguring condition caused by chronic infection with filarial worms Wuchereria bancrofti and Brugia malayi.

Elimination The reverse of an addition reaction or transformation.

In an elimination two groups (called eliminands) are lost most often from two different centres (1/2/elimination or 1/3/elimination, etc.) with concomitant formation of an unsaturation in the molecule (double bond, triple bond) or formation of a new ring.

If the groups are lost from a single centre (α-elimination, 1/1/elimination) the resulting product is a carbene or a "carbene analogue".

Elongation strain The ratio of the increase in length to the unstretched length.

Embryo The early developmental stage that, through embryological development, ultimately becomes an adult individual. In plants, that portion of a seed that will form the growing seedling following germination; it has a radicle, apical meristem, and embryonic leaf or leaves.

Embryo sac Female gametophyte of angiosperms; it is retained within the ovule.

Embryogenesis The origin and development of the embryo; embryogeny.

Embryology The study of development from the egg to the point that all major organ systems have formed.

Embryonic axis The main root/shoot body of a seedling.

Embryonic diapause The arresting of early development to allow young to hatch, or be born, when environmental conditions favor survival.

Embryophore In reference to the eggshell of many cestodes, that portion contributed by the inner envelope, derived from embryonic blastomeres.

Emergence The appearance of properties in a biological system (at the molecular, cellular, organismal, or species levels) that cannot be deduced from knowledge of the component parts taken separately or in partial combinations; such properties are termed emergent properties.

Emergent plant community Marine plant community in which most leafy growth occurs above sea level.

Emission spectrum A spectrum formed by the dispersion of light from an incandescent solid, liquid, and gas

Emission standards Regulations for restricting the amounts of air pollutants that can be released from specific point sources.

Emphysema An irreversible, obstructive lung disease in which airways become permanently constricted and alveoli are damaged or destroyed.

Empirical law A law strictly based on experiment, which may lack theoretical foundation.

Emulsion A colloidal system in which both phases are liquids.

Enantiomer One of a pair of molecular entities which are mirror images of each other and non-superimposable.

Encephalitis Infection of the brain, especially by viruses or amebas.

Encounter complex A complex of molecular entities produced at an encounter-controlled rate, and which occurs as an intermediate in a reaction mechanism. When the complex is formed from two molecular entities it is called an "encounter pair". A distinction between encounter pairs and (larger) encounter complexes may be relevant in some cases, e.g. for mechanisms involving pre-association.

Encounter-controlled rate A rate of reaction corresponding to the rate of encounter of the reacting molecular entities. This is also known as "diffusion-controlled rate" since rates of encounter are themselves controlled by diffusion rates (which in turn depend on the viscosity of the medium and the dimensions of the reactant molecular entities).

For a bimolecular reaction between solutes in water at 25°C an encounter-controlled rate is calculated to have a second-order rate constant of about 10^{10} dm^3 mol^{-1} s^{-1}.

Encrusting Describing an organism that grows as a crust over rocks and other hard surfaces.

Encystment The secretion of a protective outer covering that permits some small invertebrates to withstand exposure to extreme environmental stresses, such as desiccation and overcrowding.

End bulb A tiny swelling on the terminal end of telodendria at the distal end of an axon; also called the synaptic bouton.

Endangered species A living species that is in danger of becoming extinct because of small population sizes, poor reproduction, reduced available habitat, or a combination of these factors.

Endemic Peculiar to a certain region or country; native to a restricted area; not introduced.

Endemism A state in which species are restricted to a single region.

Endemnicity Amount or severity of a disease in a particular geographic area.

Endergonic Used in reference to a chemical reaction that requires energy; energy absorbing.

Endergonic reaction Chemical reaction that requires energy.

Endochondral Occurring with the substance of cartilage, especially bone formation.

Endocommensal Commensal symbiont that lives inside its host.

Endocrine A system of ductless glands secreting hormones into the bloodstream and targeting another organ to produce a response.

Endocrine glands A series of ductless glands in both invertebrates and vertebrates that release hormones into the body through the blood or lymph.

Endocrinology The study of the endocrine system and its role in the physiology of an animal.

Endocytosis Ingestion of particulate matter or fluid by phagocytosis or pinocytosis; that is, bringing material into a cell by invagination of its surface membrane and then pinching off the invaginated portion of a vacuole.

Endoderm Deep primary germ layer of the embryo. It gives rise to the linings of the pharynx, respiratory tree, digestive tract, urinary bladder, and the urethra.

Endodermis Innermost layer of the root cortex surrounding the stele; many of the endodermal cells have Casparian strips.

Endodyogeny Same as endopolyogeny except that only two daughter cells are formed.

Endogenous clock mechanism Any internal processes that are genetically based and that play a role in setting or regulating biological rhythms.

Endogenous factors Internal factors regulating an organism's growth and development.

Endogenous Processes within the animal; used here with particular reference to the internal, genetically based, components of biological rhythms.

Endognathous Ancestral character in insects, found in orders Diplura, Collembola, and Protura, in which the mandibles and maxillae are located in pouches.

Endolecithal Yolk for nutrition of the embryo incorporated into the egg cell itself.

Endolithic alga An alga that burrows into calcareous rocks or corals.

Endolymph Fluid that fills most of the membranous labyrinth of the vertebrate ear.

Endometrium The lining of the uterus; during the reproductive years, this layer is sloughed each month during menstruation.

Endomycorrhizae Association of a fungus with a vascular plant root or rhizome system in which the fungus penetrates the living cells of the root.

Endoparasite Parasite that lives inside its host.

Endophyte A plant that lives within the tissues of another plant.

Endoplasm The portion of cytoplasm that immediately surrounds the nucleus.

Endoplasmic reticulum Cytoplasmic organelle composed of a system of interconnected membranous tubules and vesicles; ER; rough ER has ribosomes attached to the side of the membrane facing the cytoplasm and smooth ER does not. Rough ER functions in protein synthesis while smooth ER functions in lipid synthesis.

Endopod, endopodite Medial branch of a biramous crustacean appendage(phylum Arthropoda).

Endopolyogeny Formation of daughter cells, each surrounded by its own membrane, while still in the mother cell.

Endopterygote Condition of internal wing bud development in an insect. Also, an insect in which the wing buds develop externally or any insect secondarily wingless but derived from such an ancestor; associated with holometabolous insects.

Endorphin (contraction of endogenous morphine). Group of opiate-like brain neuropeptides that modulate pain perception and are implicated in many other functions.

Endoskeleton A skeleton that lies beneath the surface of the body (e.g., the bony skeleton of vertebrates and the calcium carbonate skeleton of echinoderms); contrasts with exoskeleton.

Endosome Nucleolus in nucleus of some protozoa that retains its identity through mitosis.

Endosperm A triploid nutritive tissue resulting from the fusion of a haploid sperm nucleus with the two haploid polar nuclei in the ovule of angiosperms.

Endostyle Ciliated groove(s) in the floor of the pharynx of tunicates, cephalochordates, and larval jawless fishes useful for accumulating and moving food particles to the stomach.

Endosymbiont hypothesis The idea whereby the evolution of the eukaryotic cell might have occurred when a large anaerobic amoeboid prokaryote ingested small aerobic bacteria and stabilized them instead of digesting them.

Endosymbiosis The theory that some cellular organelles arose by the incorporation of a prokaryote into the cytoplasm of a eukaryote.

Endothelium A layer of epithelial cells forming a tissue that lines the circulatory system and body cavities; derived from the mesoderm of the embryo.

Endotherm An organism that can regulate its internal temperature. Compare ectotherm.

Endothermic Having a body temperature determined by heat derived from the animal's own oxidative metabolism; contrasts with ectothermic.

End-product (feedback) inhibition The inhibition of the first enzyme in a pathway by the end product of that pathway.

Ene reaction The addition of a compound with a double bond having an allylic hydrogen (the "ene") to a compound with a multiple bond (the "enophile") with transfer of the allylic hydrogen and a concomitant reorganization of the bonding, as illustrated below for propene (the "ene") and ethene (the "enophile"). The reverse is a "retro-ene" reaction.

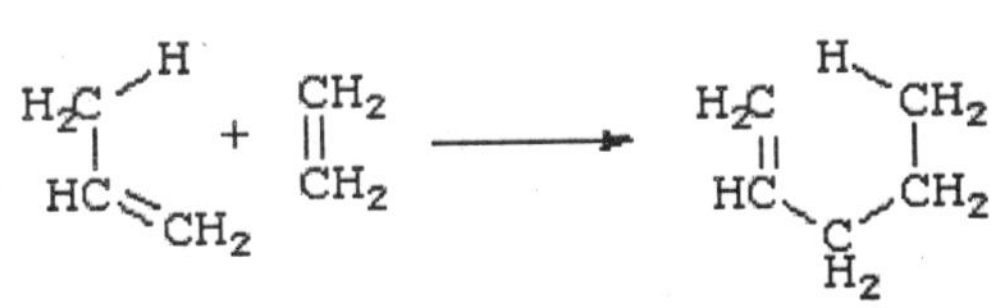

Energy Technically defined as the "ability to do work", this concept is traditionally thought of as difficult to grasp initially. The concept has to do with the colloquial idea of effort. Since it takes some effort to move an object, we say a moving object has kinetic energy. Similarly, it takes some effort to lift an object; we would like to define an energy to associate with lifting or dropping an object — this is the potential energy. The "ability to do work," then is our ability to get energy out of an object. An object loses energy by giving it to other objects. If a moving truck hits a parked car, the car gains some energy and the truck loses some; the truck has "done some work" on the car. (Energy is expressed in units of joule which were named for James Prescott Joule.)

Since it is also possible to have work done by objects which are warm, heat can be said to be a form of energy. In fact, Einstein's famous equation, $E=mc^2$, expresses that mass can also be considered a form of energy, although it is less obvious that this is the case.

Energy budget An accounting of the way in which energy coming into an ecosystem from the sun is lost or processed by organisms of the ecosystem.

Energy efficiency A measure of energy produced compared to energy consumed.

Energy gap The energy gap is the amount of energy necessary for an electron in the valence band to jump to the conduction band.

Energy level One of a series of discrete energy values that characterize a physical system governed by quantum rules.

Energy pyramid A representation of the loss of useful energy at each step in a food chain.

Energy recovery Incineration of solid waste to produce useful energy.

Enforced concerted mechanism Variation of reaction parameters in a series of reactions proceeding in non-concerted steps may lead to a situation, where the putative intermediate will possess a lifetime shorter than a bond vibration, so that the steps become concerted. The transition state structure will lie on the coordinate of the More O'Ferrall-Jencks diagram leading to that of the putative intermediate.

Engram (memory trace) The physical manifestation of learning; neural mechanism for storage of information that involves structural synaptic changes and/or chemical changes.

Enkephalin Group of small brain neuropeptides with opiate-like qualities.

Enrichment experiments Experimental manipulations that involve providing organisms with particular types of stimuli (e.g., social, sensory, motor) to ascertain the effects later in development.

Entering group An atom or group that forms a bond to what is considered to be the main part of the substrate during a reaction. For example: the attacking nucleophile in a bimolecular nucleophilic substitution reaction.

Enterocoel A type of coelom formed by the outpouching of a mesodermal sac from the endoderm of the primitive gut.

Enterocoelic mesoderm formation Embryonic formation of mesoderm by a pouchlike outfolding from the archenteron, which then expands and obliterates the blastocoel, thus forming a large cavity, the coelom, lined with mesoderm.

Enterocoelomate An animal having an enterocoel, such as an echinoderm or a vertebrate.

Enterocoely Formation of a coelom through outpocketing of the inner portion of the archenteron in some animals (deuterostomes).

Enteron The digestive cavity.

Enteropneusta A class of hemichordates whose members live in burrows in shallow marine water. Their bodies are divided into three regions: proboscis, collar, and trunk. Acorn worms.

Enthalpy The sum of the internal energy of a body and the product of its volume multiplied by the pressure.

Entine The inner layer of a pollen grain shell.

Entire A smooth leaf margin without teeth, lobes, or undulations.

Entomology Study of insects.

Entoprocta A phylum of aschelminths commonly called entoprocts.

Entozoic Living within another animal; internally parasitic (chiefly parasitic worms).

Entrainment The process by which a biological clock is set or reset by synchronizing with the period of some external, environmental stimulus.

Entropy A measure of the degree of disorganization of a system; how much energy in a system has become so dispersed (usually as heat) so that it is no longer available to do work. The higher the entropy, the more the disorder.

Environment The circumstances or conditions that surround an organism or group of organisms as well as the complex of social or cultural conditions that affect an individual or community.

Environmental ethics A search for moral values and ethical principles in human relations with the natural world.

Environmental hormones Chemical pollutants that substitute for, or interfere with, naturally-occurring hormones in our bodies; these chemicals may trigger reproductive failure, developmental abnormalities, or tumor promotion.

Environmental insults Any factor in the physical environment that inhibits the growth and/or development of an organism.

Environmental justice Combines civil rights with environmental protection to demand a safe, healthy, life-giving environment for everyone.

Environmental literacy Fluency in the principles of ecology that gives us a working knowledge of the basic grammar and underlying syntax of environmental wisdom.

Environmental resistance All the limiting factors that tend to reduce population growth rates and set the maximum allowable population size or carrying capacity of an ecosystem.

Environmental resources Anything an organism needs that can be taken from the environment.

Environmental science The systematic, scientific study of our environment as well as our role in it.

Environmental terrorism Conducting acts of terrorism through the destruction or pollution of the environment.

Environmentalism Active participation in attempts to solve environmental pollution and resource problems.

Enzyme A substance, produced by living cells, that is capable of speeding up specific chemical transformations, such as hydrolysis, oxidation, or reduction, but is unaltered itself in the process; a biological catalyst.

Enzyme-linked immunosorbent assay (ELISA) Immunodiagnostic test designed to detect the presence of fixed antibody through linkage with an enzymatic reaction.

Enzymes Molecules, usually proteins or nucleic acids, that act as catalysts in biochemical reactions.

Enzyme-substrate complex The binding of a substrate molecule to the active site of an enzyme.

Eosinophil White blood cells characterized by the presence of cytoplasmic granules that become stained by an acid dye.

Eotvos law of capillarity The surface tension γ of a liquid is related to its temperature t, the liquid's critical temperature, t*, and its density ς by

$$\gamma = 2.12\ (t^* - t)/\varsigma 3/2.$$

Ephemeral Temporary, such as vegetation that completes its life cycle in a short time.

Ephyra Refers to castlelike appearance.Miniature medusae produced by asexual budding of a scyphistoma (class Scyphoza, phylum Cnidaria). Ephyrae mature into sexually mature medusae.

Epi A prefix meaning on, upon, or over.

Epiblast An outer layer of cells in the embryo of an amniote that forms from the proliferation and movement of cells of the blastoderm.

Epiboly A spreading and thinning of ectoderm from the animal pole of an amphibian gastrula toward the vegetal pole.

Epicaridium First larval stage of the isopod suborder Epicaridea; attaches to a free-living copepod.

Epicondyle epi = above; condyl a knuckle.

Epicotyl Portion of the shoot of an angiosperm embryo or seedling above the cotyledons.

Epicuticle Thin, outermost layer of arthropod cuticle; contains sclerotin but not chitin.

Epicycle In the Ptolemaic system of astronomy, the epicycle (literally: *on the cycle* in Greek) was a geometric model to explain the variations in speed and direction of the apparent motion of the Moon, Sun, and planets. It was designed by Apollonius of Perga at the end of the 3rd century BC. In particular it explained retrograde motion.

In the Ptolemaic system, the planets are assumed to move in a small circle, called an epicycle, which in turn moves along a larger circle called a deferent. Both circles rotate counterclockwise and are roughly parallel to the Earth's plane of orbit (ecliptic). The orbits of planets in this system are epitrochoids, and the point around which an epicyclic path revolves is the equant.

The deferent would be considered to be centered on the Earth. As viewed from Earth, the planets were seen as mostly moving eastward along the deferent. Half of the time, the added motion along the epicycle was eastward, in parallel with the eastward movement of the epicycle on the deference. However, at times the planet would move along the epicycle in an opposite direction to the motion of the epicycle along the deferent. This would cause the planet to slow down and reverse course, ie. retrogradation.

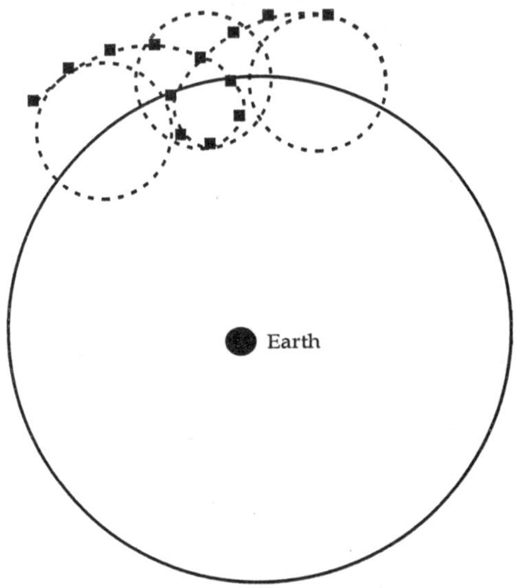

Fig. Epicycle

Over time more levels of epicycles (circles within circles) were added to the models, to match more accurately the observed planetary motions. Yet more levels of circles were not found sufficient in some cases, and "offsets" or "bars" were added to the models to compensate. This gradual process of going from a simple model to a complex model is sometimes used as an allegory for some modern technical "boondoggles". To be fair, however, the actual proliferation of epicycles that occurred has sometimes been exaggerated in modern times. During the time when the Ptolemaic system was in use people had neither the observational precision nor the mathematical methods necessary to recognize and "correct" the deviations of the planets' motions from the predictions of relatively simple epicyclic models. Epsilon 1. Dielectric coefficient or capacitivity (å).

2. Permittivity (ε).

3. Permittivity of free space (ε0).

4. Relative permittivity (εr).

5. Emittance (ε).

6. Emissivity (ε)

7. Total emissivity (ε or εt).

8. Spectral emissivity (ε).

9. Extinction coefficient (ε).

10. Self-energy (E). Luminous energy (ε)

11. Natural logarithmic base (ε but e is preferred).

12. Epoch angle (ε).

13. Eccentricity (ε or e).

Epidemic Sharp rise in the incidence of an infection or disease.

Epidemic hemorrhagic fever Virus disease transmitted by mosquitoes. Also called dengue.

Epidemiology Study concerned with all ecological aspects of a disease to explain its transmission, distribution, prevalence, and incidence.

Epidermis The outer, nonvascular layer of skin of ectodermal origin; in invertebrates, a single layer of ectodermal epithelium.

Epididymis Part of the sperm duct that is coiled and lying near the testis.

Epifauna Animals that live on the surface of the substrate. Compare infauna.

Epigenesis The embryological (and generally accepted) view that an embryo is a new creation that develops and differentiates step by step from an initial stage; the progressive production of new parts that were nonexistent as such in the original zygote.

Epigynous Floral parts (sepals, petals, and stamens) appear to arise from the top of an ovary; the ovary is said to be inferior.

Epimastigote Trypanosomatid flagellate similar to a promastigote but with a short undulating membrane, such as in Blastocrithidia.

Epimer A diastereoisomer that has the opposite configuration at only one of two or more tetrahedral "stereogenic" centres present in the respective molecular entity.

Epimerization Interconversion of epimers by reversal of the configuration at one of the "stereogenic" centres.

Epinasty The unequal growth of petioles causing the leaf blade to curve downward.

Epipelagic zone The pelagic environment from the surface to a depth of 100 to 200 m (350 to 650 ft).

Epiphysealdisk A cartilaginous growth line at the ends of long bones.

Epiphyte Plant that grows on top of another plant for support and position, but derives no nutrition from the host.

Epipod, epipodite A lateral process on the protopod of a crustacean appendage, often modified as a gill.

Epistasis Prevention of expression of an allele at one locus by an allele at another locus.

Epistome Flap over the mouth in some lophophorates bearing the protocoel.

Epitheca the larger portion of a diatom frustule

Epithelial tissue The cellular covering of internal and external surfaces of the body; consists of cells joined by small amounts of cementing substances. Epithelium is classified into types based on the number of layers deep and the shape of the superficial cells.

Epithelium A cellular tissue covering a free surface or lining a tube or cavity.

Epitoke Posterior part of a marine polychaete when swollen with developing gonads during the breeding season; contrast with atoke.

Epitoky The formation of a reproductive individual (epitoke) that differs from the nonreproductive (atoke) form of that species.

Epitope Antigenic determinant; the portion of the antigen molecule displayed on the surface of an antigen-presenting cell (APC).

Epizootic Massive infection rate among animals other than humans; identical to an epidemic in humans.

Equation With respect to comparative learning, the attempts to match situations and procedures for examining learning behavior in different species.

Equation of continuity An equation which states that a fluid flowing through a pipe flows at a rate which is inversely proportional to the cross-sectional area of the pipe. That is, if the pipe constricts, the fluid flows faster; if it widens, the fluid flows slower. It is in essence a restatement of the consevation of mass during constant flow.

Equatorial currents Major ocean currents that move parallel to the Equator.

Equilibrium The state of a body in which there is no change in its motion

Equilibrant force The force that produces equilibrium.

Equilibrium community Also called a disclimax community; a community subject to periodic disruptions, usually by fire, that prevent it from reaching a climax stage.

Equilibrium position Midpoint of the path of an object describing simple harmonic motion.

Equilibrium vapor pressure The pressure exerted by vapor molecules in equilibrium with a liquid.

Equilibrium, chemical Reversible processes (processes which may be made to proceed in the forward or reverse direction by the (infinitesimal) change of one variable, ultimately reach a point where the rates in both directions are identical, so that the system gives the appearance of having a static composition at which the gibbs energy, g, is a minimum.

Equinox As the orientation of the Earth (relative to the sun) varies through the year, the sun appears to move across the sky a little higher in the sky or a little lower. During the shorter days of winter, the sun is lower; during the longer days of summer, the sun is higher. Halfway between the extremes are the times when the night is as long as the day. The word equinox is rooted in the Latin aequinoctium, from aequi- (equal) and nox (night). The equinox marks the half-way point between the longest day of summer and the shortest day of winter.

Equivalence principle The basic postulate of a. Einstein's general theory of relativity, which posits that an acceleration is fundamentally indistinguishable from a gravitational field. In other words, if you are in an elevator which is utterly sealed and protected from the outside, so that you cannot "peek outside," then if you feel a force (weight), it is fundamentally impossible for you to say whether the elevator is present in a gravitational field, or whether the elevator has rockets attached to it and is accelerating "upward." he equivalence principle predicts interesting general relativistic effects because not only are the two indistinguishable to human observers, but also to the universe as well — any effect that takes place when an observer is accelerating should also take place in a gravitational field, and vice versa.

Ergosphere The region around a rotating black hole, between the event horizon and the static limit, where rotational energy can be extracted from the black hole.

Ergosterol The most common sterol; can be converted to vitamin D in the skin of animals exposed to sunlight.

Ergot The spore-producing reproductive body of a fungus that infects grain crops; ergot contains lysergic acid and ergonovine.

Error analysis The mathematical analysis done to show quantitatively how uncertainties in data produce uncertainty in calculated results, and to find the sizes of the uncertainty in the results.

Erythroblastosis fetalis A disease of newborn infants caused when Rh-negative mothers develop antibodies against the Rh-positive blood of the fetus.

Erythrocyte Red blood cell; has hemoglobin to carry oxygen from lungs or gills to tissues; during formation in mammals, erythrocytes lose their nuclei, those of other vertebrates retain the nuclei.

Esophagus The passage extending from the pharynx to the stomach.

Espundia Disease caused by Leishmania braziliensis. Also called chiclero ulcer, uta, pian bois, or mucocutaneous leishmaniasis.

Essential amino acid Amino acid that cannot be synthesized by an organism but must be obtained ready-made in the diet for proper health.

Essential nutrient Nutrient that cannot be synthesized by an organism but must be obtained ready-made in the diet for proper health.

Essential oils Highly volatile and aromatic oils formed in glands or special cells by some plants; probably involved in pollinator attraction or repulsion of herbivores; used in perfumes, soaps, medicine, and food.

Esthete Light sensory receptor on a shell of a chiton (phylum Mollusca).

Estrogen The female sex hormone in vertebrates.

Estrone An animal steroid found in some plants in low concentration.

Estrous The period of highest sexual receptivity (or "heat") in some female mammals that coincides with the time of egg release by the ovarymarked by intense sexual urges.

Estrus cycle A recurrent series of changes in the reproductive physiology of female mammals other than primates; females are receptive, physiologically and behaviorally, to the male only at certain times in this cycle.

Estuary A bay or drowned valley where a river empties into the sea. Fresh water mingling with salt water brings in sediment and nutrients and creates a gradient of salinity that makes estuaries among the most diverse and biologically productive ecosystems on earth.

Ethanol Ethyl alcohol is the end product of anaerobic respiration (fermentation) in plants.

Ethmoid Like a sieve.

Ethogram An inventory of all of the behavior patterns of a species.

Ethologist A person who studies the whole patterns of animal behavior in natural environments, stressing the analysis of adaptation and the evolution of the patterns.

Ethology The study of whole patterns of animal behavior in natural environments, stressing the analysis of adaptation and the evolution of the patterns.

Ethylene Gaseous plant hormone involved in fruit ripening and other aspects of plant growth and development.

Etiolation The abnormal elongation of stems caused by insufficient light or unbalanced hormonal relationships. Etiolated stems often lack chlorophyll.

Eucaryotes cells characterized by an organized nucleus and other membrane-bound subcellular structures

Euchromatic regions Less densely staining regions of chromosomes that contain active genes.

Euchromatin Part of the chromatin that takes up stain less than heterochromatin, contains active genes.

Euglenophyta A group of unicellular protists, usually green or colorless with two flagella per cell.

Eukaryote Having a true nucleus; a cell that has membranous organelles, most notably the nucleus.

Eukaryotic cell A cell containing a membrane-bounded nucleus and membrane-bounded organelles.Compare prokaryotic cell.

Euploidy Change in chromosome number from one generation to the next in which there is an addition or deletion of a complete set of chromosomes in the progeny; the most common type is polyploidy.

Euryhaline Able to tolerate wide ranges of saltwater concentrations.

Euryphagous Eating a large variety of foods.

Eurytopic Refers to an organism with a wide environmental range.

Eusocial A social system involving reproductive division of labor (i.e., castes) and cooperative rearing of young by members of previous generations.

Eutectic chemicals Phase-changing chemicals used in heat storage systems to store a large amount of energy in a small volume.

Eutely Condition where the body is composed of a constant number of cells or nuclei in all adult members of a species (e.g., rotifers, some nematodes, and acanthocephalans).

Eutrophic Rivers and lakes rich in organisms and organic material (eu 5 truly; trophic 5 nutritious).

Eutrophication Succession in a lake. The gradual accumulation of nutrients and plant and animal growth that eventually transforms a lake into a marsh and then a forest.

Evagination An outpocketing from a hollow structure.

Evaporation The escape of molecules from the liquid phase into the gaseous phase, or vapor.

Evaporative cooling The lower speed and hence the lower temperature of molecules remaining in the liquid phase after evaporation of the fastest molecules.

Evapotranspiration The combined water loss from both leaf surfaces and from the soil surface; the sum of transpiration and evaporation.

Event horizon The radius that a spherical mass must be compressed to in order to transform it into a black hole, or the radius at which time and space switch responsibilities. Once inside the event horizon, it is fundamentally impossible to escape to the outside. Furthermore, nothing can prevent a particle from hitting the singularity in a very short amount of proper time once it has entered the horizon. In this sense, the event horizon is a "point of no return."

The radius of the event horizon, R, for generalized black holes (in geometrized units) is

$R = m + (m^2 - q^2 - s/m^2)1/2$,

Where m is the mass of the hole, q is its electric charge, and s is its angular momentum.

Evergreens Coniferous trees and broad-leaved plants that retain their leaves year-round.

Evisceration The expulsion of internal organs in sea cucumbers.

Evolution Change over time. Organic or biological evolution is a series of changes in the genetic composition of a population over time.

Evolutionarily stable strategy A strategy that when employed by most individuals in a population cannot be outcompeted by some alternative strategy.

Evolutionary adaptation The changes occurring in a population of individuals over many generations by processes of natural selection.

Evolutionary duration The length of time that a species or higher taxon exists in geological time.

Evolutionary species concept A single lineage of ancestral-descendant populations that maintains its identity from other such lineages and has its own evolutionary tendencies and historical fate; differs from the biological species concept by explicitly including a time dimension and including asexual lineages.

Evolutionary systematics The study of the classification of, and evolutionary relationships among, animals; evolutionary systematists attempt to reconstruct evolutionary pathways based on resemblances between animals that result from common ancestry.

Evolutionary taxonomy A system of classification, formalized by George Gaylord Simpson, that groups species into Linnean higher taxa representing a hierarchy of distinct adaptive zones; such taxa may be monophyletic or paraphyletic but not polyphyletic.

Excimer An excited dimer, "non-bonding" in the ground state. For example, a complex formed by the interaction of an excited molecular entity with a ground state counterpart of this entity.

Exciplex An electronically excited complex of definite stoichiometry, "non-bonding" in the ground state. For example, a complex formed by the interaction of an excited molecular entity with a ground state counterpart of a different structure.

Excision repair Means by which cells are able to repair certain kinds of damage (dimerized pyrimidines) in their DNA.

Excitation Electron movement caused by light striking the chlorophyll molecule.

Excited state State of a system with energy higher than that of the ground state. This term is most commonly used to characterize a molecule in one of its electronically excited states, but can also refer to vibrational and/or rotational excitation in the electronic ground state.

Exclusion principle No two electrons in an atom can have the same set of quantum numbers

Exclusive economic zone (EEZ) A zone 200 nautical miles (370 km) wide along the coast where nations have exclusive rights to any resource.

Excretion The act, process, or function of excreting. The elimination of metabolic waste products from an animal's body.

Exergonic Characterized or accompanied by the release of energy; said of chemical reactions that release energy, so that the products have a lower free energy than the reactants.

Exergonic reaction Chemical reactions that release energy.

Exflagellation Rapid formation of microgametes from a microgametocyte of Plasmodium and related genera.

Exhaustible resources Generally considered the earth's geologic endowment: minerals, nonmineral resources, fossil fuels, and other materials present in fixed amounts in the environment.

Exine The hard outer coat of a pollen grain.

Existence value The importance we place on just knowing that a particular species or a specific organism exists.

Exite Lateral process or joint from the protopod, sometimes referred to as an epipod.

Exobiology The study of evidence relative to the possibility of life on other planets.

Exocarp Outermost layer in the pericarp (fruit wall).

Exocrine A type of gland that releases its secretion through a duct; contrasts with endocrine.

Exocrine gland A gland that secretes its product to an epithelial surface, directly or through ducts.

Exocytosis The process by which substances are moved out of a cell; the substances are transported in the cytoplasmic vesicles, the surrounding membrane of which merges with the plasma membrane in such a way that the substances are dumped outside.

Exogenous factors External factors regulating an organism's growth and development.

Exon Part of the mRNA as transcribed from the DNA that contains a portion of the information necessary for final gene product.

Exoskeleton A supporting structure secreted by ectoderm or epidermis; external, not enveloped by living tissue, as opposed to endoskeleton.

Exothermic Referring to a process that liberates energy.

Experience All of the interactions between an organism and its environment, beginning at conception and including both external and internal influences.

Experiment An artificially created situation that is used to test a hypothesis. In a controlled experiment, additional variables that might affect the experiment are prevented from doing so.

Experimental design The prescription of the treatment groups, dependent and independent variables, and sample sizes for the testing of a particular hypothesis.

Experimental error The uncertainty in the value of a quantity. This may be found from 1. statistical analysis of the scatter of data, or

2. mathematical analysis showing how data uncertainties affect the uncertainty of calculated results.

Experimental hypotheses The biological ideas, both broad and specific, developed from knowledge about previous investigations and our own imagination and creativity, that lead to tests of statistical null hypotheses using experimental designs.

Exploitation A type of competition in which organisms passively use up limited resources; also called scramble competition. Contrast with interference or contest competition, in which organisms defend or otherwise control resources.

Exploratory behavior The spontaneous search for and active investigation of objects, situations, or other organisms in the absence of any homeostatic need.

Exponential growth Growth at a constant rate of increase per unit of time; can be expressed as a constant fraction or exponent.

Extended Hammett equation This term applies in a general way to any multiparametric extension of the Hammett equation. It is sometimes used specifically for a form of dual substituent-parameter equation in which the actual value of the correlated property p under the influence of the substituent x is used, rather than the value relative to that for x = h. An intercept term h corresponding to the value of p for x = h is introduced, e.g.

$$P = \alpha \sigma_i + \beta_r + h$$

The equation may be applied to systems for which the inclusion of further terms to represent other effects, e.g. Steric, is appropriate.

Extensive property A measurable property of a thermodynamic system is extensive if, when two identical systems are combined into one, the value of that property of the combined system is double its original value in each system. Examples: mass, volume, number of moles.

Extermination The extinction of a species that is caused by humans.

External auditory canal The sound channel connecting the external and middle ears.

External combustion engine A heat engine in which the fuel burns outside the cylinder or turbine chamber.

External costs Expenses, monetary or otherwise, borne by someone other than the individuals or groups who use a resource.

Exteroception Perception of information provided by the exteroceptors.

Extinction 1. The irrevocable elimination of species; can be a normal process of the natural world as species out-compete or kill off others or as environmental conditions change.

2. In learning, the decrease of response rate or magnitude of response with lack of reinforcement in a learning situation.

Extirpate To destroy totally; extinction caused by direct human action, such as hunting, trapping, etc.

Extracellular digestion The digestion that takes place outside cells, usually in a gut, or digestive cavity. Compare intracellular digestion.

Extrinsic On the outside; related to some region other than the organ with which it is associated.

Extrusion transformation A transformation in which an atom or group Y connected to two other atoms or groups X and Z is lost from a molecule, leading to a product in which X is bonded to Z.

$$X{-}Y{-}Z \longrightarrow X{-}Z + Y$$

or

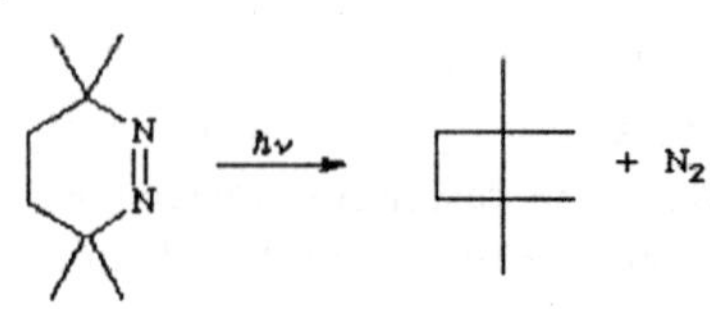

The reverse of an extrusion is called insertion.

F

F1 generation First filial generation; offspring from a genetic cross.

F2 generation Second filial generation of a genetic cross.

Facet A small, facelike articular surface.

Facette Funnel-shaped opening through the inner membrane complex of the egg of a pentastomid. It receives the product of the dorsal organ.

Facilitated diffusion Mediated transport in which a permease makes possible diffusion of a molecule across a cell membrane in the direction of a concentration gradient; contrast with active transport.

Factor One of several things multiplied together.

Facultative anaerobes Microorganisms capable of switching pathways of respiration, depending on the presence or absence

Facultative symbiont Opportunistic symbiont, establishing a relationship with a host only if the opportunity presents itself but not physiologically dependent on doing so.

FAD Abbreviation for flavine adenine dinucleotide, an electron acceptor in the respiratory chain.

Faint, Young sun paradox Theories of stellar evolution indicate that as stars mature on the main sequence, they grow steadily hotter and brighter; calculations suggest that at about the time of the formation of earth, the sun was roughly two-thirds the brightness that it is now. However, there is no geological evidence on earth (or on mars) for the sun being fainter in the past. At present there is no clear resolution for this paradox.

Family The level of classification between order and genus.

Family planning Controlling reproduction; planning the timing of birth and having as many babies as are wanted and can be supported.

Famines Acute food shortages characterized by large-scale loss of life, social disruption, and economic chaos.

Farad The derived SI unit of capacitance, defined as the capacitance in a capacitor that, if charged to 1 C, has a potential difference of 1 V; thus, it has units of C/V.

Faraday constant The electric charge carried by one mole of electrons (or singly-ionized ions). It is equal to the product of the avogadro constant and the (absolute value of the) charge on an electron; it is 9.648670×10^4 C/mol.

Faraday's first law The mass of an element deposited during electrolysis is proportional to the quantity of charge that passes through the electrolytic cell.

Faraday's first law of electrolysis The amount of chemical change during electrolysis is proportional to the charge passed.

Faraday's first law of electromagnetic induction An electromotive force is induced in a conductor when the magnetic field surrounding it changes.

Faraday's second law The mass of an element deposited during electrolysis is proportional to the chemical equivalent of that element.

Faraday's second law of electrolysis The charge q required to deposit or liberate a mass m is proportional to the charge z of the ion, the mass, and inversely proprtional to the relative ionic mass m.

Faraday's second law of electromagnetic induction The magnitude of the electromotive force is proportional to the rate of change of the field.

Faraday's third law of electromagnetic induction The sense of the induced electromotive force depends on the direction of the rate of the change of the field.

Fascicle 1. Stylet bundle or combination of mouthparts used to pierce the skin in a blood-feeding arthropod. Composition of a fascicle varies according to group.

2. A small bundle, usually referring to a collection of muscle fibers or nerve axons.

Fascicular cambium The layer of cambium that develops between the xylem and phloem within a vascular bundle.

Fast-atom bombardment (FAB) mass spectroscopy A method in which ions are produced in a mass spectrometer from nonvolatile or thermally fragile organic molecules by bombarding the compound in the condensed phase with energy-rich neutral particles.

Fat Triglyceride that is solid at room temperature; usually of animal origin. Organic molecules containing high levels of carbon and hydrogen, but little oxygen. Oils are merely fats in liquid state.

Fatigue Loss of efficiency in the performance of a motor act when that act is repeated in rapid succession.

Fat-soluble vitamin Vitamins that can be stored in fatty tissues of the body; vitamins A, D, E, and K.

Fatty acid Any of a series of saturated organic acids having the general formula $CnH_{2n}O_2$, occurs in natural fats of animals and plants.

Fault A crack in the earth's crust usually formed when two pieces of crust are moving past each other.

Fauna All of the animals present in a given region.

Fecundity The physical ability to reproduce.

Feedback loops A sequence of events in which the level of a hormone or related endocrine product circulating in the blood leads to alterations in the rate of production and release of other hormones from one or more endocrine glands. Often, feedback loops involve the hypothalamic portion of the brain where specialized sensory cells monitor circulating blood levels of numerous compounds, including both hormones and products from hormone actions at various body locations.

Feedback mechanisms control mechanisms in organisms and communities in which a change in a given factor either inhibits or stimulates processes controlling the production, release, or use of that factor

Female defense polygyny Males controlling access to females directly by competing with other males.

Femur Podomere of an insect or acarine leg fixed to the trochanter proximally and articulating with the tibia distally in insects and with the patella in acarines.

Fen An area of waterlogged soil that tends to be peaty; fed mainly by upwelling water; low productivity.

Feral A domestic animal that has taken up a wild existence.

Fermat's principle; principle of least time The principle, put forth by De fermat, that states the path taken by a ray of light between any two points in a system is always the path that takes the least time.

Fermentation Degradative pathway that begins with glycolysis and ends with the electrons being transferred back to one of the breakdown products or intermediates such as alcohols, acids, and carbon dioxide; does not require molecular oxygen.

Fermi paradox E. Fermi's conjecture, simplified with the phrase, "where are they?" Questioning that if the galaxy is filled with intelligent and technological civilizations, why haven't they come to us yet? There are several possible answers to this question, but since we only have the vaguest idea what the right conditions for life and intelligence in our galaxy, it and Fermi's paradox are no more than speculation.

Fermion Particles which have half-integralspin angular momentum are shown to behave differently than particles with integral spin angular momentum. Statistical mechanics can be used to describe the particles. The statistical rules which express how fermions interact are called Fermi-Dirac statistics. Fermions are further categorized as leptons and hadrons. Hadrons are further categorized as mesons and baryons.

Ferromagnetism The property of a substance by which it is strongly attracted by a magnet.

Fertile Crescent Area between the Tigris and Euphrates Rivers in the Near East; some of the earliest documented sites of agriculture.

Fertility Measurement of actual number of offspring produced through sexual reproduction; usually described in terms of number of offspring of females, since paternity can be difficult to determine.

Fertilization membrane A membrane that raises off the surface of an egg after sperm penetration; prevents multiple fertilization.

Fertilizer burn When the concentration of solutes in the soil is too high due to over fertilization, plasmolysis at the roots occurs, damaging the plant.

Festoons Sclerites on the posterior margin of the opisthosoma of certain hard ticks.

Fetch The span of the sea surface over which the wind blows to form wind-driven waves.

Fetus Postembryonic unborn. From the seventh or eighth week of gestation.

Feynman diagram A diagram showing the production and exchange of particles during a subatomic interaction.

Fiber Long and narrow sclerenchymous cell; functions in support; an important dietary component that provides bulk; component of fabrics, ropes, and paper.

Fibril A strand of protoplasm produced by a cell and lying within the cell.

Fibrillar Composed of or pertaining to fibrils or fibers.

Fibrillar flight muscle Insect flight muscle responsible for indirect flight. A single nerve impulse results in many cycles of flight muscle contraction and relaxation.

Fibrin Protein that forms a meshwork, trapping erythrocytes, to become blood clot. Precursor is fibrinogen.

Fibrocartilage The type of cartilage made up of parallel, thick, compact bundles, separated by narrow clefts containing typical cartilage cells (chondrocytes).

Fibrosis Deposition of fibrous connective tissue in a localized site, during process of tissue repair or to wall off a source of antigen. The general name for accumulation of scar tissue in the lung.

Fibrous connective tissue The tissue that is made up of fibers that are very densely packed (e.g., tendons and ligaments).

Fibrous roots A root system with many equally sized roots forming a mat, as in grasses. There is no primary taproot.

Fibula fibul = buckle. A leg bone.

Fiddlehead The curled fern frond prior to unrolling and elongation; also known as a crozier.

Fidelity A principle that forbids misleading or deceiving any creature capable of being mislead or deceived. We are to be truthful in our dealings with others.

Field 1. The study of physics is categorized into fields which group related associations between physical phenomenon. The broadest classification splits Classical Physics from Modern Physics, making a division between the focus of study prior to the decades surrounding the turn of the 18th to 19th century and the focus of study since the decades surrounding the turn of the 18th to 19th century. Classical Physics vaguely refers to Classical Mechanics, Electricity & Magnetism, Thermodynamics, and Optics. Modern Physics vaguely refers to Atomic, Nuclear, Particle, and Condensed Matter (formerly known as solid state). There is no hard-and-fast rule about which topic covers which phenomenon; however, any topic which requires the use of quantum mechanics is generally considered modern; anything else is classical. Indeed, many fields are beginning to overlap as our computational prowess allows more and more complicated models. While the majority of research physicists focus on sub-field of modern physics, there is still work being done in classical physics.

Field 2. With the observation that the gravitational interaction (for example) is able to tug on massive objects without physical contact, Sir Isaac Newton invoked the principle of action-at-a-distance. While he was not happy with "an invisible hand" reaching across the cosmos, there was no physical connection to be made. Further, there is obviously a force at work, since a dropped object speeds up (accelerates) and thereby gains kinetic energy. Since people must do work to impart energy (it takes some effort to move objects), we need to explain where the gravitational interaction stores its energy which it gives to a dropped object as it falls. (People get their energy from the chemical reactions of digesting food.) The idea is that the fundamental interactions (not just gravity) store energy in the region between the charges. This energy storage is the interaction field. When a charge is placed in the field of another charge, it feels a force and is either attracted or repelled by that other charge.

Field capacity The soil-water storage capacity; the saturated soil profile after gravitational percolation ceases to flow.

Field effect An experimentally observable effect symbolized by F (on reaction rates, etc.) of intramolecular coulombic interaction between the centre of interest and a remote unipole or dipole, by direct action through space rather than through bonds. The magnitude of the field effect (or "direct effect") depends on the unipolar charge/dipole moment, orientation of dipole, shortest distance between the centre of interest and the remote unipole or dipole, and on the effective dielectric constant. An approach to its calculation was described by Kirkwood and Westheimer (1938) and has been elaborated in more recent years.

Filament Part of the stamen in a flower that supports the anther.

Filamentous A slender or threadlike structure or organism, in the latter consisting of a chain of interconnected cells as in filamentous algae.

Filial An offspring generation, for example, F1, the first filial generation.

Filial imprinting The process by which young animals form a social attachment for a particular stimulus, often the mother.

Filiform Thread-like.

Filipodium A type of pseudopodium that is very slender and may branch but does not rejoin to form a mesh.

Filoplume feather A small thin feather that probably has sensory functions in birds (pinfeather).

Filters A porous mesh of cotton cloth, spun glass fibers, or asbestos-cellulose that allows air or liquid to pass through but holds back solid particles.

Filter feeding Any feeding process by which particulate food is filtered from water in which it is suspended.

Filtration Movement of material through a membrane as a result of hydrostatic pressure.

Fin ray Each of the bony spines in the fins of bony fishes.

Finite resources Resources that have a limit to their availability; not boundless.

Finlet small median fin on the dorsal and ventral sides of the rear parts of tuna and similar fish

Fire ecology The study of the environmental effects of fire.

Fire-climax community An equilibrium community maintained by periodic fires; examples include grasslands, chaparral shrubland, and some pine forests.

First law of photoelectric emission The rate of emission of photoelectrons is directly proportional to the intensity of the incident light.

First law of thermodynamics The total amount of energy in the universe remains constant; more energy cannot be created and existing energy cannot be destroyed; energy can only undergo conversion from one form to another.

First World The industrialized capitalist or market-economy countries of Western Europe, North America, Japan, Australia, and New Zealand.

Fish meal A fish-protein supplement used in animal feeds.

Fission Asexual reproduction in which the cell divides into two (binary fission) or more (multiple fission) daughter parts, each of which becomes an individual organism.

Fitness Degree of adjustment and suitability for a particular environment. Genetic fitness is relative contribution of one genetically distinct organism to the next generation; organisms with high genetic fitness are naturally selected and become prevalent in a population.

Fixed action pattern (FAP) An innate behavior pattern that is stereotyped, spontaneous, and independent of immediate control, genetically encoded, and independent of individual learning.

Fizeau method One of the first truly relativistic experiments, intended to measure the speed of light. Light is passed through a spinning cogwheel driven by running water, is reflected off a distant mirror, and then passed back through the spinning cogwheel. When the rate of running water (and thus the spinning of the cogwheel) is synchronized so that the returning pulses are eclipsed, c can be calculated.

Fjord A deep coastal embayment caused by glacial erosion

Flabellum Recurved process often found on the first two thoracic exopods of branchiuran crustaceans.

Flaccid Pertaining to a cell or tissue with less than full

Flagella Elongated appendages of certain cells used in locomotion.

Flagellar pocket Depression, sometimes long and deep, from which a flagellum arises.

Flagellum Whiplike cellular structure of motility; eukaryotic flagella are composed of microtubules.

Flagging behavior Alarm signaling, as with the use of the tail.

Flame cell Specialized hollow excretory or osmoregulatory structure of one or several small cells containing a tuft of flagella (the "flame") and situated at the end of a minute tubule; connected tubules ultimately open to the outside.

Flash tube The ionization tube that emits the light in a chemical laser

Flash vacuum pyrolysis (FVP) Thermal reaction of a molecule by exposing it to a short thermal shock at high temperature, usually in the gas phase.

Flatworms (phylum Platyhelminthes) Invertebrates that are dorsoventrally flattened and have an incomplete digestive tract, true organs, and organ systems.

Flavin adenine dinucleotide A coenzyme that is a condensation product of riboflavin phosphate and adenylic acid; it forms the prosthetic group of certain enzymes; FAD.

Flavin mononucleotide An electron acceptor in the electron transport scheme or aerobic respiration.

Flavonoids A group of secondary compounds produced by plants and important in chemical identification of those plants; believed to be contained in petals that reflect ultraviolet patterns and thus an element in pollinator attraction.

Fleming valve The first vacuum-tube diode.

Fleshy fruit Fruit in which the cells of the pericarp are alive at maturity.

Flexion flect = bend.

Flocculate A process by which dissolved substances come out of solution and aggregate together to form macroscopic particles.

Flood An overflow of water onto land that normally is dry.

Floodplains Low lands along riverbanks, lakes, and coastlines subjected to periodic inundation.

Flora All of the plants present in a given region.

Floret One of the small flowers that make up the inflorescences in the composite and grass families.

Florigen The hypothetical flowering hormone. This compound (or compounds) has never been identified chemically.

Flower The reproductive structure of the anthophyta or angiosperms.

Flowering plants Plants that have flowers, seeds, and true leaves, stems, and roots.

Flue-gas scrubbing Treating combustion exhaust gases with chemical agents to remove pollutants. Spraying crushed limestone and water into the exhaust gas stream to remove sulfur is a common scrubbing technique.

Fluid feeding The process by which an animal feeds on fluid.

Fluid mosaic Refers to the model for the structure of membranes, consisting of a bilayer of lipids in which globular proteins are embedded and can move laterally.

Fluid mosaic model Model of cell membrane structure composed of a lipid bilayer with scattered proteins; often described as a sea of lipids with protein icebergs.

Fluidized bed combustion High pressure air is forced through a mixture of crushed coal and limestone particles, lifting the burning fuel and causing it to move like a boiling fluid. Fresh coal and limestone are added continuously to the top of the combustion bed while ash and slag are drawn off below.

Fluke A member of class Trematoda or class Monogenea. Also, certain of the flatfishes (order Pleuronectiformes).

Fluorescence Longer wavelength emissions which can result form excited electron energy, results in a glowing light.

Fluxional A chemical species is said to be fluxional if it undergoes rapid degenerate rearrangements (generally detectable by methods which allow the observation of the behaviour of individual nuclei in a rearranged chemical species, e.g. NMR, X-ray).

Example: Bullvalene (1 209 600 interconvertible arrangements of the ten CH groups).

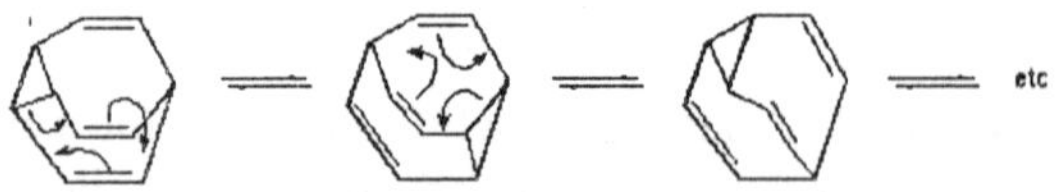

The term is also used to designate positional change among ligands of complex compounds and organometallics. In these cases, the change is not necessarily degenerate.

FMN. Abbreviation for flavin mononucleotide, the prosthetic group of a protein (flavoprotein) and a carrier in the electron transport chain in respiration.

F-number The ratio of the focal length of a lens to the effective aperture

Focal length The distance between the principal focus of a lens or mirror and its optical center or vertex.

Focal plane The plane perpendicular to the principal axis of a converging lens or mirror and containing the principal focus.

Focal point The focal point of a lens is defined by considering a narrow beam of light incident upon the lens, parallel to the optic (symmetry) axis of the lens and centered on that axis. The focal point is that point to which the rays converge or from which they diverge after passing through the lens. The convergent case defines a converging (positive) lens. The second case defines a diverging (negative) lens.

Focal-animal sampling A technique for recording the behavior of individual animals in which the observer concentrates on the activities of one individual for a set time period and then switches to watch another animal for a set time period.

Foci Groups of cells in an embryo that differentiate into specific structures and organs.

Focus A point at which light rays meet or from which rays of light appear to diverge.

Follicle Single, dry, dehiscent fruit that splits along one seam; example: milkweed.

Food aid Financial assistance intended to boost less-developed countries' standards of living.

Food chain A sequence of organisms in which plants are the primary food source for herbivores, which are in turn the food source for carnivores, etc., until the top carnivore level is reached.

Food security The ability of individuals to obtain sufficient food on a day-to-day basis.

Food surpluses Excess food supplies.

Food vacuole An organelle in the cell that functions in intracellular digestion.

Food web A community food chain depicting which species feed on each other and how many interrelationships are involved.

Foot 1. The basal portion of a moss sporophyte, embedded in the gametophyte.

2.The muscular locomotory structure of molluscs.

Forage crops Crops that are grown as food for domesticated herbivores.

Forager Hunter-gatherer group.

Foraging behavior The process animals use in locating food resources.

Foramen foram = an opening.

Foraminiferan A member of the class Granuloreticulosea (phylum Sarcomastigophora) bearing a test with many openings.

Force That which causes matter to change its velocity. Studied by many people, including Galileo, it was not until Newton formulated his three laws that the concept of force was placed on solid mathematical footing. While forces may be due to an explicit "push" or a collision, they may also be due to the interaction between objects with some property which can in general be called charge. The latter type of force is referred to as a fundamental force or more generally as an interaction between charges.

Force, fundamental That which causes matter to change its velocity via the interaction of chargedparticles. For example, the Earth pulls on any object with mass via the gravitational interaction. However, this so-called action-at-a-distance (Newton used the word "action" to describe what we now call "force"; we now have a different meaning for the word action.) Has been cause for much consternation, even for Newton. The idea of causing an acceleration without direct contact (action at a distance) eventually led to the development of field theory in which all interactions are due to the exchange of particles called gauge bosons.

Forced vibration Vibration that is due to the application of a periodic force, and not to the natural vibrations of the system.

Fore reef The outer part of a barrier reefor atoll.

Forebrain The forebrain consists of the diencephalon and telencephalon.

Forest management Scientific planning and administration of forest resources for sustainable harvest, multiple use, regeneration, and maintenance of a healthy biological community.

Form drag Hydrodynamic drag on an organism caused by its cross-sectional area.

Formed-element fraction The cellular component of vertebrate blood.

Forward bias Voltage applied to a semiconductor P-N junction that increases the electron current across the junction.

Fossa foss = depression or trench.

Fossil Any evidence of pre-existing life which is preserved in the earth's crust. In addition to the remains of entire, or parts of, organisms, fossils also include indirect evidence such as animal tracks and footprints, as well as the impression of leaves.

Fossil fuels Organic molecules derived from partially decayed plant and animal matter, produced primarily during the Carboniferous period; includes oil, gas, and coal.

Fossorial Characterized by digging or burrowing.

Fouling. Contamination of feeding or respiratory areas of an organism by excrement, sediment, or other matter. Also, accumulation of sessile marine organisms on the hull of a boat or ship so as to impede its progress through the water.

Fouling organisms Organisms that live attached to submerged surfaces such as boats and pilings.

Founder effect Changes in gene frequency that occur when a few individuals from a parental population colonize new habitats; the change is a result of founding individuals not having a representative sample of the parental population's genes.

Fourth World A political/economic category describing very poor nations that have neither market economies nor central planning and are either not developing or are developing very slowly. Also used to describe indigenous communities within wealthier nations.

Fovea A small pit or depression; especially the fovea centralis, a small rodless pit in the retina of some vertebrates, a point of acute vision.

FPS The system of units based on the fundamental units of the 'English system': foot, pound and second.

Fraction I protein Equivalent to RUBP carboxylase; the primary leaf protein in many green plants.

Fractional distillation The process of separating the components of a liquid mixture by means of differences in their boiling points.

Frame-shift mutation A mutation caused by the insertion or deletion of nucleotides (less than three or a number not a multiple of three) resulting in the improper grouping into codons.

Free energy. The energy available for doing work in a chemical system.

Free-running rhythm The activity cycle that an animal exhibits when placed in a constant environment; its period is different from any known cyclic environmental variable.

Free-threshing grain Grain that separates easily from enclosing bracts.

Freezing condensation A process that occurs in the clouds when ice crystals trap water vapor. As the ice crystals become larger and heavier, they begin to fall as rain or snow.

Freezing point The temperature at which a liquid changes to a solid.

Frequency How often an event occurs; esp. In the context of light, for which the frequency is the reciprocal of the amount of time it takes for a single wavelength to pass a point (i.e. The rate at which the waves pass).

Fresh water Water other than seawater; covers only about 2 percent of earth's surface, including streams, rivers, lakes, ponds, and water associated with several kinds of wetlands.

Freshwater ecosystems Ecosystems in which the fresh (nonsalty) water of streams, rivers, ponds, or lakes plays a defining role.

Friction A force that resists the relative motion of objects that are in contact with each other.

Frictional drag the resistance created by an animal's body surface when it moves through a fluid medium

Fringing reef a large coral-reef formation that closely borders the shoreline

Frond The photosynthetic leaf blade of a fern.

Front The boundary between two air masses of different temperature and density.

Frontal frons = brow, foliage.

Frontal plane. A plane parallel to the main axis of the body and at right angles to the sagittal plane.

Frontier An unexploited natural area at the leading edge of human settlement.

Frontier mentality The idea that the world has an unlimited supply of resources for human use regardless of the consequences to natural ecosystems and the biosphere.

Frontier orbitals The highest-energy occupied molecular orbital (HOMO) (filled or partly filled) and lowest-energy unoccupied molecular orbital (LUMO) (completely or partly vacant) of a molecular entity. Examination of the mixing of frontier molecular orbitals of reacting molecular entities affords an approach to the interpretation of reaction behaviour; this constitutes a simplified perturbation molecular orbital theory of chemical behaviour.

Fructose Six-carbon monosaccharide; often referred to as fruit sugar.

Fruit A mature ovary.

Fruiting body The reproductive structure of certain fungi.

Frustule The siliceous, box-like cell wall of diatoms.

Fucoxanthin An accessory brown pigment found in brown algae and some protists.

Fuel assembly A bundle of hollow metal rods containing uranium oxide pellets; used to fuel a nuclear reactor.

Fuel cell An electrochemical cell in which the chemical energy of continuously supplied fuel is converted into electric energy.

Fuel-switching A change from one fuel to another.

Fuelwood Branches, twigs, logs, wood chips, and other wood products harvested for use as fuel.

Fugitive emissions Substances that enter the air without going through a smokestack, such as dust from soil erosion, strip mining, rock crushing, construction, and building demolition.

Function A relation between the elements of one set, X (the domain), and the elements of another set, Y (the range), such that for each element in the domain X there's only one corresponding element in the range Y. When a function is written in the form of an equation relating values of variables, y = y(x), y must be single-valued, that is each value of x corresponds to only one value of y. While y = x2 is a function, x = y1/2 is not. Both equations express relations, however. Science deals with mathematical relations between measurements, not functions. Scientists often use the word function colloquially in the sense of "depends on" as in "Pressure is a function of volume and temperature", when they really mean "Pressure depends on volume and temperature."

Functional group Organic compounds are thought of as consisting of a relatively unreactive backbone, for example a chain of sp^3 hybridized carbon atoms, and one or several functional groups. The functional group is an atom, or a group of atoms that have similar chemical properties whenever it occurs in different compounds. It defines the characteristic physical and chemical properties of families of organic compounds.

Functional neuroanatomy The study of the size, structure, and arrangement of cells within the nervous system, particularly the brain.

Functionalists People who conduct studies involving attempts to discern how the mind works, as opposed to studies of its structure.

Fundamental niche The multidimensional space that an animal would occupy under optimal conditions in the absence of competitors; also called preferred niche. Contrast with realized niche.

Fundamental The lowest frequency produced by a musical tone source That harmonic component of a wave which has the lowest frequency.

Fundamental unit Any one of seven basic units of measure.

Fungi One of the five kingdom classifications; consists of nonphotosynthetic, eukaryotic organisms with cell walls, filamentous bodies, and absorptive nutrition.

Fungicidal A compound that is able to kill a fungus and stop an infection.

Fungicide A chemical that kills fungi.

Fungistatic A compound that slows or stops the growth of a fungus.

Funiculus The stalklike structure connecting an ovule to its placental surface within an ovary.

Funnel The siphon of cephalopods.

Fusiform Spindle shaped; tapering toward each end.

Fusion 1. The change of phase from a solid to a liquid; melting

2. A reaction in which light nuclei combine, forming a nucleus with a larger mass number.

G

G1 Part of interphase known as Gap 1; time of active metabolism in the cell cycle.

G2 Part of interphase after the synthesis of DNA and before the start of nuclear division and known as Gap 2.

Gaia hypothesis The idea that the earth as a whole should be regarded as a living organism and that biological processes stabilize the environment.

Galactose A 6-carbon sugar.

Galapagos Islands An archipelago on the equator in the Pacific Ocean about 1,000 km west of Ecuador. Charles Darwin's observations of the plant and animal life of these islands were important in the formulation of the theory of evolution by natural selection.

Gallbladder The pear-shaped reservoir for bile.

Galvanometer An instrument used to measure minute electric currents.

Gametangia Any cell or organ in which gametes are formed.

Gametangium Structure in which gametes are produced.

Gamete A haploid reproductive cell that develops into a new individual after its union with another gamete.

Gametic meiosis. Meiosis that occurs during formation of the gametes, as in humans and other metazoa.

Gametocyst Cyst produced by some apicomplexan parasites. Sexual reproduction and spore formation occurs within this cyst.

Gametocyte The mother cell of a gamete, that is, immature gamete.

Gametogenesis The formation of gametes by way of meiosis.

Gametogony Multiple fission that forms gametes that fuse to form a zygote. Also called gamogony. Occurs in the class Sporozoea.

Gametophyte The haploid, gamete-producing plant in the alternation of generations; undergoes mitosis to produce the haploid gametes, which fuse to form the diploid zygote of the sporophyte.

Gamma rays That portion of the sun's total range of radiation in which rays are shorter than x rays; below 0.1 nm in length.

Gamont Apicomplexan life cycle stage that is committed to undergoing gametogenesis.

Ganglion An aggregation of nerve tissue containing nerve cells.

Ganoid scales Thick, bony, rhombic scales of some primitive bony fishes; not overlapping.

Gap analysis A biogeographical technique of mapping biological diversity and endemic species to find gaps between protected areas that leave endangered habitats vulnerable to disruption.

Gap junction An area of tiny canals communicating the cytoplasm between two cells.

Garden city A new town with special emphasis on landscaping and rural ambience.

Gas (respiratory) exchange The movement of oxygen and other gases between the atmosphere and the ocean or between the water or atmosphere and living organisms.

Gasohol A mixture of gasoline and ethanol.

Gastric shield A chitinized plate in the stomach of a bivalve (phylum Mollusca) on which the crystalline style is rotated.

Gastrodermis Lining of the digestive cavity of Cnidarians.

Gastrolith Calcareous body in the wall of the cardiac stomach of crayfish and other Malacostraca, preceding the molt.

Gastropoda The class of molluscs characterized by torsion. A shell, when present, is usually coiled. Snails.

Gastrotricha A small phylum of marine and freshwater species of gastrotrichs that inhabit the spaces between bottom sediments.306

Gastrovascular canals Fluid-filled canals opening at the mouth of Cnidarians and ctenophores that function in gas exchange and in the distribution of nutrients.

Gastrovascular cavity Body cavity in certain lower invertebrates that functions in both digestion and circulation and has a single opening serving as both mouth and anus.

Gastrozooid The feeding polyp of a hydroid, a hydranth.

Gastrula Embryonic stage, usually cap or sac shaped, with walls of two layers of cells surrounding a cavity (archenteron) with one opening (blastopore).

Gastrulation The embryological process that results in the formation of the gastrula; results in the formation of the embryonic gut, ectoderm, and endoderm.

Gauss' law The electric flux through a closed surface is proportional to the algebraic sum of electric charges contained within that closed surface; in differential form,

Div e = rho,

Where rho is the charge density.

Geiger tube Ion sensitive instrument used for the detection of subatomic particles.

Gel That state of a colloidal system in which the solid particles form the continuous phase and the fluid medium the discontinuous phase.

Geminate pair Pair of molecular entities in close proximity in solution within a solvent cage and resulting from reaction (e.g. bond scission, electron transfer, group transfer) of a precursor that constitutes a single kinetic entity.

Geminate recombination The reaction with each other of two transient molecular entities produced from a common precursor in solution. If reaction occurs before any separation by diffusion has occurred, this is termed "primary geminate recombination". If the mutually reactive entities have been separated, and come together by diffusion, this is termed "secondary geminate recombination".

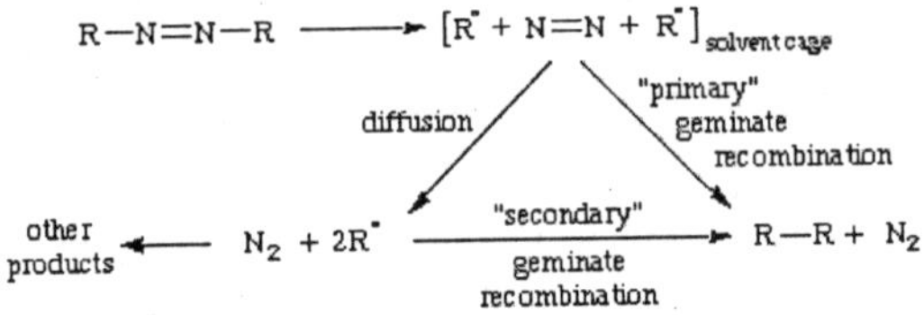

Gemmae cup A specialized structure on the gametophyte of certain liverworts that produces asexual plantlets (gemmae) capable of starting new gametophytes.

Gemmule Resistant, overwintering capsule formed by freshwater, and some marine, sponges that contains masses of mesenchyme cells; amoeboid mesenchyme cells are released and organize themselves into a sponge.

Gene A unit of heredity; a segment of DNA nucleus of the cell that contains genetic information

Gene activators Hormones that can turn a given gene on or off.

Gene banks Storage for seed varieties for future breeding experiments.

Gene flow An exchange of genes between two populations of a species, or in extreme cases, between populations of two species (hybridization).

Gene insertion The process by which one or more genes from one organism are incorporated into the genetic makeup of a second organism.

Gene pool A collection of all of the alleles of all of the genes in a population.

Gene regulation The process by which genes are turned on and off to regulate growth and development of an organism.

Gene synthesizer A machine that can produce a functional gene segment in a short period of time.

General acid catalysis The catalysis of a chemical reaction by a series of Brønsted acids (which may include the solvated hydrogen ion) so that the rate of the catalysed part of the reaction is given by $\Sigma k_{ha}[ha]$ multiplied by some function of substrate concentrations. (The acids are unchanged by the overall reaction.) General catalysis by acids can be experimentally distinguished from specific catalysis by hydrogen cations (hydrons) by observation of the rate of reaction as a function of buffer concentration.

General fertility rate Representation of population age structure and fecundity; crude birth rate multiplied by the percentage of fecund women (between approximately fifteen and forty-four years of age) by 1000.

Generalists Animals that are capable of eating a variety of foods or living in a variety of places.

Generalization In learning, a principle that states if an animal has been conditioned to respond to a certain stimulus, the response will usually also occur to stimuli used during the original acquisition trials.

Generalized niche Niche that has broad requirements; tolerates a range of conditions.

Generation time The length of time between when an organism is born and when it first reproduces.

Generative nucleus Produced by the haploid microspore nucleus of a pollen grain, the generative nucleus divides mitotically to form two sperm nuclei.

Generator potential A graded potential that travels only a short distance along the plasma membrane of a sensory cell.

Genetic The study of the mechanisms of transmission of genes from parents to offspring.

Genetic assimilation The disappearance of a species as its genes are diluted through crossbreeding with a closely related species.

Genetic drift Random change in allelic frequencies in a population occurring by chance. In small populations, genetic variation at a locus may be lost by chance fixation or a single allelic variant.

Genetic engineering Modifying the genetic structure of one organism by splicing in selected genetic information from another organism.

Genetic erosion Irreversible loss of genetic diversity due to extinction of traditional varieties and wild ancestors of crop plants.

Genetic recombination Crossing-over; a major source of genetic variation in a population or a given species.

Genetic screening The use of a number of mutant forms of a species to test for a specific type of change or deficit, as with respect to biological clocks.

Genetic variability Produced by sexual reproduction, all organisms are genetically different from each other.

Genetically engineered microorganisms (gems) Bacteria that have been genetically engineered by the insertion or deletion of DNA segments.

Genital atrium Cavity in the body wall of a flatworm into which male and female genital ducts open.

Genital slit The genital opening of cetaceans.

Genitointestinal canal Duct connecting the oviduct and intestine of some polyopisthocotylean Monogenea.

Genome All the DNA in a haploid set of chromosomes (nuclear genome), organelle (mitochondrial genome, chloroplast genome) or virus (viral genome, which in some viruses consists of RNA rather than DNA).

Genotype The genetic constitution, expressed and latent, of an organism; the total set of genes present in the cells of an organism; contrasts with phenotype.

Genotypic ratio The relative numbers of progeny in each genotypic category produced by a genetic cross.

Genus A group of related species with taxonomic rank between family and species.

Geometric growth Growth that follows a geometric pattern of increase, such as 2, 4, 8, 16, etc.

Geometrized units A system of units whereby certain fundamental constants (g, c, k, and h) are set to unity. This makes calculations in certain theories, such as general relativity, much easier to deal with, since these constants appear frequently.

As a result of converting to geometrized units, all quantities are expressed in terms of a unit of distance, traditionally the cm.

Georeceptor A specialized nerve ending that responds to the force of gravity.

Geothermal energy Energy drawn from the internal heat of the earth, either through geysers, fumaroles, hot springs, or other natural geothermal features, or through deep wells that pump heated groundwater.

Geotropism The bending responses of a plant to the forces

Germ Embryo of a cereal grain.

Germ layer In the animal embryo, one of three basic layers (ectoderm, endoderm, mesoderm) from which the various organs and tissues arise in the multicellular animal.

Germ plasm Genetic material that may be preserved for future agricultural, commercial, and ecological values (plant seeds or parts or animal eggs, sperm, and embryos).

Germinate To resume growth and increase metabolic activity,

Gerontology The scientific problems of aging in all their aspects, including clinical, biological, and sociological.

Gestation Period of development of the young in viviparous animals, from the time of fertilization of the ovum until birth.

Gestation period The portion of the reproductive cycle in a female mammal extending from fertilization to birth of its offspring.

Giardiasis A common infection of the lumen of the small intestine with the flagellated protozoan Giardia lamblia, and spread via contaminated food and water and by direct person-to-person contact.

Gibberellin A group of related compounds which cause single-gene dwarf mutants of corn and peas to elongate normally.

Gibbs energy diagram A diagram showing the relative standard gibbs energies of reactants, transition states, reaction intermediates, and products, in the same sequence as they occur in a chemical reaction.

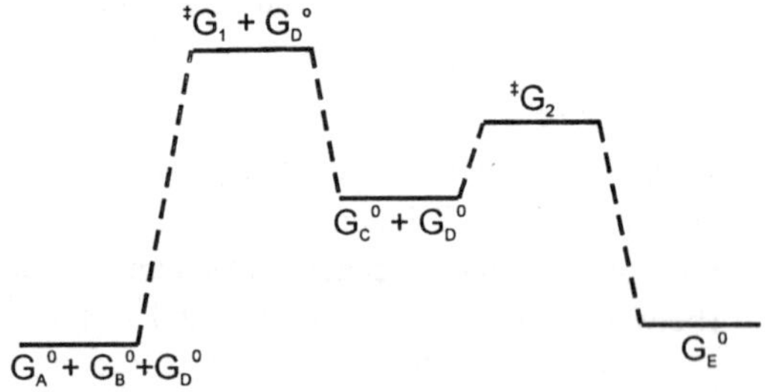

These points are often connected by a smooth curve (a "gibbs energy profile", commonly still referred to as a "free energy profile") but experimental observation can provide information on relative standard gibbs energies only at the maxima and minima and not at the configurations between them. The abscissa expresses the sequence of reactants, products, reaction intermediates and transition states and is usually undefined or only vaguely defined by the reaction coordinate (extent of bond breaking or bond making). In some adaptations the abscissas are however explicitly defined as bond orders, brønsted exponents, etc.

Contrary to statements in many text books, the highest point on a gibbs energy diagram does not necessarily correspond to the transition state of the rate-limiting step. For example, in a stepwise reaction consisting of two reaction steps

1. a + b $\longrightarrow$ c

2. c + d $\longrightarrow$ e

One of the transition states of the two reaction steps must (in general) have a higher standard gibbs energy than the other, whatever the concentration of d in the system. However, the value of that concentration will determine which of the reaction steps is rate-limiting. If the particular concentrations of interest, which may vary, are chosen as the standard state, then the rate-limiting step is the one of highest gibbs energy.

Gill An aquatic respiratory organ for obtaining oxygen and getting rid of carbon dioxide.

Gill arches Bony or cartilaginous gill supports of some vertebrates; also called visceral arches.

Gill cover (operculum) The flap of bony plates that covers the gills of bony fishes.

Gill filaments A thin-walled, fleshy extension of a gill arch that contains vessels carrying blood to and from gas exchange surfaces.

Gill lamellae Thin plates of tissue on gill filaments that contain the capillary beds across which gases are exchanged.

Gill raker Each of the projections along the inner surface of fish gills.

Gill slit One of several openings in the pharyngeal region of chordates. Gill slits allow water to pass from the pharynx to the outside of the body. In the process, water passes over gills or suspended food is removed in a filter-feeding mechanism.

Ginkgo A decidious gymnosperm tree (Ginkgo biloba) with distinctive fan-shaped leaves, native to eastern China.

Girth Circumference, as of a tree trunk.

Glass sponges Deep-water sponges with a skeleton of fused silica spicules.

Glia' cells Supportive cells that are closely associated with neurons.

Global environmental citizenship A shift in our attention from pollution in a specific place to a concern about the life-support systems of the whole planet.

Global warming Some experts contend our earth's annual average temperature is gradually increasing, partly due to the greenhouse effect.

Globular protein A coiled polymer of amino acids forming a compact macromolecule.

Globulins A large group of compact proteins with high molecular weight; includes immunoglobulins (antibodies).

Glochidium A larval stage of freshwater bivalves in the family Unionidae; it lives as a parasite on the gills or fins of fishes.

Glomerulus A tuft of capillaries projecting into a renal corpuscle in a kidney. Also, a small spongy mass of tissue in the proboscis of hemichordates, presumed to have an excretory function. Also, a concentration of nerve fibers situated in the olfactory bulb.

Glossa Tonguelike mouthpart in Hymenoptera.

Gluconeogenesis Synthesis of glucose from protein or lipid precursors.

Glucose Six-carbon monosaccharide; one of the most abundant simple sugars. It is the building block of both cellulose and starch and is important to several metabolic pathways.

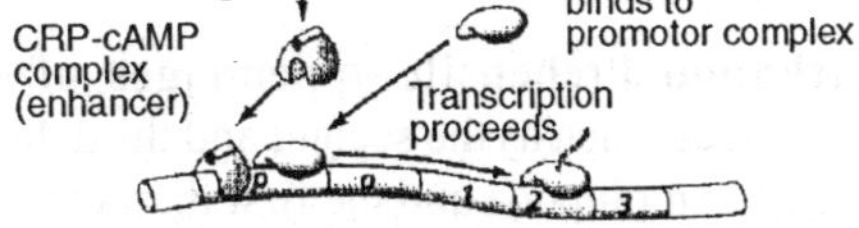

Fig. Glucose

Glumes Pair of bracts at the base of a spikelet in a grass flower.

Gluon Subatomic particle that binds quarks together in the nucleus of an atom.16

Gluten Protein complex in endosperm of wheat and some other cereals that is essential in making a leavened bread.

Glycerol An organic molecule to which fatty acids are attached to form a fat.

Glycocalyx Finely filamentous layer containing carbohydrate, found on the outer surface of many cells, from 7.5 to 200.0 nm thick.

Glycogen A polysaccharide constituting the principal form in which carbohydrate is stored in animals; animal starch.

Glycolysis The conversion of glucose to pyruvic acid (pyruvate) with the release of some energy in the form of ATP. Occurs in the cytosol; literally "sugar splitting."

Glycoprotein A protein with attached sugars.

Glycoside Physiologically active compound in plants that always contains a sugar group, although the active part of the molecule may differ.

Glycosomes Organelles found in Trypanosoma that contain enzymes of glycolysis and for oxidizing reduced NAD.

Glyoxylate cycle Metabolic pathway that functions to convert fatty acids or acetate to carbohydrate.

Glyoxysomes A subcellular microbody present in the cytoplasm of many oil seeds. Enzymes packaged in the glyoxysome convert lipids to carbohydrates during the germination process.

Gnathobase A median basic process on certain appendages in some arthropods, usually for biting or crushing food.

Gnathopod Prehensile appendages of some Crustacea, such as the second and third thoracic legs of Amphipoda and the first thoracic legs of some Isopoda.

Gnathosoma Anterior of two basic regions of the body of a mite or tick. Also called a capitulum.

Gnathostomata A superclass of vertebrates whose members possess hinged jaws and paired appendages. Notochord may be replaced by the vertebral column.

Gnetophytes Taxa of the division Gnetophyta, a group of gymnosperms.

Goblets Markings on the stigmatal plates of certain hard ticks.

Golgi body (apparatus) Organelle of membranous, hollow sacs arranged in a stack; functions in modification, storage, and packaging of secretion materials; may be called dictyosome in plants.

Golgi complex An arrangement of sacs and membranes in many eukaryotic cells that is involved in collecting and transporting molecules.

Gonad An organ that produces gametes (ovary in the female and testis in the male).

Gonadotropin A hormone that stimulates activity in the gonads.

Gonads Glands responsible for the production of gametes and where certain gonadal hormones are produced. These consist of the ovaries in females and the testes in males.

Gonangium Reproductive zooid of hydroid colony (Cnidaria).

Gondwana One of two large continents, the southern one, that formed when the supercontinent Pangaea broke up about 180 million years ago.

Gonochoristic Having separate sexes; that is, an individual is either male or female, but never both. "Dioecious" means the same thing.

Gonoduct Duct leading from a gonad to the exterior.

Gonopore A genital pore found in many invertebrates.

Gonotyl Muscular sucker or other perigenital specialization surrounding or associated with the genital atrium of a digenetic trematode.

Gonozooid A polyp of a hydrozoan Cnidarian that produces medusae.

Gorgonians Colonial anthozoans that secrete a skeleton made of protein.

Grade A level of organismal complexity or adaptive zone characteristic of a group of evolutionarily related organisms.

Graded signal A form of communication that is analog, or varying in intensity or frequency, providing quantitative information about the strength of the stimulus.

Gradualism A component of Darwin's evolutionary theory postulating that evolution occurs by the temporal accumulation of small, incremental changes, usually across very long periods of geological time; it opposes claims that evolution can occur by large, discontinuous or macromutational changes.

Grafting A method of asexual propagation characterized by placing a shoot (the scion) onto the rootstock (the stock) of another plant.

Grain Single, dry indehiscent fruit of a single seed that is fused to the ovary wall.

Gram-negative Denoting bacterial cells that do not stain with crystal violet and iodine.

Gram-positive Denoting bacterial cells that are readily stained with crystal violet and iodine.

Grana flattened saclike structures inside chloroplasts containing chlorophyll and other photosynthetic enzymes

Granite An igneous rock overlying most of the landmasses

Granulocytes White blood cells (neutrophils, eosinophils, and basophils) bearing "granules" (vacuoles) in their cytoplasm that stain deeply.

Granuloma, granulomatous tissue Repaired area of a body marked by fibrous connective tissue (fibrosis). Also, fibrous connective tissue surrounding an antigen source.

Granum Structures within chloroplasts, seen as green granules with the light microscope and as a series of stacked disk-shaped membranes with an electron microscope; the grana contain the chlorophylls and carotenoids and are the sites of the light reactions of photosynthesis.

Grasslands A biome characterized by grasses and a few scattered trees. Grassland animals include numerous insects and other herbivores such as bison.

Gravel Inorganic soil particles larger than 2.0 mm in diameter.

Gravitational field Region of space in which each point is associated with a value of gravitational acceleration.

Gravitational force The mutual force of attraction between particles of matter

Gravitational interaction The interaction between particles of matter that has no known distance limitations, but is the weakest interaction of all.

Gravitational potential energy Potential energy acquired by an object when it is moved against gravity

Graviton The carrier for the gravitational interaction.

Gravity The force of gravitation on an object on or near the surface of a celestial body.

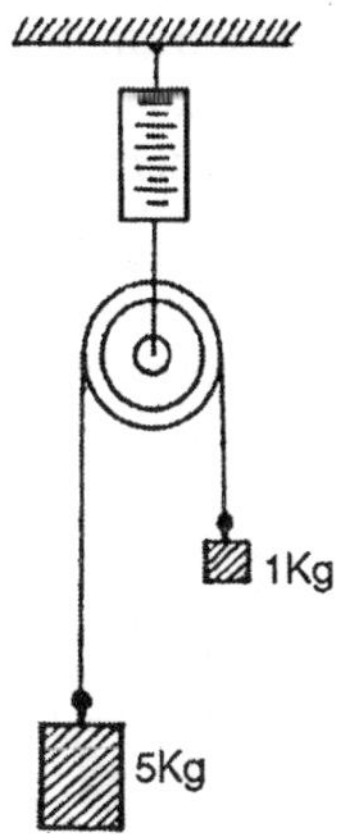

Fig. Gravity

Gray crescent A dark arching band that forms on the surface of the amphibian zygote opposite the point of sperm penetration; forms in the region where gastrulation will occur.

Gray; gy The derived SI unit of absorbed dose, defined as the absorbed dose in which the energy per unit mass imparted to the matter by ionizing radiation is 1 J/kg; it thus has units of J/kg.

Great ocean conveyor A global circulation pattern in which water cycles throughout the ocean basins.

Great whales Large whales: the sperm and the baleen whales.

Green algae Seaweeds in which chlorophyll is not green beard effecta mechanism for recognizing kin that does not require previous experience. Recognition genes enable an organism to identify and behave altruistically towards other organisms bearing that same gene.

Green gland Excretory gland of certain Crustacea; the antennal gland.

Green plans Integrated national environmental plans for reducing pollution and resource consumption while achieving sustainable development and environmental restoration.

Green political parties Political organizations based on environmental protection, participatory democracy, grassroots organization, and sustainable development.

Green revolution Dramatically increased agricultural production brought about by "miracle" strains of grain; usually requires high inputs of water, plant nutrients, and pesticides.

Greenhouse effect The warming of the Earth due to the atmospheric accumulation of carbon dioxide and other gases, which trap heat and reradiate it back to the Earth's surface.

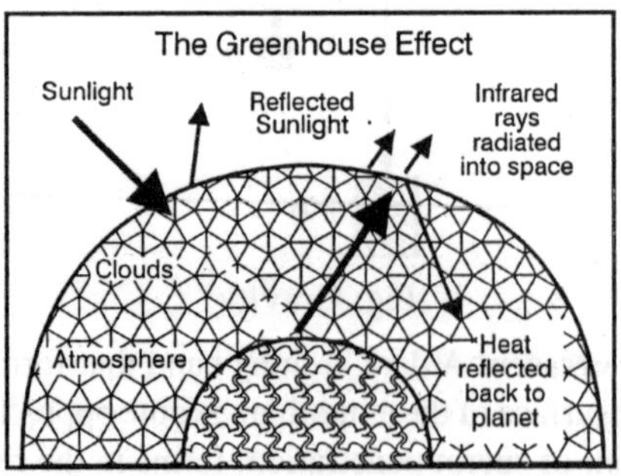

Fig. Greenhouse effect

Gregarious Living in groups or flocks.

Grid An element of an electronic tube An electrode used to control the flow of electrons from the cathode to the plate.

Grid bias The grid-to-cathode voltage.

Gross national product (GNP) The sum total of all goods and services produced in a national economy. Gross domestic product (GDP) is used to distinguish economic activity within a country from that of off-shore corporations.

Gross primary production The total amount of photosynthesis accomplished in a given period of time.

Ground itch Skin rash caused by bacteria introduced by invasive hookworm larvae.

Ground meristem The basic or fundamental tissue of the apical meristem; dermal tissues surround the ground meristem, and the provascular strands are embedded in it.

Ground parenchyma The basic ground tissue consisting of living parenchymal cells.

Ground state The state of lowest Gibbs energy of a system.

Ground tissue Includes primary tissues of parenchyma, sclerenchyma, and collenchyma that make up much of the bulk of the primary plant body; function in support, photosynthesis, and storage; also known as fundamental tissue.

Groundwater Water held in gravel deposits or porous rock below the earth's surface; does not include water or crystallization held by chemical bonds in rocks or moisture in upper soil layers.

Group A defined linked collection of atoms or a single atom within a molecular entity. This use of the term in physical organic and general chemistry is less restrictive than the definition adopted for the purpose of nomenclature of organic compounds.

Group selection Selection that operates on two or more genetic lineages (groups) as units; broadly defined, this includes kin and interdemic selection.

Growth An irreversible increase in size or volume of a cell, tissue, organ, or organism.

Growth ring A concentric layer of wood (secondary xylem) formed in one growing season through cell divisions of the vascular cambium. Also termed annual ring.

Guanine A white crystalline purine base, $C_5H_5N_5O$, occurring in various animal tissues and in guano and other animal excrements.

Guano The accumulation of the excrement of seabirds.

Guard cell One of a pair of specialized cells in the epidermis that regulates the opening and closing of a stoma.

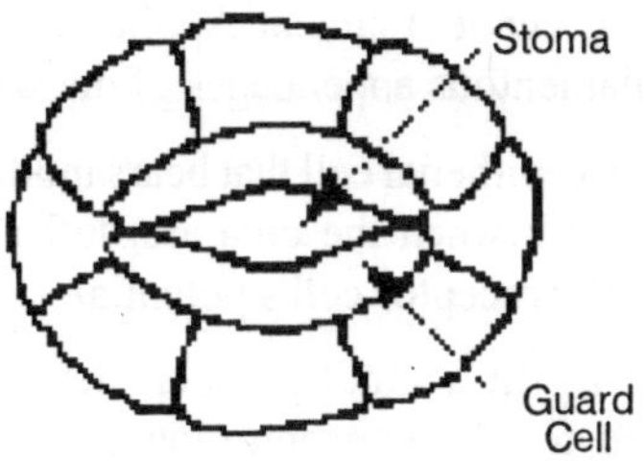

Fig. Guard cell

Guest An organic or inorganic ion or molecule that occupies a cavity, cleft or pocket within the molecular structure of a host molecular entity and forms a complex with it or that is trapped in a cavity within the crystal structure of a host.

Guild In ecology, a group of species that exploit the same class of environment in a similar way.

Gular flutter The type of breathing experienced by some birds. Rapid movement of the throat region promotes evaporative water loss.

Gully erosion Removal of layers of soil, creating channels or ravines too large to be removed by normal tillage operations.

Gustation The act of tasting or the sense of taste.

Gustatory Having to do with the sensation of taste, the chemoreception of molecules in solution by specialized epithelial receptor cells.

Guttation Root pressure forcing water out of the leaves, usually overnight when stomates are closed.

Guyot A flat-topped seamount.

Gymnosperm Seed-producing plants in which the seed is not enclosed in an ovary as in angiospems. Ginkgos, cycads, and conifers are examples.

Gynandromorph An abnormal individual exhibiting characteristics of both sexes in different parts of the body; for example the left side of a bilateral organism may show characteristics of one sex and the right side those of the other sex.

Gynandry In a hermaphroditic organism, maturation first of the female gonads and then of the male organs. Also called protogyny.

Gynocophoric canal Groove in male schistosomes (certain trematodes) that carries the female.

Gynoecium The collective term referring to the carpels or ovule-bearing structures, within a flower.

Gyre A large, nearly circular system of wind-driven surface currents that center around latitude 30in both hemispheres.

H

Habitat imprinting The tendency of an animal to prefer a place to live based on early experience in that habitat.

Habitat restoration The recovery of stressed or destroyed habitats.

Habitat selection The choice of an animal's place to live. Habitat selection involves the interaction of physiological and psychological factors.

Habitat The place where an organism normally lives or where individuals of a population live.

Habituation The relatively persistent waning of a response that results from repeated presentations not followed by any form of reinforcement.

Hadal zone The bottom below 6,000 m (20,000 ft).

Hadley cells Circulation patterns of atmospheric convection currents as they sink and rise in several intermediate bands.

Hadron This type of particle is affected by the strong interaction. They may or may not also be affected by the weak interaction. This is to be compared against leptons. According to the current theory of the standard model hadrons are made up of quarks which can be divided into families. The hadrons are further categorized depending on which constituent quarks are bound together. These categories are the mesons and the baryons.

Hair A long slender filament. Applied especially to such filamentous appendages of the skin.

Hair cell An epithelial cell that bears modified cilia on one end; when the cilia are deflected, they change the receptor cell's potential.

Half-life, $t_{1/2}$. In a kinetic experiment, the time required for the concentration of a particular reacting species to fall to one-half of its initial value. (Its dependence on initial concentration depends upon the order of reaction. It is independent of initial concentration only for a first-order process.)

Hall effect When charged particles flow through a tube which has both an electric field and a magnetic field (perpendicular to the electric field) present in it, only certain velocities of the charged particles are preferred, and will make it undeviated through the tube; the rest will be deflected into the sides. This effect is exploited in such devices as the mass spectrometer and in the thompson experiment.

Haller's organ Depression on the first tarsi of ticks; functions as an olfactory and humidity receptor.

Hallucinogen Psychoactive drug capable of altering moods and perceptions of time and space.

Hallux L. Digit number one of the foot.

Halochromism Halochromism means the colour change which occurs on addition of acid (or base, or a salt) to a solution of a compound. A chemical reaction (e.g. ion formation) transforms a colourless compound into a coloured one.

Halophilic Pertaining to microorganisms that tolerate high concentrations of saline media.

Halophyte A salt-tolerant terrestrial plant.

Halter In Diptera, small club-shaped structure on each side of the metathorax representing the hindwings; believed to be sense organs for balancing; also called balancer.

Halzoun Disease resulting from blockage of the nasopharynx by a parasite. Also called marrara.

Hammond principle (or Hammond postulate) The hypothesis that, when a transition state leading to an unstable reaction intermediate (or product) has nearly the same energy as that intermediate, the two are interconverted with only a small reorganization of molecular structure. Essentially the same idea is sometimes referred to as "Leffler's assumption", namely, that the transition state bears the greater resemblance to the less stable species (reactant or reaction intermediate/product). Many text books and physical organic chemists, however, express the idea in Leffler's form, but attribute it to Hammond.

As a corollary, it follows that a factor stabilising a reaction intermediate will also stabilize the transition state leading to that intermediate.

The acronym "Bemahapothle" is sometimes used in recognition of the principal contributors towards expansion of the original idea of the Hammond postulate.

Hamuli Large hooks on the opisthaptor of a monogenetic trematode; referred to as anchors by American authors.

Handicap hypothesis The hypothesis that apparently deleterious sexual "ornaments" possessed by males in many species are attractive to females because they indicate that the males have such vigor (good genes) that they can survive even with the handicap.

Hansch constant A measure of the capability of a solute for hydrophobic (lipophilic) interaction based on the partition coefficient P for distribution of the solute between octan-1-ol and water. The most general way of applying P in correlation analysis, QSAR, etc. is as log P, but the behaviour of substituted benzene derivatives may be quantified by a substituent constant scale,, which is defined in a way analogous to the Hammett scale. There are various scales, depending on the substrate series used as reference.

Haplodiploidy The mode of sex determination by which males develop from haploid (unfertilized) eggs and females from diploid (fertilized) eggs; characteristic of the insect order Hymenoptera.

Haploid The reduced, or N, number of chromosomes, typical of gametes, as opposed to the diploid, or 2N, number found in somatic cells. In certain groups, mature organisms may have a haploid number of chromosomes.

Haploid cell a cell containing only one of each type of chromosome characteristic of its species

Haptens Molecules of small molecular weight (usually) that are immunogenic only when attached to carrier molecules, usually proteins.

Haptera Short, sturdy rootlike structures that form the holdfast of seaweeds.

Hapto The hapto symbol, with numerical superscript, provides a topological description for the bonding of hydrocarbons and other pi-electron systems to metals, by indicating the connectivity between the ligand and the central atom. For example, 3 indicates that three atoms of the ligand are bonded to the central atom.

Hard acid A Lewis acid with an acceptor centre of low polarizability. Other things being approximately equal, complexes of hard acids and bases or soft acids and bases have an added stabilization (sometimes called "HSAB" rule). For example the hard O- (or N-) bases are preferred to their S- (or P-) analogues by hard acids. Conversely a "soft acid" possesses an acceptor centre of high polarizability and exhibits the reverse preference for coordination of a soft base. These preferences are not defined in a quantitative sense.

Hard base A Lewis base with a donor centre (e.g. an oxygen atom) of low polarizability; the converse applies to "soft bases".

Hardwood A term often used in reference to all woody dicots; more accurately, wood having a high specific gravity.

Hardy weinberg equilibrium Mathematical demonstration that the Mendelian hereditary process does not change the populational frequencies of alleles or genotypes across generations, and that changes in allelic or genotypic frequencies requires factors such as natural selection, genetic drift in finite populations, recurring mutation, migration of individuals among populations, and nonrandom mating.

Hardy weinberg formula The application of the binomial expansion to determine the frequency of two alleles in the absence of any evolutionary forces.

Harem A large group of females herded by a large male pinniped for the purpose of mating; harems also occur in some fishes.

Harmonics The fundamental and the tones whose frequencies are whole number multiples of the fundamental

Hawking radiation The theory that black holes emit radiation like any other hot body. Virtual particle-antiparticle pairs are constantly being created in supposedly empty space. Occasionally, a pair will be created just outside the event horizon of a black hole. There are three possibilities: 1.Both particles are captured by the hole; 2.Both particles escape the hole; 3.One particle escapes while the other is captured. The first two cases are straightforward; the virtual particle-antiparticle pair recombine and return their energy back to the void via the uncertainty principle.

It is the third case that interests us. In this case, one of the particles has escaped (and is speeding away to infinity), while the other has been captured by the hole. The escapee becomes real and can now be detected by distant observers. But the captured particle is still virtual; because of this, it has to restore conservation of energy by assigning itself a negative mass-energy. Since the hole has absorbed it, the hole loses mass and thus appears to shrink. From a distance, it appears as if the hole has emitted a particle and reduced in mass.

The rate of power emission is proportional to the inverse square of the hole's mass; thus, the smaller a hole gets, the faster and faster it emits hawking radiation. This leads to a runaway process; what happens when the hole gets very small is unclear; quantum theory seems to indicate that some kind of "remnant" might be left behind after the hole has emitted away all its mass-energy.

Hawking temperature The temperature of a black hole caused by the emission of hawking radiation.

Hazardous Describes chemicals that are dangerous, including flammables, explosives, irritants, sensitizers, acids, and caustics; may be relatively harmless in diluted concentrations.

Hazardous waste Any discarded material containing substances known to be toxic, mutagenic, carcinogenic, or teratogenic to humans or other life-forms; ignitable, corrosive, explosive, or highly reactive alone or with other materials.

HDL (high density lipoprotein) Transport molecule that removes excess cholesterol from the body's tissues to the liver for degradation and elimination.

Head Many individual flowers tightly compressed into the shape of a single large flower.

Head-foot The body region of a mollusc that contains the head and is responsible for locomotion as well as retracting the visceral mass into the shell.

Health A state of physical and emotional well-being; the absence of disease or ailment.

Heap-Leach extraction A technique for separating gold from extremely low-grade ores. Crushed ore is piled in huge heaps and sprayed with a dilute alkaline-cyanide solution, which percolates through the pile to extract the gold, which is separated from the effluent in a processing plant. This process has a high potential for water pollution.

Heart urchins Burrowing sea urchins with a flattened test and short spines.

Heartwood The wood found in the center of a tree trunk; often a darker color due to the accumulation of resins, oils, gums, and other metabolic by-products, which prevent water movement through this tissue.

Heartworm disease A parasitic infection in dogs caused by the nematode Dirofilaria immitis.

Heat Heat, like work, is a measure of the amount of energy transferred from one body to another because of the temperature difference between those bodies. Heat is not energy possessed by a body. We should not speak of the "heat in a body." The energy a body possesses due to its temperature is a different thing, called internal thermal energy. The misuse of this word probably dates back to the 18th century when it was still thought that bodies undergoing thermal processes exchanged a substance, called caloric or phlogiston, a substance later called heat.

Heat capacity The amount of heat energy that must be added or subtracted to change the temperature of a body; water has a high heat capacity.

Heat engine Any device that converts heat energy into mechanical work.

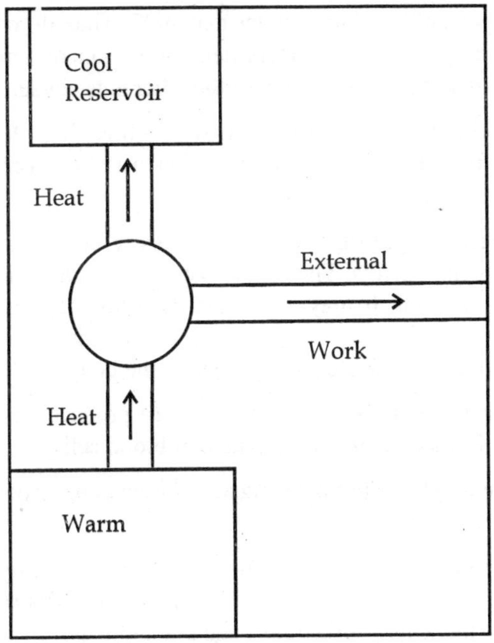

Fig. Heat engine

Heat of fusion The heat required per unit mass to change a substance from solid to liquid at its melting point.

Heat of vaporization The heat required per unit mass to change a substance from liquid to vapor at its boiling point.

Heat pump A device that absorbs heat from a cool environment and gives it off to a region of higher temperature.

Heat sink A reservoir that absorbs heat without a significant increase in temperature.

Heavy metals A group of toxic metals: mercury, lead, and others.

Hectocotylus Specialized, and sometimes autonomous, arm that serves as a male copulatory organ in cephalopods.

Heisenberg's Uncertainty Principle Pairs of measurable quantities whose product has dimensions of energy×time are called conjugate quantities in quantum mechanics, and have a special relation to each other, expressed in Heisenberg's uncertainty principle. It says that the product of the uncertainties of the two quantities is no smaller than h/2?. Thus if you improve the measurement precision of one quantity the precision of the other gets worse.

Helpers-at-the-nest Individuals other than the parents that forego reproduction to assist rearing the young of others.

Hemal system Strands of tissue found in echinoderms. The hemal system is of uncertain function. It may aid in the transport of large molecules or coelomocytes, which engulf and transport waste particles within the body.

Hematopoiesis hema = blood; poie = produce. The formation and development of blood cells.

Hemelytra The front wing of an insect of the order Hemiptera.

Hemerythrin A red, iron-containing respiratory pigment found in the blood of some polychaetes, sipunculids, priapulids, and brachiopods.

Hemicellulose Polysaccharide in plant cell walls that cross-link cellulose fibrils.

Hemichordata The phylum of marine, wormlike animals whose members have an epidermal nervous system and pharyngeal gill slits. Acorn worms and pterobranchs.

Hemimetabolous metamorphosis A type of insect metamorphosis in which immature insects are different in form and habitats from the adult. It is different from holometabolous metamorphosis in that there is a gradual series of changes in form during the transition from immature to adult.

Hemizygous An individual having one member of a pair of genes.

Hemocoel Main body cavity of arthropods, the embryonic development of which differs from that of a true coelom but that includes a vestige of a true coelom.

Hemocyanin A nonheme, blue respiratory pigment that is found in the plasma of many molluscs and arthropods and is composed of monomers, each of which contains two atoms of copper and can bind one molecule of oxygen.

Hemocyte Any blood corpuscle or formed element of the blood.

Hemoglobin An iron-containing respiratory pigment occurring in vertebrate red blood cells and in blood plasma of many invertebrates; a compound of an iron porphyrin heme and globin proteins.

Hemoglobinuria Bloody urine.

Hemolymph Fluid in the coelom or hemocoel of some invertebrates that represents the blood and lymph of vertebrates.

Hemozoin Insoluble digestion product of malaria parasites produced from hemoglobin.

Hemp Cannabis sativa; source for fibers used in making hemp rope.

Henderson-Hasselbach equation An equation of the form

$$pH = pKa - lg([HA]/[A^-])$$

for the calculation of the pH of solutions where the ratio $[HA]/[A^-]$ is known.

Henry The derived SI unit of inductance, defined as the inductance of a closed circuit in which an electromotive force of 1 V is produced when the electric current varies uniformly at a rate of 1 a/s; it thus has units of v s/a.

Hepatic Pertaining to the liver.

Hepatosplenomegaly Swollen liver and spleen.

Herb Nonwoody plant; aromatic plants whose leaves are used in seasoning.

Herbaceous Without woody tissues; typical of most annuals and biennials; herbaceous perennials die back to the soil level each year.

Herbal A text that describes plants that are useful medicinally and in other ways.

Herbalism The use of plants for treating ailments.

Herbarium A permanent collection of dried and pressed plants that provides information on the location and identification of the local flora.

Herbicide A plant growth regulator that inhibits growth or kills a plant when applied at relatively low concentrations.

Herbivore Animals that feed only on plants.

Herbivory The process of existing by eating macroscopic plants.

Heredity The faithful transmission of biological traits from parents to their offspring.

Heritability (in the broad sense)The amount of the variance in a trait in a population that is due to genetic factors; defined as the ratio of the variance due to genetic factors to the sum of variance due to both genetic and environmental factors; also called the degree of genetic determination.

Heritable Capable of being inherited, such as a physical trait.

Hermaphrodite An organism with both male and female functional reproductive organs. Hermaphroditism may refer to an aberration in unisexual animals; monoecy implies that this is the normal condition for the species.

Hermaphroditism A state characterized by the presence of both male and female reproductive organs in the same animal.

Hermatypic Relating to reef-forming corals.

Hermatypic coral A reef-building coral.

Hertz; Hz The derived SI unit of frequency, defined as a frequency of 1 cycle per s; it thus has units of s^{-1}.

Hesperidium A berry with a thick leathery peel (exocarp and mesocarp) and a fleshy endocarp arranged in sections.

Heterobimetallic complex A metal complex having two different metal atoms.

Heterocercal In some fishes, a tail with the upper lobe larger than the lower, and the end of the vertebral column somewhat upturned in the upper lobe, as in sharks.

Heterochromatic regions Having inactive genes. Inactive regions of chromosomes are said to be heterochromatic.

Heterochromatin Chromatin that stains intensely and appears to represent inactive genetic areas.

Heterochrony Evolutionary change in the relative time of appearance or rate of development of characteristics from ancestor to descendant.

Heteroconjugation Association between a base and the conjugate acid of a different base through a hydrogen bond. The term has its origin in the conjugate acid-base pair and is in no way related to conjugation of orbitals. Heteroassociation is a more appropriate term. Some authors refer to conjugated systems containing a heteroatom, e.g. pyridine, as "heteroconjugated systems".

Heterocyst A specialized cell of certain cyanobacteria; usually larger and thick-walled, this cell excludes oxygen and allows for anaerobic nitrogenase activity.

Heterodont Having teeth differentiated into incisors, canines, and molars for different purposes.

Heterogonic life cycle Life cycle involving alternation of parasitic and free-living generations.

Heteroleptic Transition metal or Main Group compounds having more than one type of ligand.

Heterolysis, heterolytic The cleavage of a covalent bond so that both bonding electrons remain with one of the two fragments between which the bond is broken.

$$A-B \longrightarrow A^+ + B^-$$

Heterolytic bond fission is a feature of many bimolecular reactions in solution (e.g., electrophilic substitution, nucleophilic substitution).

Heterolytic bond-dissociation energy The energy required to break a given bond of some specific compound by heterolysis. For the dissociation of a neutral molecule AB in the gas phase into A^+ and B^- the heterolytic bond-dissociation energy $D(A^+B^-)$ is the sum of the bond dissociation energy, D(A-B), and the adiabatic ionization energy of the radical A. minus the electron affinity of the radical B.

Heterophile reaction Antigen-antibody reaction in which the antibody was not specifically elicited by the antigen to which it binds.

Heterosporous Producing two distinct types of spores.

Heterotherm An animal whose body temperature fluctuates markedly; "cold-blooded."

Heterotroph An organism that is incapable of synthesizing its own food and, therefore, must feed upon organic compounds produced by other organisms.

Heterotrophic theory A theory of the origin of life which proposes that the first organisms obtained nutrition from the spontaneous formation of organic molecules derived from primordial gases.

Heteroxenous Describes a parasite that lives within more than one host during its life cycle.

Heterozygote An organism in which homologous chromosomes contain different allelic forms (often dominant and recessive) of a locus; derived from a zygote formed by union of gametes of dissimilar allelic constitution.

Heterozygous A genotype for a given phenotypic expression containing a dominant and a recessive allele for that trait.

Hexacanth Oncosphere; a six-hooked larva hatching from the egg of a eucestode.

Hexactinellida The class of sponges whose members are characterized by triaxon siliceous spicules, which are sometimes formed into an intricate lattice. Cup or vase shaped. Scyconoid body form. Glass sponges.

Hexamerous Six parts, specifically, symmetry based on six or multiples thereof.

Hexapoda The class of mandibulate arthropods whose members are characterized by having three pairs of legs. Commonly called insects. Hexapods often have wings and a body divided into head, thorax, and abdomen. Insecta has been used as an alternate class name.

Hexose A 6-carbon sugar.

Hibernation Condition, especially of mammals, of passing the winter in a torpid state in which the body temperature drops nearly to freezing and the metabolism drops close to zero.

Hierarchical system. A scheme arranging organisms into a series of taxa of increasing inclusiveness, as illustrated by Linnean classification.

High tide The highest level reached by the rising tide.

High-level waste repository A place where intensely radioactive wastes can be buried deep in the ground and remain unexposed to groundwater and earthquakes for tens of thousands of years. The first such site chosen in the United States is Yucca Mountain, Nevada, although it is not certain that a waste repository will ever be built there.

High-quality energy Intense, concentrated, and high-temperature energy that is considered high-quality because of its usefulness in carrying out work.

Hildebrand parameter A parameter measuring the cohesion of a solvent (energy required to create a cavity in the solvent).

Hill reaction The splitting of a molecule of water during the light reactions of photosynthesis; the photolysis of water.

Hilum Scar on a seed indicating where it was attached to the ovary.

Hindbrain Includes the medulla oblongata, cerebellum, and pons.

Hirudinea The class of annelids whose members are characterized by bodies with 34 segments, each of which is subdivided into annuli. Anterior and posterior suckers present. Leeches.

Histogenesis Formation and development of tissue.

Histology Study of tissues.

Histone Any of several simple proteins found in cell nuclei and complexed at one time or another with DNA. Histones yield a high proportion of basic amino acids on hydrolysis; characteristic of eukaryotes.

Histozoic Dwelling within the tissues of a host.

Hofmann rule "The principal alkene formed in the decomposition of quaternary ammonium hydroxides that contain different primary alkyl groups is always ethylene, if an ethyl group is present." Originally given in this limited form by A.W. Hofmann, the rule has since been extended and modified as follows: "When two or more alkenes can be produced in a -elimination reaction, the alkene having the smallest number of alkyl groups attached to the double bond carbon atoms will be the predominant product." This orientation described by the Hofmann rule is observed in elimination reactions of quaternary ammonium salts and tertiary sulfonium salts, and in certain other cases.

Holdfast An organ at the base of macroalgae that attaches the stalk to a rocky surface.

Hole Generally, the atoms in a semicondutor will form bonds with neighboring atoms in such a manner that each atom sees a "closed shell". Since the acceptor atom introduces fewer electrons, the shell is not closed; all of the electrons are tightly bound. The lack of a closed shell in the context of a semiconducting material is referred to as a hole. If an additional electron can be found, it will tend to fill the hole (close the shell). Occaisionally, it is possible for an electron in a neighboring closed shell to fill this shell leaving a new unclosed shell behind. So, the flow of electrons to the right (each moving right to fill a hole and leaving a new hole behind on the left) is equivolent to a hole moving to the left.

Holistic medicine The treatment of the whole entity, the human body, spirit, and mind_especially accepted in Eastern and primitive cultures.

Holoblastic Division of a zygote that results in separate blastomeres.

Holoblastic cleavage Complete and approximately equal division of cells in early embryo. Found in mammals, amphioxus, and many aquatic invertebrates that have eggs with a small amount of yolk.

Holometabolous metamorphosis A type of insect metamorphosis in which immatures, called larvae, are different in form and habitats from the adult; the last larval molt results in the formation of a pupa; radical cellular changes in the pupal stage end in adult emergence.

Holophytic nutrition Occurs in green plants and certain protozoa and involves synthesis of carbohydrates from carbon dioxide and water in the presence of light, chlorophyll, and certain enzymes.

Holoplankton species of zooplankton that remain in a planktonic stage throughout their lives

Holothuroidea The class of echinoderms whose members are elongate along the oral-aboral axis, have microscopic ossicles embedded in a muscular body wall, and have circumoral tentacles. Lack rays. Sea cucumbers.

Holozoic nutrition Type of nutrition involving ingestion of liquid or solid organic food particles.

Home range The area over which an animal ranges in its activities. Unlike territories, home ranges are not defended.

Homeobox A highly conserved 180-base pair sequence found in regulatory sequences of protein-coding genes that regulate development.

Homeostasis Tendency of living organisms to maintain a steady state in their internal environmental conditions, including body temperature, blood sugar level, and metabolic rate.

Homeotherm An animal, such as a bird or mammal, that maintains precisely controlled internal body temperatures using its own heating and cooling mechanisms.

Homeothermic Having a nearly uniform body temperature, regulated independent of the environmental temperature; "warm blooded."

Homeotic genes Genes, identified through mutations, that give developmental identity to specific body segments.

Homestead Act Legislation passed in 1862 allowing any citizen or applicant for citizenship over twenty-one years old and head of a family to acquire 160 acres of public land by living on it and cultivating it for five years.

Homing The ability of an animal to return to its home site after being displaced.

Hominid A member of the family Hominidae, now represented by one living species, Homo sapiens.

Hominoid Relating to the Hominoidea, a superfamily of primates to which the great apes and humans are assigned.

HOMO 1. An acronym for highest occupied molecular orbital. s

2. A prefix (consisting of lower case letters, homo,) used to indicate a higher homologue of a compound.

Homoaromatic Whereas in an aromatic molecule there is continuous overlap of p-orbitals over a cyclic array of atoms, in a homoaromatic molecule there is a formal discontinuity in this overlap resulting from the presence of a single sp3 hybridized atom at one or several positions within the ring; p-orbital overlap apparently bridges these sp3 centres, and features associated with aromaticity are manifest in the properties of the compound. Pronounced homoaromaticity is not normally associated with neutral molecules, but mainly with species bearing an electrical charge, e.g., the "homotropylium" cation, $C_8H_9^+$.

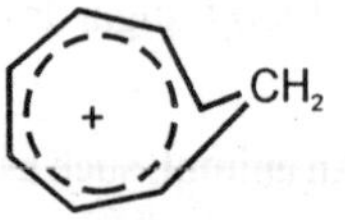

In bis, tris, (etc.) homoaromatic species, two, three, (etc.) single sp^3 centres separately interrupt the pi-electron system.

Homocercal A tail with the upper and lower lobes symmetrical and the vertebral column ending near the middle of the base, as in most teleost fishes.

Homoconjugation 1. Association between a base and its conjugate acid through a hydrogen bond ($B..HB^+$ or $AH..A^-$). Homoassociation is a more appropriate term for this phenomenon.

2. The orbital overlap of two pi systems separated by a non-conjugating group, such as CH_2.

Homodont Having a series of similar, unspecialized teeth.

Homogeneous Consistent and similar in composition.

Homogonic life cycle Life cycle in which all generations are parasitic or all are free living. There is no (or little) alternation of the two.

Homoleptic Transition metal or Main Group compounds having only one type of ligand are said to be homoleptic.

Homologous Structures that have a common evolutionary origin; the wing of a bat and the arm of a human are homologous; each can be traced back to a common ancestral appendage.

Homologous chromosomes Chromosomes that exist in pairs; each homologue possesses the same genes or loci, but the homologues may have different alleles at the same locus; one member of each pair comes from each parent.

Homologous pairs Two chromosomes that are morphologically and structurally identical and that pair during prophase of meiosis.

Homology Similarity of parts or organs of different organisms caused by evolutionary derivation from a corresponding part or organ in a remote ancestor, and usually having a similar embryonic origin. May also refer to a matching pair of chromosomes. Serial homology is the correspondence in the same individual of repeated structures having the same origin and development, such as the appendages of arthropods.

Homolysis, homolytic The cleavage of a bond ("homolytic cleavage" or "homolytic fission") so that each of the molecular fragments between which the bond is broken retains one of the bonding electrons. A unimolecular reaction involving homolysis of a bond (not forming part of a cyclic structure) in a molecular entity containing an even number of (paired) electrons results in the formation of two radicals:

It is the reverse of colligation. Homolysis is also commonly a feature of bimolecular substitution reactions (and of other reactions) involving radicals and molecules.

Homoplasy Phenotypic similarity among characteristics of different species or populations (including molecular, morphological, behavioral or other features) that does not accurately represent patterns of common evolutionary descent (= nonhomologous similarity); it is produced by evolutionary parallelism, convergence and/or reversal, and is revealed by incongruence among different characters on a cladogram or phylogenetic tree.

Homosporous Producing only one kind of spore.

Homothetogenic fission Mitotic fission across the rows of cilia of a protozoan.

Homozygote An organism in which the pair of alleles for a trait is composed of the same genes (either dominant or recessive but not both).

Homozygous Having two identical alleles for a given trait.

Hood Dorsal wall of the camerostome that extends over the capitulum.

Hooke's law The stress applied to any solid is proportional to the strain it produces within the elastic limit for that solid. The constant of that proportionality is the young modulus of elasticity for that substance.

Hoop conjecture The conjecture (as yet unproven, though there is substantial evidence to support it) that a nonspherical object, nonspherically compressed, will only form a black hole when all parts of the object lie within its event horizon; that is, when a "hoop" of the event horizon circumference can be rotated in all directions and will completely enclose the object in question.

Horizons Soil zones from the surface downward that reveal visible horizontal layers of soil.

Hormone 1. An organic molecule synthesized by a plant that exerts, even in low concentrations, profound regulation of growth and/or development.

2. Chemical products of ductless glands that are carried by the circulatory system and that influence various physiological processes in the body.

Horn The paired growths on the head of certain ungulate animals; the median growth of hair on the snout of the rhinoceros.

Horsehair worms Pseudocoelomate animals that belong to the phylum Nematomorpha. Also known as Gordian worms or hairworms (Gordius is the name for an ancient king who tied an intricate knot).

Horseshoe crabs Arthropods with a large horseshoe-shaped carapace.

Horticultural therapy The activities of propagating, growing, and planting plants is therapeutic for a variety of people including the elderly and mentally challenged.

Host A molecular entity that forms complexes with organic or inorganic guests, or a chemical species that can accommodate guests within cavities of its crystal structure. Examples include cryptands and crowns (where there are ion-dipole attractions between heteroatoms and positive ions), hydrogen-bonded molecules that form "clathrates" (e.g. hydroquinone and water), and host molecules of inclusion compounds (e.g. urea or thiourea). van der Waals forces and hydrophobic interactions bind the guest to the host molecule in clathrates and inclusion compounds.

Host organism An organism that provides lodging for a parasite.

Host specificity Degree to which a parasite is able to mature in more than one host species.

Hot desert Deserts of the American Southwest and Mexico; characterized by extreme summer heat and cacti.

Hubble constant The constant which determines the relationship between the distance to a galaxy and its velocity of recession due to the expansion of the universe. Since the universe is self-gravitating, it is not truly constant. In cosmology, it is defined as

H = (da/dt)/a,

Where a is the 4-radius of the universe. When evaluated for the present, it is written

H0 == h(t = now).

The hubble constant is not known to great accuracy (only within about a factor of 2), but is believed to lie somewhere between 50 and 100 km/s/mpc.

Hubble's law A relationship discovered between distance and radial velocity. The further away a galaxy is away from is, the faster it is receding away from us. The constant of proportionality is the hubble constant, h0. The cause is interpreted as the expansion of spacetime itself.

Hückel (4n + 2) rule Monocyclic planar (or almost planar) systems of trigonally (or sometimes digonally) hybridized atoms that contain (4n + 2) electrons (where n is a non-negative integer) will exhibit aromatic character. The rule is generally limited to n = 0-5.

This rule is derived from the Hückel MO calculation on planar monocyclic conjugated hydrocarbons $(CH)_m$ where m is an integer equal to or greater than 3 according to which (4n + 2) π electrons are contained in a closed-shell system. Examples of systems that obey the Hückel rule include:

Cyclopropenyl cation ($m = 3$, $n = 0$) Cyclopentadienide ($m = 5$, $n = 1$) Benzene ($m = 6$, $n = 1$)

Systems containing 4n π electrons (such as cyclobutadiene and the cyclopentadienyl cation) are "antiaromatic".

Human ecology The study of the interactions of humans with the environment.

Human resources Human wisdom, experience, skill, labor, and enterprise.

Humerus humer = shoulder.

Humidity Water vapor in the atmosphere.

Humoral immune response Binding of antigen with soluble antibody in blood serum. Also, the entire process by which the body responds to an antigen by producing antibody to that antigen.

Humoral Pertaining to an endocrine secretion.

Humus Sticky, brown, insoluble residue from the bodies of dead plants and animals; gives soil its structure, coating mineral particles and holding them together; serves as a major source of plant nutrients.

Hunter-gatherers Human social group secures food sources from wild resources such as hunting animal prey or collecting edible plants from the wild.

Huygens' construction; huygens' principle The mechanical propagation of a wave (specifically, of light) is equivalent to assuming that every point on the wavefront acts as point source of wave emission.

Hyaline A clear, glassy, structureless material occurring, for example, in cartilage, vitreous body, mucin, and glycogen.

Hybrid Offspring of a cross between two species or between alternate homozygous conditions.

Hybridization Linear combination of atomic orbitals on an atom. Hybrid orbitals are often used in organic chemistry to describe the bonding molecules containing tetrahedral (sp^3), trigonal (sp^2), and digonal (sp) atoms.

Hybridoma Fused product of a normal and a myeloma (cancer) cell, which has some of the characteristics of the normal cell.

Hydatid cyst A type of cyst formed by juveniles of certain tapeworms (Echinococcus) in their vertebrate hosts.

Hydatid sand Sediment in a hydatid cyst formed by free protoscolices.

Hydranth Nutritive zooid of hydroid colony.

Hydration Addition of water or of the elements of water (i.e. H and OH) to a molecular entity. For example, hydration of ethene:

$$CH_2{=}CH_2 + H_2O \rightarrow CH_3CH_2OH$$

The term is also used in a more restricted sense for the process.

Hydraulic skeleton The use of body fluids in open circulatory systems to give support and facilitate movement; muscles contracting in one part of the body force body fluids into some distant tissue space, thus causing a part of the body to extend or become turgid.

Hydrocarbon An organic molecule that contains only carbon and hydrogen and has its carbons bonded in a linear fashion.

Hydrocaulus Stalks or "stems" of a hydroid colony, the parts between the hydrorhiza and the hydranths.

Hydrocoel Second or middle coelomic compartment in echinoderms; left hydrocoel gives rise to water vascular system.

Hydrogen An element that is one of the constituents of water, organic matter, and many other chemicals. Hydrogen gas (H_2) is composed of two hydrogen atoms.

Hydrogen bond The hydrogen bond is a form of association between an electronegative atom and a hydrogen atom attached to a second, relatively electronegative atom. It is best considered as an electrostatic interaction, heightened by the small size of hydrogen, which permits proximity of the interacting dipoles or charges. Both electronegative atoms are usually (but not necessarily) from the first row of the Periodic Table, i.e., N, O, or F. Hydrogen bonds may be intermolecular or intramolecular. With a few exceptions, usually involving fluorine, the associated energies are less than 20-25 KJ mol^{-1} (5-6 KCal mol^{-1}).

Hydrogen potential Same as ph; refers to the number of charged hydrogen atoms or protons present, acidity of the soil.

Hydrogen sulfide (H_2S) The gas that is produced in anoxic sediments.

Hydrogenation Addition of one or more hydrogens to monunsaturated and polyunsaturated fatty acid chains.

Hydrogenosomes Small organelles in certain anaerobic protozoa that produce molecular hydrogen as an end product of energy metabolism.

Hydroid The polyp form of a Cnidarian as distinguished from the medusa form. Any Cnidarian of the class Hydrozoa, order Hydroida.

Hydrological cycle The cycling of water between reservoirs in oceans, lakes, and groundwater, and the atmosphere.

Hydrolysis The decomposition of a chemical compound by the addition of water; the splitting of a molecule into its groupings so that the split products acquire hydrogen and hydroxyl groups.

Hydron General name for the ion H^+ either in natural abundance, or where it is not desired to distinguish between the isotopes, as opposed to proton for $1H^+$, deuteron for $2H^+$ and triton for $3H^+$.

Hydrophilic "Water loving". The capacity of a molecular entity or of a substituent to interact with polar solvents, in particular with water, or with other polar groups.

Hydrophobic The property of a substance that has a tendency to repel water.

Hydrophobic interaction The tendency of hydrocarbons (or of lipophilic hydrocarbon-like groups in solutes) to form intermolecular aggregates in an aqueous medium, and analogous intramolecular interactions. The name arises from the attribution of the phenomenon to the apparent repulsion between water and hydrocarbons. However, the phenomenon ought to be attributed to the effect of the hydrocarbon-like groups on the water-water interaction. The misleading alternative term "hydrophobic bond" is discouraged.

Hydroponics A system of growing plants in a liquid nutrient solution without soil.

Hydrorhiza Rootlike stolon that attaches a hydroid to its substrate.

Hydrosphere Aqueous envelope of the earth.

Hydrostatic pressure The pressure exerted by a fluid (gas or liquid), defined as force per unit area. For example, the hydrostatic pressure of one atmosphere (1 atm) is 14.7 lb/in^2.

Hydrostatic skeleton. A mass of fluid or plastic parenchyma enclosed within a muscular wall to provide the support necessary for antagonistic muscle action; for example, parenchyma in acoelomates and perivisceral fluids in pseudocoelomates serve as hydrostatic skeletons.

Hydrothermal vents Deep, oceanic regions where the tectonic plates of the earth's crust are moving apart. They are characterized by occasional lava flows and hot water springs.

Hydroxyl Containing an OH_ group, a negatively charged ion formed by alkalies in water.

Hydrozoa The class of Cnidarians whose members have epidermally derived gametes, mesoglea without wandering amoeboid cells, and gastrodermis without nematocysts. Medusae, when present, with a velum. Hydra, Obelia, Physalia.

Hydrozoans Cnidarians whose life cycle typically includes a polyp, which is often colonial, and a medusa.

Hygroreceptor A receptor found in insects that detects the water content of air.

Hymenium The layer of fertile cells that produces spores in a fungal fruiting body.

Hyoid Y-shaped.

Hyomandibular Bone derived from the hyoid gill arch, forming part of articulation of the lower jaw of fishes, and forming the stapes of the ear of amniotic vertebrates.

Hyperapolysis Detachment of a tapeworm proglottid while still immature, before eggs are formed.

Hyperbola Graph of an inverse proportion.

Hypercharge A property of some baryons and leptons that is conserved in strong and electromagnetic interactions but not in weak interactions.

Hyperconjugation In the formalism that separates bonds into σ and π types, hyperconjugation is the interaction of σ-bonds (e.g. C-H, C-C, etc.) with a π network. This interaction is customarily illustrated by contributing structures, e.g. for toluene (below), sometimes said to be an example of "heterovalent" or "sacrificial hyperconjugation", so named because the contributing structure contains one two-electron bond less than the normal Lewis formula for toluene.

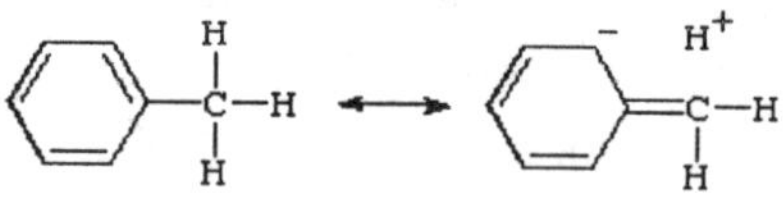

At present, there is no evidence for sacrificial hyperconjugation in neutral hydrocarbons.

The concept of hyperconjugation is also applied to carbenium ions and radicals, where the interaction is now between σ-bonds and an unfilled or partially filled π or p-orbital. A contributing structure illustrating this for the tert-butyl cation is:

This latter example is sometimes called an example of "isovalent hyperconjugation" (the contributing structure containing the same number of two-electron bonds as the normal Lewis formula).

Both structures shown on the right hand side are also examples of "double bond- no-bond resonance".

The interaction between filled π or p orbitals and adjacent antibonding σ^* orbitals is referred to as "negative hyperconjugation", as for example in the fluoroethyl anion.

Hyperinfection Condition in Strongyloides infections in which filariform juveniles repenetrate mucosa of the small intestine and proceed with migration.

Hypermetamorphosis Type of metamorphic development in which different larval instars have markedly dissimilar body forms.

Hyperosmotic Refers to a solution whose osmotic pressure is greater than that of another solution to which it is compared; contains a greater concentration of dissolved particles and gains water through a selectively permeable membrane from a solution containing fewer particles; contrasts with hypoosmotic.

Hyperparasitism Parasitism of a parasite by another parasite.

Hyperphagia A condition in which an animal does not stop eating when it normally would.

Hyperpolarized A description for a membrane whose polarity is greater than its typical resting potential.

Hypertonic A solution having a greater number of solute particles than another solution to which it is compared.

Hypertrophy Abnormal increase in size of a part or organ.

Hypha Microscopic threads that make up the body of most fungi.

Hypnozoite Dormant exoerythrocytic form found in certain Plasmodium species.

Hypoblast An inner layer of cells that results from the proliferation and movement of cells in the blastoderm of an avian or reptilian embryo.

Hypocotyl The shoot portion of a seedling below the cotyledonary node.

Hypocotyl hook The hooked portion of a hypocotyl formed at germination that assists a dicot shoot in pulling itself through the soil crust.

Hypodermis The cellular layer lying beneath and secreting the cuticle of annelids, arthropods, and certain other invertebrates.

Hypogynous Floral whorls (sepals, petals, stamens) inserted below the ovary of a flower.

Hypoosmotic Refers to a solution whose osmotic pressure is less than that of another solution with which it is compared or taken as a standard; contains a lesser concentration of dissolved particles and loses water during osmosis; contrasts with hyperosmotic.

Hypophagia A condition in which an animal does not eat as much as it normally would.

Hypopharynx Tonguelike lobe arising from the floor of the mouth in insects; variously modified for feeding in many groups.

Hypophysis Pituitary body.

Hypostome Name applied to structure in various invertebrates (such as mites and ticks), located at posterior or ventral area of mouth.

Hypothalamus A ventral part of the forebrain beneath the thalamus; one of the centers of the autonomic nervous system.

Hypotheca the smaller portion of a diatom frustule

Hypothesis An untested statement about nature; a scientific conjecture, or educated guess. Formally, a hypothesis is made prior to doing experiments designed to test it. Compare: law and theory.

Hypothetico-deductive Scientific process of making a conjecture and then seeking empirical tests that potentially lead to its rejection.

Hypothyroidism Listlessness and other metabolic symptoms caused by low thyroid hormone levels.

Hypotonic A solution having a lesser number of solute particles than another solution to which it is compared.

Hypsochromic shift Shift of a spectral band to higher frequency or shorter wavelength upon substitution or change in medium. It is informally referred to as blue shift.

Hysterosoma Combination of the metapodosoma and opisthosoma of the body of a tick or mite.

I

Ice age A period of time when significant amounts of ice form on the continents, inducing a fall in sea level.

Ice point The melting point of ice when in equilibrium with water saturated with air at standard atmospheric pressure.

Ick Serious disease of freshwater fishes, caused by the ciliate protozoan Ichthyophthirius multifiliis.

Icterus Yellowing of the skin and other organs because of bile pigments in the blood.

Ideal free distribution The distribution of individuals among resource patches of different quality that equalizes the net rate of gain of each individual. Assumes that organisms are free to move and have ideal knowledge about patch quality.

Ideal gas A theoretical gas consisting of infinitely small molecules that exert no forces on each other; also called perfect gas.

Ideal gas constant; universal molar gas constant; r The constant that appears in the ideal gas equation. It is equal to 8.314 34 j/k/mol.

Ideal-lens equation $1/p + 1/q = 1/f$, where p is the distance from object to lens, q is the distance from lens to image, and f is the focal length of the lens. This equation has important limitations, being only valid for thin lenses, and for paraxial rays. Thin lenses have thickness small compared to p, q, and f. Paraxial rays are those which make angles small enough with the optic axis that the approximation (angle in radian measure) = sin(angle) may be used.

Identity reaction A chemical reaction whose products are chemically identical with the reactants, for example the bimolecular self exchange reaction of CH_3I with I^-.

Idiosoma Posterior of the two basic parts of the body of a mite or tick, bearing the legs and most internal organs.

Ige Immunoglobulin E; specific class of antibodies involved in allergic reactions; individuals who suffer from allergies have elevated levels of these antibodies.

Igneous Most of the molten mass of the earth, igneous material cooled to form the earth's crust.

Igneous rocks Crystalline minerals solidified from molten magma from deep in the earth's interior; basalt, rhyolite, andesite, lava, and granite are examples.

Illumination The luminous flux per unit area of a surface.

Image A point mapping of luminous points of an object located in one region of space to points in another region of space, formed by refraction or reflection of light in a manner which causes light from each point of the object to converge to or diverge from a point somewhere else (on the image). The images which are useful generally have the character that adjacent points of the object map to adjacent points of the image without discontinuity, and is a recognizable (though perhaps somewhat distorted) mapping of the object.

Imago The adult and sexually mature insect.

Imagochrysalis Quiescent stage between the nymph and adult in the life cycle of a chigger mite.

Imbalance The situation in which reaction parameters that characterize different bond forming or bond breaking processes in the same reaction have developed to different extents as the transition state is approached along some arbitrarily defined reaction coordinate. For example, in the nitroalkane anomaly, the Brønsted β exponent for proton removal is smaller than the Brønsted α for the nitroalkane, because of imbalance between the amount of bond breaking and resonance delocalization in the transition state. Imbalance is common in reactions such as elimination, addition and other complex reactions that involve proton (hydron) transfer.

Imbibition The process of taking up water physically.

Imitation An act that occurs when an animal immediately copies the actions of another animal while they are in each other's presence.

Immediate hypersensitivity Biological manifestation of an antigen-antibody reaction in which the maximal response is reached in a few minutes or hours. Intradermal injection of the antigen produces local swelling and redness with heavy infiltration of polymorphonuclear leukocytes. Intravenous injection may produce anaphylactic shock and death.

Immune cross reaction Binding of an antibody or cell receptor site with an antigen other than the one that would provide an exact "fit"; that is, an antigen-antibody reaction in which the antigen is not the same one that stimulated the production of that antibody.

Immunity State in which a host is more or less resistant to an infective agent; preferably used in reference to resistance arising from tissues that are capable of recognizing and protecting the animal against "nonself."

Immunization The process of making one immune. The induction of protective immunity by administration of either 1. a vaccine or toxoid (active immunization) or

2. preformed antibodies (passive immunization).

Immunogenic Refers to any substance that is antigenic; that is, stimulates production of antibody or cell-mediated immunity.

Immunoglobulin Any of a group of plasma proteins, produced by plasma cells, that participates in the immune response by combining with the antigen that stimulated its production. Antibody.

Immunological Pertaining to the immune response, in which a protein antibody is synthesized by an organism to counteract some pathogenic factor.

Immunology The study of the immune system. This study reveals the many phenomena that are responsible for both acquired and innate immunity. It also includes the use of antibody-antigen reactions in other laboratory work (serology and immunochemistry).

Impedance 1. The ratio of applied wave-producing force to resulting displacement velocity of a wave-transmitting medium

2. The ratio of sound pressure to volume displacement at a given surface in a sound-transmitting medium

3. The ratio of the effective voltage to the effective current in an a-c circuit.

Impedance matching A technique used to ensure maximum transfer of energy from the output of one circuit to the input of another.

Imperfect flower Unisexual flower; either staminate or pistillate.

Imperfect stage The asexual phase in a fungus life cycle characterized by the production of asexual spores.

Impermeable Having the property of restricting the passage of substances.

Implantation im = into; plantare = to set. Attachment of the blastocyst to the endometrium.

Imprinting A process that occurs when an animal learns to make a particular response to only one type of animal or object. The sensory modes used for establishing such a connection can be visual, auditory, olfactory, or some combination of these, depending upon the animal.

Impulse The product of a force and the time interval during which it acts.In moving particles along the cell surface or, in ciliate protozoans, for locomotion.

Inbreeding depression In a small population, an accumulation of harmful genetic traits (through random mutations and natural selection) that lowers viability and reproductive success of enough individuals to affect the whole population.

Inbreeding The tendency among members of a population to mate preferentially with close relatives.

Incidence In epidemiology, the number of new cases of a disease per unit time; that is, a rate measurement. Contrast with prevalence.

Incidental parasite Accidental parasite.

Incisive Cut into.

Inclusion compound (inclusion complex) A complex in which one component (the host) forms a cavity or, in the case of a crystal, a crystal lattice containing spaces in the shape of long tunnels or channels in which molecular entities of a second chemical species (the guest) are located. There is no covalent bonding between guest and host, the attraction being generally due to van der Waals forces. If the spaces in the host lattice are enclosed on all sides so that the guest species is "trapped" as in a cage, such compounds are known as "clathrates" or "cage" compounds".

Incomplete dominance An interaction between alleles in which both alleles are expressed more or less equally, and the phenotype of the heterozygote is different from either homozygote.

Incomplete flower Lacking one or more of the four floral parts (sepals, petals, stamens, pistil).

Incomplete linkage Two genes that are carried on the same chromosome but are far enough apart that crossing-over between them occurs.

Incomplete protein Protein that lacks the full complement of essential amino acids in the correct proportions.

Incus The middle of a chain of three bones of the mammalian middle ear.

Indehiscent Denoting mature fruit that do not split open to release their seed.

Indehiscent fruit Dry fruit that does not split open on maturity.

Independent assortment The random alignment of homologous chromosomes in meiosis.

Independent data points A reference to the fact that each of the data points collected to use in a statistical test of a hypothesis must be independent of every other data point.

Independent variables The variables that define the treatment conditions in an experiment; literally, the variables that we manipulate to test their effects on behavior.

Indeterminate cleavage A type of embryonic development in which the fate of the blastomeres is not determined very early as to tissues or organs, for example, in echinoderms and vertebrates; regulative cleavage.

Indeterminate Pertaining to a stem that produces unrestricted vegetative growth; the stem does not terminate in flowering.

Index of refraction The ratio of the speed of light in a vacuum to its speed in a given matter medium.

Indigenous Pertains to organisms that are native to a particular region; not introduced.

Indirect development In arthropods, refers to development in which larva or nymph hatches from an egg and is distinctly different in body form from the adult; that is, development with metamorphosis.

Indirect fitness A measure of the potential for an individual to contribute copies of the genes it possesses to future generations through its influence on the reproduction of nondescendant relatives. Only the portion of the nondescendant kins' reproduction due to helping by the relative is included in the measure of its inclusive fitness.

Indirect flight Insect flight accomplished by flight muscles acting on the body wall. Changes in shape of the thorax cause wing movements. A single nerve impulse results in many cycles of the wings; also called asynchronous flight.

Individual distance The fixed minimal distance an animal keeps between itself and other members of its species.

Indoleacetic acid (IAA) The naturally occurring auxin.

Indo-west Pacific region The tropical Indian and west and central Pacific oceans.

Induced fit The precise fit between an enzyme and its substrate.

Induced magnetism Magnetism produced in a ferromagnetic substance by the influence of a magnetic field

Induced ovulators Species in which copulation is necessary to effect ovulation.

Inducible enzyme An enzyme present only when induced by some particular substrate.

Inductance The Property of an electric circuit by which a varying current induces a back emf in that circuit or a neighboring circuit.

Induction Reasoning from the particular to the general, that is, deriving a general statement (hypothesis) based on individual observations. In embryology, the alteration of cell fates as the result of interaction with neighboring cells.

Induction period The initial slow phase of a chemical reaction which later accelerates. Induction periods are often observed with radical reactions, but they may also occur in other systems (for example before steady-state concentration of the reactants is reached).

Inductive effect In strict definition, an experimentally observable effect (on rates of reaction, etc.) of the transmission of charge through a chain of atoms by electrostatic induction. A theoretical distinction may be made between the field effect, and the inductive effect as models for the Coulomb interaction between a given site within a molecular entity and a remote unipole or dipole within the same entity. The experimental distinction between the two effects has proved difficult, except for molecules of peculiar geometry, which may exhibit "reversed field effects". Ordinarily the inductive effect and the field effect are influenced in the same direction by structural changes in the molecule and the distinction between them is not clear. This situation has led many authors to include the field effect in the term "inductive effect". Thus the separation of values into inductive and resonance components does not imply the exclusive operation of a through-bonds route for the transmission of the non-conjugative part of the substituent effect. To indicate the all-inclusive use of the term inductive, the phrase "so-called inductive effect" is sometimes used. Certain modern theoretical approaches suggest that the "so-called inductive effect" reflects a field effect rather than through-bonds transmission.

Inductive reactance Reactance in an a-c circuit containing inductance, which causes a lagging current.

Inductomeric effect A molecular polarizability effect occurring by the inductive mechanism of electron displacement. The consideration of such an effect and the descriptive term have been regarded as obsolescent or even obsolete, but in recent years theoretical approaches have reintroduced substituent polarizability as a factor governing reactivity, etc. and its parametrization has been proposed.

Indusium A thin membrane that covers the sori of many ferns; the membrane often breaks at the time of spore maturation.

Industrial fishery A fishery in which the catch is used for purposes other than direct human consumption.

Industrial revolution Advances in science and technology that have given us power to understand and change our world.

Industrial sewage Wastewater from industries. Compare domestic sewage.

Industrial timber Trees used for lumber, plywood, veneer, particleboard, chipboard, and paper; also called roundwood.

Inelastic collision A collision in which the colliding objects stick together after impact.

Inert Stable and unreactive under specified conditions.

Inertia A descriptive term for that property of a body which resists change in its motion. Two kinds of changes of motion are recognized: changes in translational motion, and changes in rotational motion.

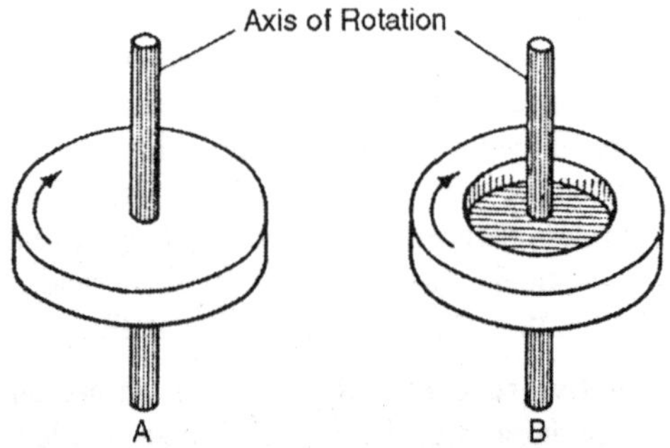

Fig. Moment of inertia

In modern usage, the measure of translational inertia is mass. Newton's first law of motion is sometimes called the "Law of Inertia", a label which adds nothing to the meaning of the first law. Newton's first and second laws together are required for a full description of the consequences of a body's inertia. The measure of a body's resistance to rotation is its Moment of Inertia.

Inertial confinement A nuclear fusion process in which a small pellet of nuclear fuel is bombarded with extremely high-intensity laser light.

Inertial frame A non-accelerating coordinate system. One in which F = ma holds, where F is the sum of all real forces acting on a body of mass m whose acceleration is a. In classical mechanics, the real forces on a body are those which are due to the influence of another body. Contact forces, gravitational, electric, and magnetic forces are real. Fictitious forces are those which arise solely from formulating a problem in a non-inertial system, in which ma = F

Infauna Animals that burrow in the substrate. Compare epifauna.

Inferior ovary Ovary lies below the attachment of the sepals, petals, and stamens; an epigynous flower.

Infiltration The process of water percolation into the soil and pores and hollows of permeable rocks.

Inflammation The complicated physiological process in mobilization of body defenses against foreign substances and infectious agents and repair of damage from such agents.

Inflammatory response A complex series of interactions between fragments of damaged cells, surrounding tissues, circulating blood cells, and specific antibodies; typical of infections.

Inflorescence A flower cluster with a definite arrangement

Informal economy Small-scale family businesses in temporary locations outside the control of normal regulatory agencies.

Information In communication, a measure of the uncertainty of the behavioral response of an individual. In learning, refers to the fact that for successful classical conditioning to occur there must be some intuitive knowledge of the relationship between the CS and US.

Infra Below.

Infraciliature All cilia, basal bodies, and their associated fibrils in a ciliate protozoan.

Infrapopulation All individuals of a single parasite species in one host.

Infrared That portion of the sun's total range of radiation having wavelengths immediately longer than the longest of the visible spectrum (red); between approximately 750 nm and 1m in length.

Infrared light Electromagnetic waves longer than those of visible light and shorter than radio waves.

Infrasonic range Vibrations in matter below 20 cycles/second.

Infundibulum Stalk of the neurohypophysis linking the pituitary to the diencephalon.

Infusoriform larva Ciliated larva produced by an infusorigen within a dicyemid mesozoan.

Infusorigen Mass of reproductive cells within a rhombogen.

Ingroup Taxon being studied in a cladistic analysis of evolutionary history.

Inhibition The decrease in rate of reaction brought about by the addition of a substance (inhibitor), by virtue of its effect on the concentration of a reactant, catalyst or reaction intermediate. For example, molecular oxygen and p-benzoquinone can react as "inhibitors" in many reactions involving radicals as intermediates by virtue of their ability to act as scavengers toward these radicals.

If the rate of a reaction in the absence of inhibitor is V_0 and that in the presence of a certain amount of inhibitor is V, the degree of inhibition (i) is given by

$$i = (V_0 - V)/V_0$$

Initiation A reaction or process generating free radicals which then induce a chain reaction. For example, in the chlorination of alkanes by a radical mechanism the initiation step is the dissociation of molecular chlorine.

Ink sac A gland found in some cephalopods that secretes a dark fluid

Innate A characteristic based partly or wholly on genetic or epigenetic constitution.

Innate releasing mechanism (IRM) A neural process, triggered by the sign stimulus, which preprograms an animal for receiving the sign stimulus and mediates specific behavioral responses.

Inner ear A paired, sound-sensitive organ in vertebrates.

Inner-sphere (electron transfer) Historically an electron transfer between two metal centres sharing a ligand or atom in their respective coordination shells. The definition has more recently been extended to any situation in which the interaction between the donor and acceptor centres in the transition state is significant (>20 kJ mol^{-1}).

Inorganic molecules (compounds) Pertaining to compounds that are not hydrocarbons or their derivatives. Compounds other than organic compounds.

Insecticide A chemical that kills insects.

Insectivore An insectivorous animal or plant. Any mammal of the order Insectivora, comprising the moles, shrews, and Old World hedgehogs.

Insectivorous plants Plants which actively or passively capture insects as an adaptation to acquire nitrogen. Examples include the venus flytrap, sundew, and pitcher plant.

Insects Arthropods with three pairs of legs and one pair of antennae. Few are marine, an exception being the water strider.

Insertion A chemical reaction or transformation of the general type

$$X\text{-}Z + Y \longrightarrow X\text{-}Y\text{-}Z$$

in which the connecting atom or group Y replaces the bond joining the parts X and Z of the reactant XZ. An example is the carbene insertion reaction

$$R_3C\text{-}H + H_2C: \longrightarrow R_3C\text{-}CH_3$$

The reverse of an insertion is called an extrusion.

Insertion mutant A type of mutation in which an extra base pair is inserted and shifts the sequence out of phase by one pair; also called a frame shift.

Insight learning The use of cognitive or mental processes to associate experiences and solve problems.

Insolation Incoming solar radiation.

Instantaneous current The magnitude of a varying current at any instant of time.

Instantaneous velocity Short displacement divided by elapsed time Slope of the line that is tangent to a velocity graph at a given point.

Instantaneous voltage The magnitude of a varying voltage at any instant of time.

Instar Stage in the life of an insect or other arthropod between molts.

Instinct Stereotyped, predictable, genetically programmed behavior. Learning may or may not be involved.

Instrumental conditioning Trial-and-error learning. The reinforcement of certain behaviors in animals leads to an animal repeating the behavior.

Instrumental value Value or worth of objects that satisfy the needs and wants of moral agents. Objects that can be used as a means to some desirable end.

Insulator A material which has a large energy gap is an insulating material. An insulator may conduct electricity; but only with a lot of effort — notice that wood is a good insulator, but when lightning strikes a tree, this is electricity flowing through wood. Both semi-conductors and conductors conduct better than insulators.

Intangible resources Abstract commodities, such as open space, beauty, serenity, genius, information, diversity, and satisfaction.

Integral Any quantity which can only have values equal to some integer (1, 2, 3..)

Integral protein Protein that spans or penetrates the lipid bilayer of cell membranes.

Integral, half Any quantity which can only have values equal to half of some odd integer (1/2, 3/2, 5/2..)

Integument Outermost layers of an ovule that typically develop into the seed coat.

Integumentary exchange The exchange of gases through the integument. Also cutaneous exchange.

Intelligence A collection of mental capacities including imagination, problem-solving ability, memory, the ability to use information gained from past experiences, perceptiveness, and behavioral flexibility. The processes by which animals obtain information about their environment, retain it and use the information to make decisions during the course of their behavioral activities.

Intensity Number of parasites in an infected host (infrapopulation). Mean intensity is the average number of parasites per infected host.

Intensity level The logarithm of the ratio of the intensity of a sound to the intensity of the threshold of hearing

Intensive variable A measurable property of a thermodynamic system is intensive if when two identical systems are combined into one, the variable of the combined system is the same as the original value in each system. Examples: temperature, pressure.

Intention movement The preparatory movements that an animal makes prior to a complete behavior pattern, as with wing movements before taking flight.

Interaction Any change in the amount or quantum numbers of particles that are near each other

Intercalary meristem A type of meristem present at the base of the blade and/or sheath of many monocots.

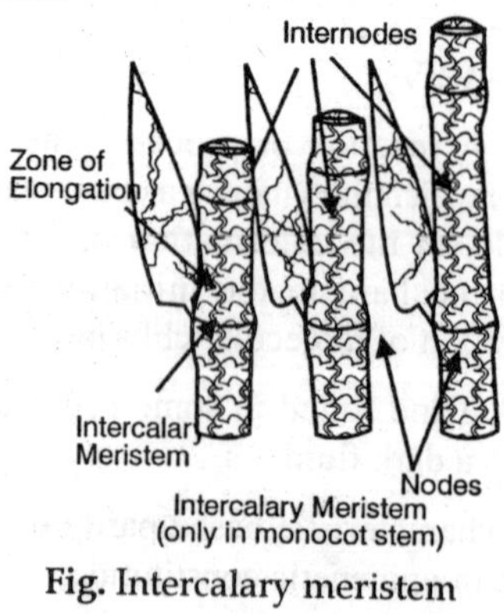

Fig. Intercalary meristem

Intercalated disk Membranous boundary between adjacent cardiac muscle cells.

Intercellular Occurring between body cells.

Interdemic selection The selection of entire breeding populations as the basic units.

Interface A surface that forms the boundary between two phases or systems.

Interfascicular cambium The layer of cambium that develops between vascular bundles and connects with the fascicular cambium to form the vascular cambium of woody tissues.

Interference 1. The superposing of one wave on another 2. The mutual effect of two beams of light, resulting in a reduction of energy in certain areas and an increase of energy in others.

Interferons Several cytokines encoded by different genes, important in mediation of natural immunity and inflammation.

Interkinesis The activities occurring between meiosis I and meiosis II; similar to interphase but without chromosome replication.

Interleukin Cytokines produced by white blood cells and mediating their own activities or those of other white blood cells.

Interleukin-1 A cytokine produced by macrophages that stimulates T helper lymphocytes.

Interleukin-2 A lymphokine produced by T helper lymphocytes that leads to proliferation of T helper cells and other T lymphocytes.

Intermediary meiosis Meiosis that occurs neither during gamete formation nor immediately after zygote formation, resulting in both haploid and diploid generations, such as in foraminiferan protozoa.

Intermediate A molecular entity with a lifetime appreciably longer than a molecular vibration (corresponding to a local potential energy minimum of depth greater than RT) that is formed (directly or indirectly) from the reactants and reacts further to give (either directly or indirectly) the products of a chemical reaction; also the corresponding chemical species.

Intermediate filament The chemically heterogeneous group of protein fibers, the specific proteins of which can vary with cell type. One of the three most prominent types of cytoskeletal filaments.

Intermediate host A host in which some development of a symbiont occurs, but in which maturation and sexual reproduction do not take place.

Intermediate inheritance. Neither of alternate alleles of a gene are completely dominant, and heterozygote shows a condition intermediate between or different from homozygotes for each allele.

Intermediate layer Of the three main layers of the ocean, the one below the surface, or mixed, layer. It includes the main thermocline.

Intermediate lobe The area in the pituitary gland between the anterior and posterior lobes.

Intermittent parasite Temporary parasite.

Intermittent reinforcement Schedules of reinforcement that involve providing rewards, not on every trial, but on usually one of four other types of schedules: fixed ratio, fixed interval, variable ratio, or variable interval.

Intermolecular 1. Descriptive of any process that involves a transfer (of atoms, groups, electrons, etc.) or interactions between two or more molecular entities.

2. Relating to a comparison between different molecular entities.

Internal combustion engine A heat engine in which the fuel burns inside the cylinder or turbine chamber.

Internal costs The expenses, monetary or otherwise, borne by those who use a resource.

Internal energy Total potential and kinetic energy of the molecules and atomic particles of a substance.

Internalizing costs Planning so that those who reap the benefits of resource use also bear all the external costs.

Interneuron A neuron located between a sensory neuron and a motor neuron. Interneurons function as integrating centers.

Internode Region on a stem between nodes.

Internuncial processes Cytoplasmic channels that connect one part of a cell to another, such as those linking the distal cytoplasm to the tegumental cytons in many flatworms.

Interobserver reliability A method for checking on the agreement between two or more people watching the same animals behave. This becomes important both as a check on the definitions developed for particular behavior patterns and when we have two or more individuals gathering the data in a particular investigation and we wish to know that they are taking down their records of the data in as nearly similar a manner as possible.

Interphase Period between two cell divisions when a cell is carrying on its normal functions. Long period of the cell cycle between one mitosis and the next. Includes G1 phase, S phase, and G2 phase. The replication of DNA occurs during interphase.

Interplanting The system of planting two or more crops, either mixed together or in alternating rows, in the same field; protects the soil and makes more efficient use of the land.

Intersecting storage ring An accelerator in which particles collide as they move in opposite directions

Intersexual selection Selection on characteristics of one sex (usually males) based on mate choices made by members of the other sex (usually females).

Interspecific competition In a community, competition for resources between members of different species.

Interstitial Situated in the interstices or spaces between structures such as cells, organs, or grains of sand.

Interstitial animal An animal that occupies the spaces (interstices) between sediment particles

Interstitial fauna Animals living between sediment particles.

Interstitial water The water contained between sediment particles.

Intertidal Living in the area between high and low tides and, thus, alternately exposed to the air and to the sea.

Intertidal zone The vertical extent of the shoreline between the high and low tide lines.

Intine The inner layer of the pollen wall.

Intracellular digestion Digestion that occurs inside a cell.

Intralecithal cleavage Cleavage in which the nuclei undergo several divisions within the yolk mass without concurrent cytokinesis; common in arthropods.

Intramolecular 1. Descriptive of any process that involves a transfer (of atoms, groups, electrons, etc.) or interactions between different parts of the same molecular entity.

2. Relating to a comparison between atoms or groups within the same molecular entity.

Intramolecular catalysis The acceleration of a chemical transformation at one site of a molecular entity through the involvement of another functional ("catalytic") group in the same molecular entity, without that group appearing to have undergone change in the reaction product. The use of the term should be restricted to cases for which analogous intermolecular catalysis by chemical species bearing that catalytic group is observable. Intramolecular catalysis can be detected and expressed in quantitative form by a comparison of the reaction rate with that of a comparable model compound in which the catalytic group is absent, or by measurement of the effective molarity of the catalytic group.

Intrasexual selection Selection on characteristics of the sexes based on competition among them (usually males) for access to members of the other sex (usually females).

Intraspecific competition In a community, competition for resources among members of the same species.

Intrinsic barrier (ΔG)The Gibbs energy of activation (G) in the limiting case where $\Delta G_0 = 0$, i.e. when the effect of thermodynamic driving force is eliminated. According to the Marcus equation, the intrinsic barrier is related to the reorganization energy, λ, of the reaction by the equation

$\Delta G = \lambda/4$

Intrinsic growth rate Exponential growth rate of a population, that is, the difference between the density-independent components of the birth and death rates of a natural population with stable age distribution.

Intron An intervening or noncoding segment of a gene; introns separate exons.

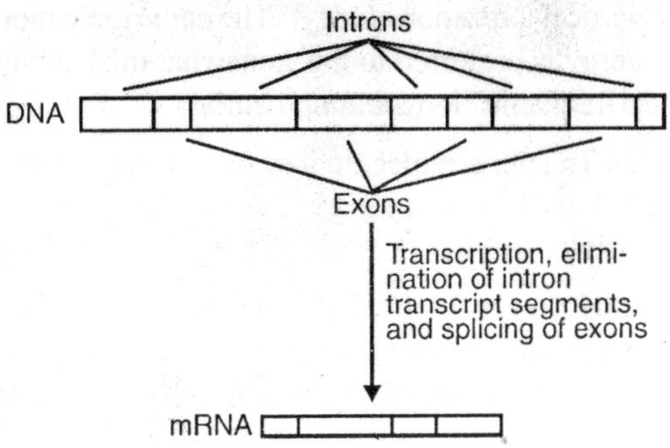

Fig. Intron

Introvert A tubular, eversible extension of the head, bearing the mouth at its tip.

Invagination An infolding of a layer of tissue to form a sac-like structure.

Inverse photoelectric effect The emission of photons of radiation from a material when bombarded with high speed electrons.

Inversion A turning inward or inside out, as in embryogenesis of sponges; also, reversal in order of genes or reversal of a chromosome segment.

Invertebrates Animals without backbones.

Involucre Whorl of bracts that subtend a flower or an inflorescence.

Involuntary Nervous system Stimulates smooth and cardiac muscle and glands of the body.

Involutions Infolding of membranes to increase surface area.

Iodinophilous vacuole Vacuole within a protozoan that stains readily with iodine.

Ion An atom or a group of atoms having an electric charge.

Ion pair A pair of oppositely charged ions held together by Coulomb attraction without formation of a covalent bond. Experimentally, an ion pair behaves as one unit in determining conductivity, kinetic behaviour, osmotic properties, etc.

Oppositely charged ions with their centres closer together than a distance

$q = 8.36 \times 10^6 Z^+Z^-/(\varepsilon_r T)$ pm

are considered to constitute an ion pair [Z+ and Z^- are the charge numbers of the ions, and r is the relative permittivity (or dielectric constant) of the medium.]

An ion pair, the constituent ions of which are in direct contact (and not separated by an intervening solvent or other neutral molecule) is designated as a "tight ion pair" (or "intimate" or "contact ion pair"). A tight ion pair of X^+ and Y^- is symbolically represented as X^+Y^-.

By contrast, an ion pair whose constituent ions are separated by one or several solvent or other neutral molecules is described as a "loose ion pair", symbolically represented as $X^+||Y^-$. The members of a loose ion pair can readily interchange with other free or loosely paired ions in the solution. This interchange may be detectable (e.g., by isotopic labelling) and thus afford an experimental distinction between tight and loose ion pairs.

A further conceptual distinction has sometimes been made between two types of loose ion pairs. In "solvent-shared ion pairs" the ionic constituents of the pair are separated by only a single solvent molecule, whereas in "solvent-separated ion pairs" more than one solvent molecule intervenes. However, the term "solvent-separated ion pair" must be used and interpreted with care since it has also widely been used as a less specific term for "loose" ion pair.

Ion pair return The recombination of a pair of ions R^+ and Z^- formed from ionization of RZ.

If the ions are paired as a tight ion pair and recombine without prior separation into a loose ion pair this is called "internal ion-pair return":

R^+Z^- (tight ion pair) $\rightleftharpoons$ RZ (covalent molecule)

It is a special case of "primary geminate recombination".

If the ions are paired as a loose ion pair and form the covalent chemical species via a tight ion pair, this is called "external ion-pair return":

$R^+||Z^-$ (loose ion pair) $\rightleftharpoons R^+Z^-$ (tight ion pair) RZ (covalent molecule)

It is a special case of "secondary geminate recombination".

When the covalent molecule RZ is reformed without direct evidence of prior partial racemization or without other direct evidence of prior formation of a tight ion pair, (e.g., without partial racemization if the group R is suitably chiral) the internal ion-pair return is sometimes called a "hidden return".

External (unimolecular) ion-pair return is to be distinguished from "external (bimolecular) ion return", the (reversible) process whereby dissociated ions are converted into loose ion pairs:

$R+ + Z^- \rightleftharpoons R^+||Z^-$

Ionic bond Chemical bond formed when ions of opposite charge attract.

Ionization The generation of one or more ions. It may occur, e.g. by loss of an electron from a neutral molecular entity, by the unimolecular heterolysis of such an entity into two or more ions, or by a heterolytic substitution reaction involving neutral molecules, such as

$CH_3CO_2H + H_2O \rightarrow H_3O+ + CH_3CO_2-$

$Ph_3CCl + AlCl_3' \rightarrow Ph_3C^+ + AlCl_4^-$ (electrophile-assisted)

$Ph_3CCl \rightarrow Ph_3C + Cl^-$(ion pair, in benzene)

The loss of an electron from a singly, doubly, etc. charged cation is called second, third, etc. ionization. This terminology is used especially in mass spectroscopy.

Ionization chamber A device used to detect the passage of charged rays or particles by their ionizing effect on a gas.

Ionization energy(SI unit kJ mol^{-1} or J per molecule) The minimum energy required to remove an electron from an isolated molecular entity (in its vibrational ground state) in the gaseous phase. If the resulting molecular entity is considered to be in its vibrational ground state, one refers to the energy as the "adiabatic ionization energy". If the molecular entity produced possesses the vibrational energy determined by the Franck-Condon principle (according to which the electron ejection takes place without an accompanying change in molecular geometry), the energy is called the "vertical ionization energy". The name ionization energy is preferred to the somewhat misleading earlier name "ionization potential".

Ionize To split a molecule into two or more parts, each part becoming an electrically charged particle.

Ionizing power A term to denote the tendency of a particular solvent to promote ionization of an uncharged or, less often, charged solute. The term has been used both in a kinetic and in a thermodynamic context.

Ionizing radiation High-energy electromagnetic radiation or energetic subatomic particles released by nuclear decay.

Ionosphere The lower part of the thermosphere.

Ions Electrically-charged atoms that have gained or lost electrons.

Ipso-attack The attachment of an entering group to a position in an aromatic compound already carrying a substituent group (other than hydrogen). The entering group may displace that substituent group but may also itself be expelled or migrate to a different position in a subsequent step. The term "ipso-substitution" is not used, since it is synonymous with substitution.

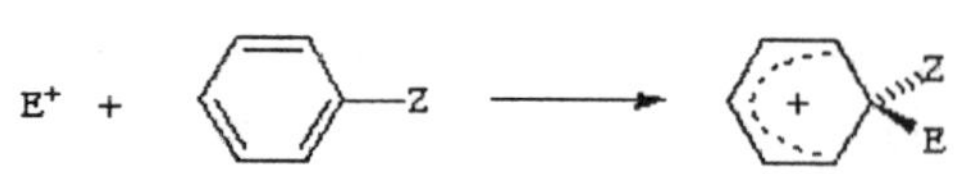

Iridocyte A fish skin cell that contains reflecting crystals made of guanine.

Iridophore A chromatophore with light-reflecting crystals.

Irregular reflection Scattering Reflection in many different directions from an irregular surface.

Irregular symmetry Bilateral symmetry; having only one plane through which the structure (flower) could be cut to result in mirror image halves; zygomorphy.

Irrigation The application of water to agricultural crops by transporting it to the crop through canals, pipes, and sprinklers.

Irritability The ability of a cell to undergo a change in membrane potential.

Irritants Corrosives (strong acids), caustics (alkaline reagents), and other substances that damage biological tissues on contact.

Ischia Hip.

Island arc A curved chain of volcanic islands that form along a trench.

Island biogeography The study of rates of colonization and extinction of species on islands or other isolated areas based on size, shape, and distance from other inhabited regions.

Island effects Reductions in species diversity caused by reduction in ecosystem area.

Island of fertility The region in a desert ecosystem directly beneath a tree or shrub. Leaf fall and accumulation of litter result in nutrient cycling directly around the plant, even though the bare soil between plants may be depleted of nutrients.

Isodesmic reaction A reaction (actual or hypothetical) in which the types of bonds that are made in forming the products are the same as those which are broken in the reactants, e.g.

$PhCOOH + p\text{-}ClC_6H_4COO^- \longrightarrow PhCOO\text{-} + p\text{-}ClC_6H_4COOH$

$ClCH{=}CH_2 + ClCH_2CH_2Cl\ CH_2 \longrightarrow CH_2 + Cl_2CHCH_2Cl$

Isoelectronic Two or more molecular entities are described as isoelectronic if they have the same number of valence electrons and the same structure, i.e. number and connectivity of atoms, but differ in some of the elements involved. Thus

CO, N_2 and NO^+ are isoelectronic

$CH_2{=}C{=}O$ and $CH_2{=}N{=}N$ are isoelectronic

CH_3COCH_3 and $CH_3N{=}NCH_3$ have the same number of electrons, but have different structures, hence they are not described as isoelectronic.

Isoentropic A reaction series is said to be isoentropic if the individual reactions of the series have the same standard entropy of activation.

Isogametes Outwardly similar male and female gametes.

Isogamy A condition in which gametes are equal in size.

Isohaline Having the same salinity.

Isokinetic relationship When a series of structurally related substrates undergo the same general reaction or when the reaction conditions for a single substrate are changed in a systematic way, the enthalpies and entropies of activation sometimes satisfy the relation

$\Delta H - \beta \Delta S = \text{constant}$

where the parameter β is independent of temperature. This equation (or some equivalent form) is said to represent an "isokinetic relationship". The temperature (at which all members of a series obeying the isokinetic relationship react at the same rate) is termed the "isokinetic temperature".

Supposed isokinetic relationships as established by direct correlation of ΔH with ΔS are often spurious and the calculated value of β is meaningless, because errors in H lead to compensating errors in ΔS. Satisfactory methods of establishing such relationships have been devised.

Isolating mechanisms Any physical or behavioral characteristic that prevents successful exchange of genes between members of different species or populations.

Isolecithal Pertaining to a zygote (or ovum) with yolk evenly distributed. Homolecithal.

Isolobal The term is used to compare molecular fragments with each other and with familiar species from organic chemistry. Two fragments are isolobal if the number, symmetry properties, approximate energy, and shape of the frontier orbitals and the number of electrons in them are similar.

Isomer One of several species (or molecular entities) that have the same atomic composition (molecular formula) but different line formulae or different stereochemical formulae and hence different physical and/or chemical properties.

Isomerization A chemical reaction, the principal product of which is isomeric with the principal reactant. An intramolecular isomerization that involves the breaking or making of bonds is a special case of a molecular rearrangement.

Isomerization does not necessarily imply molecular rearrangement (e.g. in the case of the interconversion of conformational isomers).

Isopods Small, dorsoventrally flattened crustaceans such as the sea louse.

Isoprenes Five-carbon compounds that are the basic unit of terpenes.

Isosbestic point This term is usually employed with reference to a set of absorption spectra, plotted on the same chart for a set of solutions in which the sum of the concentrations of two principal absorbing components, A and B, is constant. The curves of absorbance against wavelength (or frequency) for such a set of mixtures often all intersect at one or more points, called isosbestic points.

Isosbestic points are commonly met when electronic spectra are taken (a) on a solution in which a chemical reaction is in progress (in which case the two absorbing components concerned are a reactant and a product, A + B), or (b) on a solution in which the two absorbing components are in equilibrium and their relative proportions are controlled by the concentration of some other component, typically the concentration of hydrogen ions, e.g., an acid-base indicator equilibrium.

$$A \rightleftharpoons B + H^+_{aq}$$

The effect may also appear (c) in the spectra of a set of solutions of two unrelated non-interacting components having the same total concentration. In all these examples, A (and/or B) may be either a single chemical species or a mixture of chemical species present in invariant proportion.

If A and B are single chemical species, isosbestic points will appear at all wavelengths at which their molar absorption coefficients (formerly called extinction coefficients) are the same. (A more involved identity applies when A and B are mixtures of constant proportion.)

If absorption spectra of the types considered above intersect not at one or more isosbestic points but over progressively changing wavelength, this is prima facie evidence in case (a) for the formation of a reaction intermediate in substantial concentration ($A \rightarrow C \rightarrow B$), in case (b) for the involvement of a third absorbing species in the equilibrium, e.g.

$$A \rightleftharpoons B + H^+_{aq} \rightleftharpoons C + 2\,H^+_{aq}$$

or in case (c) for some interaction of A and B, e.g.,

$$A + B \rightleftharpoons C$$

Isoselective relationship A relationship analogous to the isokinetic relationship, but applied to selectivity data of reactions. At the isoselective temperature, the selectivities of the series of reactions following the relationship are identical.

Isothermal process A thermal process that takes place at constant temperature.

Isotonic Pertaining to solutions having the same or equal osmotic pressure; isosmotic.

Isotope Alternate form of an element with a different number of neutrons but the same number of protons and electrons.

Isotope effect The effect on the rate or equilibrium constant of two reactions that differ only in the isotopic composition of one or more of their otherwise chemically identical components is referred to as a kinetic isotope effect or a thermodynamic (or equilibrium) isotope effect respectively.

Isotope effect, heavy atom An isotope effect due to isotopes other than those of hydrogen.

Isotope effect, intramolecular A kinetic isotope effect observed when a single substrate, in which the isotopic atoms occupy equivalent reactive positions, reacts to produce a non-statistical distribution of isotopologue products. In such a case the isotope effect will favor the pathway with lower force constants for displacement of the isotopic nuclei in the transition state.

Isotope effect, kinetic The effect of isotopic substitution on a rate constant is referred to as a kinetic isotope effect.

For example in the reaction

$$A + B \longrightarrow C$$

the effect of isotopic substitution in reactant A is expressed as the ratio of rate constants kl/kh, where the superscripts l and h represent reactions in which the molecules A contain the light and heavy isotopes, respectively.

Within the framework of transition state theory in which the reaction is rewritten as

$$A + B \rightleftharpoons [TS] \longrightarrow C$$

and with neglect of isotopic mass on tunnelling and the transmission coefficient, k^l/kh can be regarded as if it were the equilibrium constant for an isotope exchange reaction between the transition state [TS] and the isotopically substituted reactant A, and calculated from their vibrational frequencies as in the case of a thermodynamic isotope effect.

Isotope effects like the above, involving a direct or indirect comparison of the rates of reaction of isotopologues, are called "intermolecular", in contrast to intramolecular isotope effects, in which a single substrate reacts to produce a non-statistical distribution of isotopologue product molecules.

Isotope effect, primary A kinetic isotope effect attributable to isotopic substitution of an atom to which a bond is made or broken in the rate-controlling step or in a pre-equilibrium step of a specified reaction is termed a primary isotope effect. The corresponding isotope effect on the equilibrium constant of a reaction in which one or more bonds to isotopic atoms are broken, is called a "primary equilibrium isotope effect".

Isotope effect, secondary A kinetic isotope effect that is attributable to isotopic substitution of an atom to which bonds are neither made nor broken in the rate-controlling step or in a pre-equilibrium step of a specified reaction, and is therefore not a primary isotope effect, is termed a secondary isotope effect. The corresponding isotope effect on the equilibrium constant of such a reaction is called a "secondary equilibrium isotope effect".

Secondary isotope effects have been discussed in terms of the conventional electronic effects of physical organic chemistry, e.g. induction, hyperconjugation, hybridization, etc., since these properties are determined by the electron distribution, that depends on vibrationally averaged bond lengths and angles which vary slightly with isotopic substitution.

Isotope effect, solvent A kinetic or equilibrium isotope effect resulting from change in the isotopic composition of the solvent.

Isotope effect, steric A secondary isotope effect attributed to the different vibrational amplitudes of isotopologues. For example, both the mean and mean-square amplitudes of vibrations associated with C-H bonds are greater than those of C-D bonds. The greater effective bulk of molecules containing the former may be manifested by a steric effect on a rate or equilibrium constant.

Isotope effect, thermodynamic The effect of isotopic substitution on an equilibrium constant is referred to as a thermodynamic (or equilibrium) isotope effect.

For example, the effect of isotopic substitution in reactant A that participates in the equilibrium:

$$A + B \rightleftharpoons C$$

is the ratio Kl/Kh of the equilibrium constant for the reaction in which A contains the light isotope to that in which it contains the heavy isotope. The ratio can be expressed as the equilibrium constant for the isotopic exchange reaction:

$$Al + Ch \rightleftharpoons Ah + Cl$$

in which reactants such as B that are not isotopically substituted do not appear.

The potential energy surfaces of isotopic molecules are identical to a high degree of approximation, so thermodynamic isotope effects can only arise from the effect of isotopic mass on the nuclear motions of the reactants and products, and can be expressed quantitatively in terms of partition function ratios for nuclear motion:

$$\frac{K^1}{K^h} = \frac{\left(Q^1_{nuc} / Q^h_{nuc}\right)C}{\left(Q^1_{nuc} / Q^h_{nuc}\right)A}$$

Although the nuclear partition function is a product of the translational, rotational and vibrational partition functions, the isotope effect is determined almost entirely by the last named, specifically by vibrational modes involving motion of isotopically different atoms. In the case of light atoms (i.e. protium vs. deuterium or tritium) at moderate temperatures, the isotope effect is dominated by zero-point energy differences.

Isotope exchange A chemical reaction in which the reactant and product chemical species are chemically identical but have different isotopic composition. In such a reaction the isotope distribution tends towards equilibrium (as expressed by fractionation factors) as a result of transfers of isotopically different atoms or groups.

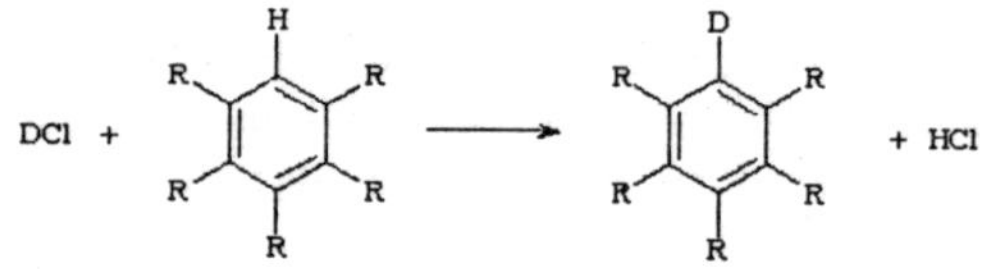

Isotopic perturbation, method of NMR shift difference measurement of the isotope effect on a fast (degenerate) equilibrium between two, except for isotopic substitution, species which are equivalent. This can be used to distinguish a rapidly equilibrating mixture with time-averaged symmetry from a single structure with higher symmetry.

Isotopic scrambling The achievement, or the process of achieving, an equilibrium distribution of isotopes within a specified set of atoms in a chemical species or group of chemical species.

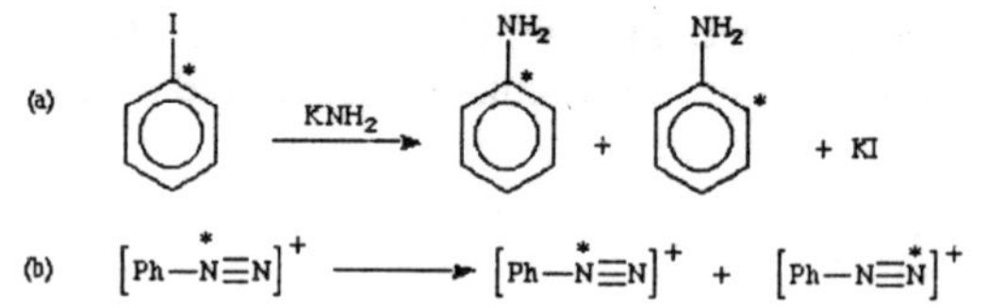

Isotopologue A molecular entity that differs only in isotopic composition (number of isotopic substitutions), e.g. CH_4, CH_3D, CH_2D_2.

Isotopomer Isomers having the same number of each isotopic atom but differing in their positions. The term is a contraction of "isotopic isomer".

Isotopomers can be either constitutional isomers (e.g. $CH_2DCH{=}O$ and $CH_3CD{=}O$) or isotopic stereoisomers (e.g. (R)- and (S)-CH_3CHDOH or (Z)- and (E)-$CH_3CH{=}CHD$).

Isozyme Different chemical forms of the same enzyme; thought to be important in adaptation to environmental extremes.

Iteroparity The production of offspring by an organism in successive bouts. Contrast with semelparity.

J

J curve A growth curve that depicts exponential growth; called a J curve because of its shape.

Jacket cells Ciliated somatoderm of an orthonectid mesozoan.

Jacobson's (vomeronasal) organ Olfactory receptor present in most reptiles; blind-ending sacs that open through the secondary palate into the mouth cavity; they are used to sample airborne chemicals.

Jawless fishes Fishes that lack jaws and paired fins: hagfishes (or slime eels) and lampreys.

Jejunum Portion of the small intestine located between the duodenum and the ileum.

Jet streams Powerful winds or currents of air that circulate in shifting flows; similar to oceanic currents in extent and effect on climate.

Johnston's organ Mechanoreceptor (auditory receptor) Found at the base of the antennae of male mosquitoes and midges.

Josephson effects Electrical effects observed when two superconducting materials are separated by a thin layer of insulating material.

Joule The derived SI unit of energy defined as the amount of work done by moving an object through a distance of 1 m by applying a force of 1 n; it thus has units of n m.

Joule's first law The heat q produced when a current i flows through a resistance r for a specified time t is given by

$$Q = i^2rt$$

Joule's second law The internal energy of an ideal gas is independent of its volume and pressure, depending only on its temperature.

Joule-Thomson effect The change in temperature that occurs when a gas expands into a region of lower pressure.

Junction detector A solid-state device based on the transistor principle that is used to detect the passage of charged particles.

Juvenile hormone Hormone produced by the corpora allata of insects; among its effects are maintenance of larval or nymphal characteristics during development.

Juxtaglomerular apparatus Complex of sensory cells located in the afferent arteriole adjacent to the glomerulus and a loop of the distal tubule, which produces the enzyme renin.

K

Kala-azar Disease caused by Leishmania donovani. Also called Dumdum fever or visceral leishmaniasis.

Kamlet-Taft solvent parameters Parameters of the Kamlet-Taft solvatochromic relationship which measure separately the hydrogen bond donor (α), hydrogen bond acceptor (β), and dipolarity/polarizability (π) properties of solvents as contributing to overall solvent polarity.

Karyogamy In sexual reproduction, the fusion of genetically distinct nuclei.

Karyotyping The determination of the number and structure of chromosomes in an individual.

Kekulé structure (for aromatic compounds) A representation of an aromatic molecular entity (such as benzene), with fixed alternating single and double bonds, in which interactions between multiple bonds are assumed to be absent.

For benzene,

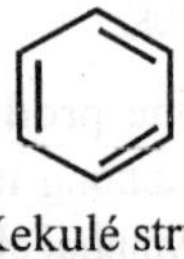 and 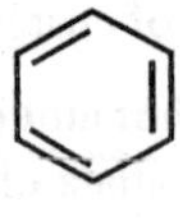

are the Kekulé structures.

Kelp A group of large brown seaweeds.

Kelvin scale The scale of temperature having a single fixed point, the temperature of the triple point of water, which is assigned the value 273.16^{o}k

Kentrogon Larva in the crustacean order Rhizocephala that is attached to its host crab; formed after the cypris larva molts and its appendages and carapace are discarded.

Kepler's first law A planet orbits the sun in an ellipse with the sun at one focus.

Kepler's second law A ray directed from the sun to a planet sweeps out equal areas in equal times.

Kepler's third law The square of the period of a planet's orbit is proportional to the cube of that planet's semimajor axis; the constant of proportionality is the same for all planets.

Keratin A tough, water-resistant protein found in the epidermal layers of the skin. Found in hair, feathers, hoofs, nails, claws, bills, etc.

Kerr effect The ability of certain substances to differently refract light waves whose vibrations are in different directions when the substance is placed in an electric field.

Keystone predator A predator that is very important in the maintenance of the structure of a community.

Keystone species A species upon which several other species depend. Removal of the keystone species leads to the death or disappearance of the dependent species.

Kilocalorie A unit of heat equal to 1,000 calories.

Kilogram; kg The fundamental SI unit of mass, which is the only SI unit still maintained by a physical artifact: a platinum-iridium bar kept in the international bureau of weights and measures at Sevres, France.

Kilowatt hour A unit of electric energy equal to 3.6×10^6 Ws.

Kin selection The selection of genes due to individuals assisting the survival and reproduction of nondescendant relatives who possess the same genes by common descent.

Kinesis Movements by an organism in random directions in response to stimulus.

Kinetic ambiguity Synonymous with kinetic equivalence.

Kinetic control (of product composition) The term characterizes conditions (including reaction times) that lead to reaction products in a proportion governed by the relative rates of the parallel (forward) reactions in which the products are formed, rather than by the respective overall equilibrium constants.

Kinetic electrolyte effect (kinetic ionic-strength effect) The general effect of an added electrolyte (i.e. an effect other than, or in addition to, that due to its possible involvement as a reactant or catalyst) on the observed rate constant of a reaction in solution. At low concentrations (when only long-range coulombic forces need to be considered) the effect on a given reaction is determined only by the ionic strength of the solution and not by the chemical identity of the ions. For practical purposes, this concentration range is roughly the same as the region of validity of the Debye-Hückel limiting law for activity coefficients. At higher concentrations, the effect of an added electrolyte depends also on the chemical identity of the ions. Such specific action can usually be interpreted as the incursion of a reaction path involving an ion of the electrolyte as reactant or catalyst, in which case the action is not properly to be regarded just as a kinetic electrolyte effect.

Kinetic electrolyte effects are usually (too restrictively and therefore incorrectly) referred to as "kinetic salt effects". A kinetic electrolyte effect ascribable solely to the influence of the ionic strength on activity coefficients of ionic reactants and transition states is called a "primary kinetic electrolyte effect". A kinetic electrolyte effect arising from the influence of the ionic strength of the solution upon the pre-equilibrium concentration of an ionic species that is involved in a subsequent rate-limiting step of a reaction is called a "secondary kinetic electrolyte effect". A common case encountered in practice is the effect on the concentration of hydrogen ion (acting as catalyst) produced from the ionization of a weak acid in a buffer solution.

Kinetic energy The energy a body has by virtue of its motion. The kinetic energy is the work done by an external force to bring the body from rest to a particular state of motion.

Kinetic equivalence Two reaction schemes are kinetically equivalent if they imply the same rate law.

For example, consider the two schemes (i) and (ii) for the formation of C from A:

(i) $$A \underset{k_{-1}\,.\,OH^-}{\overset{k_1\,.\,OH^-}{\rightleftharpoons}} B \xrightarrow{k_2} C$$

Providing that B does not accumulate as a reaction intermediate.

$$\frac{d[C]}{dt} = \frac{k_1k_2[A][OH^-]}{k_2 + k_{-1}[OH^-]} \qquad (1)$$

(ii) $$A \underset{k_{-1}\,.\,OH^-}{\overset{k_1\,.\,OH^-}{\rightleftharpoons}} B \xrightarrow[OH^-]{k_2} C$$

Providing that B does not accumulate as a reaction intermediate.

$$\frac{d[C]}{dt} = \frac{k_1k_2[A][OH^-]}{k_{-1} + k_2[OH^-]} \qquad (2)$$

Both equations for d[C]/dt are of the form

$$\frac{d[C]}{dt} = \frac{r\,[A][OH^-]}{1 + s[OH^-]} \qquad (3)$$

where r and s are constants (sometimes called "coefficients in the rate equation"). The equations are identical in their dependence on concentrations and do not distinguish whether OH^- catalyses the formation of B, and necessarily also its reversion to A, or is involved in its further transformation to C. The two schemes are therefore kinetically equivalent under conditions to which the stated provisos apply.

Kinetic theory The molecules of matter are continuously in motion and the collisions between molecules are perfectly elastic.

Kinetid Axoneme of a cilium of flagellum together with its basal fibrils and organelles. Also called a mastigont.

Kinetochore A disc of proteins located on the centromere, specialized to interact with the spindle fibers during mitosis.

Kinetodesma Fibril arising from the kinetosome of a cilium in a ciliate protozoan, and passing along the kinetosomes of cilia in that same row.

Kinetodesmose Compound fiber joining cilia into rows.

Kinetoplast Conspicuous part of a mitochondrion in a trypanosome; usually found near the kinetosome.

Kinetosome Centriole from which an axoneme arises. Also called a basal body or blepharoplast.

Kinety All the kinetosomes and kinetodesmata of a row of cilia.

Kingdom The highest level of classification of life; the most widely accepted classification system includes five kingdoms: Monera, Protista, Fungi, Plantae, and Animalia.

Kinin A type of local hormone that is released near its site of origin; also called parahormone or tissue hormone.

Kinocilium A long cilium at the apex of a hair cell.

Kinorhyncha The phylum of aschelminths that contains members called kinorhynchs; small elongate worms found exclusively in marine environments where they live in mud and sand.

Kirchhoff's first law An incandescent solid or gas under high pressure will produce a continuous spectrum.

Kirchhoff's second law A low-density gas will radiate an emission-line spectrum with an underlying emission continuum.

Kirchhoff's third law Continuous radiation viewed through a low-density gas will produce an absorption-line spectrum.

Kirkwood gaps Gaps in the asteroid belt, caused by resonance effects from jupiter. Similar gaps exist in saturn's rings, due to the resonance effects of shepherd moons.

Knockout gene A genetically engineered mutant gene that is introduced into an embryo to study the specific effects of that gene.

Knot Commonly seen as an irregular formation in a longitudinal section of wood formed when a branch becomes surrounded (buried) by formation of secondary xylem.

Known resources Minerals or other useful environmental materials or services that are identified and partially mapped; may not be environmentally or socially acceptable or economically feasible to exploit.

Koch's blue bodies Schizonts of Theileria parva in circulating lymphocytes.

Kohlrausch's law If a salt is dissolved in water, the conductivity of the solution is the sum of two values — one depending on the positive ions and the other on the negative ions.

Koppel-palm solvent parameters Parameters to measure separately the ability of a solvent to enter into non-specific solvent-solute interactions (permittivity and refractive index nD) and specific solvent-solute interaction (solvent basicity or nucleophilicity B and solvent acidity or electrophilicity E) as contributing to overall solvent polarity.

Kranz anatomy A specialized leaf anatomy found in C4 plants in which the vascular bundle is surrounded by bundle sheath cells.

Krebs Cycle The second stage of cellular respiration that occurs in the mitochondria; completes the breakdown of glucose into carbon dioxide.

Krill Any of the small, pelagic, shrimplike crustaceans. Krill are an important source of energy in antarctic food webs.

K-selected Organisms whose populations are maintained near the carrying capacity of the environment.

K-strategist Species of organism that uses a survival and reproductive "strategy" characterized by low fecundity, low mortality, and longer life and with populations approaching the carrying capacity of the environment, controlled by density-dependent factors.

Kupffer cells Phagocytic epithelial cells lining the sinusoids of the liver.

Kwashiorkor Malnutrition caused by diet high in carbohydrate and extremely low in protein.

L

Labellum Expanded tip of the labium in an insect.

Labial palp 1. Chemosensory appendage found on the labium of insects (Arthropoda).

2. Flaplike lobe surrounding the mouth of bivalve molluscs that directs food toward the mouth.

Labile The term has loosely been used to describe a relatively unstable and transient chemical species or (less commonly) a relatively stable but reactive species. It must therefore not be used without explanation of the intended meaning.

Labium The posterior mouthpart of insects. It is often referred to as the "lower lip," is chemosensory, and was derived evolutionarily from paired head appendages (Hexapoda, Arthropoda).

Labrum The upper lip of insects and crustaceans situated above or in front of the mandibles; also refers to the outer lip of a gastropod shell.

Labyrinth organ One of a pair of equilibrium organs in vertebrates that contains three fluid-filled semicircular canals.

Labyrinth Vertebrate internal ear, composed of a series of fluid-filled sacs and tubules (membranous labyrinth) suspended within bone cavities (osseous labyrinth).

Labyrinthodont A group of Paleozoic amphibians containing the temnospondyls and the anthracosaurs.

Lachrymal Secreting or relating to tears.

Lactation The production of milk by the mammary glands.

Lacteal One of the lymph vessels in the villus of the intestine.

Lactic acid The end product of anaerobic respiration in animals, causes muscle soreness.

Lactose Milk sugar; a disaccharide composed of glucose and galactose.

Lacunae Channels making up the lacunar system in Acanthocephala. Also, in developing wings of insects, canals that contain nerves, tracheae, and hemolymph.

Lacunar system System of canals in the body wall of an acanthocephalan, functioning as a circulatory system.

Lagena Portion of the primitive ear in which sound is translated into nerve impulses; evolutionary beginning of cochlea.

Lagoon A shallow, sheltered body of water separated from the open sea by coral reefs, sand bars, and/or barrier islands.

Lagrange points Points in the vicinity of two massive bodies (such as the earth and the moon) where each others' respective gravities balance. There are five, labelled 11 through 15. L1, 12, and 13 lie along the centerline between the centers of mass between the two masses; l1 is on the inward side of the secondary, l2 is on the outward side of the secondary; and l3 is on the outward side of the primary. L4 and l5, the so-called trojan points, lie along the orbit of the secondary around the primary, sixty degrees ahead and behind of the secondary.

L1 through l3 are points of unstable equilibrium; any disturbance will move a test particle there out of the lagrange point. L4 and l5 are points of stable equilibrium, provided that the mass of the secondary is less than about 1/24.96 the mass of the primary. These points are stable because centrifugal pseudoforces work against gravity to cancel it out.

Lamarckism Hypothesis, as expounded by Jean Baptiste de Lamarck, of evolution by the acquisition during an organism's lifetime of characteristics that are transmitted directly to offspring.

Lambdoidal lambda = like the Greek letter lambda.

Lambert's first law The illuminance on a surface illuminated by light falling on it perpendicularly from a point source is proportional to the inverse square of the distance between the surface and the source.

Lambert's second law If the rays meet the surface at an angle, then the illuminance is proportional to the cosine of the angle with the normal.

Lambert's third law The luminous intensity of light decreases exponentially with distance as it travels through an absorbing medium.

Lamella One of the two plates forming a gill in a bivalve mollusc. One of the thin layers of bone laid concentrically around an osteon (Haversian canal). Any thin, platelike structure.

Lamp shells Invertebrates that have

Lancelets Chordates with the three basic chordate characteristics but lacking a backbone.

Land ethic A philosophy or attitude about land that provides for its protection or more careful management.

Land races Traditional varieties of plant crops.

Land reform Democratic redistribution of landownership to recognize the rights of those who actually work the land to a fair share of the products of their labor.

Land rehabilitation A utilitarian program to repair damage and make land useful to humans.

Landauer's principle A principle which states that it doesn't explicitly take energy to compute data, but rather it takes energy to erase any data, since erasure is an important step in computation.

Landfills Land disposal sites for solid waste; operators compact refuse and cover it with a layer of dirt to minimize rodent and insect infestation, wind-blown debris, and leaching by rain.

Landscape ecology The study of the reciprocal effects of spatial pattern on ecological processes. A study of the ways in which landscape history shapes the features of the land and the organisms that inhabit it as well as our reaction to, and interpretation of, the land.

Landscape epidemiology Approach to epidemiology that employs all ecological aspects of a nidus. By recognizing certain physical conditions, the epidemiologist can anticipate whether a disease can be expected to exist.

Landslide The sudden fall of rock and earth from a hill or cliff. Often triggered by an earthquake or heavy rain.

Langmuir cells Parallel pairs of surface ocean convection cells driven by winds.

Laplace equation For steady-state heat conduction in one dimension, the temperature distribution is the solution to laplace's equation, which states that the second derivative of temperature with respect to displacement is zero.

Lappets Lobes around the margin of scyphozoan medusae (phylum Cnidaria).

Large intestine That part of the digestive system between the ileocecal valve of the small intestine and the anus; removes salt and water from undigested food and releases feces through the anus.

Larva 1. The immature, feeding stage of an insect that undergoes holometabolous metamorphosis. 2. The immature stage of any animal species in which adults and immatures are different in body form and habitat.

Larvacea The class of urochordates whose members are planktonic and whose adults retain a tail and notochord. With a gelatinous covering of the body.

Larval instars Any of the different immature feeding stages of an insect that undergoes holometabolous metamorphosis.

Larynx Modified upper portion of respiratory tract of air-breathing vertebrates, bounded by the glottis above and the trachea below; voice box.

Laser An acronym for light amplification by stimulated emission of radiation

Last glacial maximum (LGM) the time that the last major continental glacial advance in the Northern Hemisphere reached its maximum extent, about 18,000 years ago.

Latent heat of evaporation The amount of heat energy that is needed to evaporate a substance, that is, to change it from a liquid to a gas.For water, it is 540 cal/g.

Latent heat of fusion The heat that must be extracted from a liquid to freeze it to a solid at the same temperature. For water, it is 80 cal/g.

Latent heat of melting The amount of heat energy needed to melt a substance, that is, to change it from a solid to a liquid.

Latent heat of vaporization the heat energy required to convert a liquid to a gas at the same temperature.

Latent learning Associations made with neither immediate reinforcement or reward nor with particular behavior evident at the time of learning. The animal may store and use such information in appropriate situations at a later time.

Lateral Of or pertaining to the side of an animal; a bilateral animal has two sides.

Lateral bud Bud found in the axil of a leaf, also called an axillary bud.

Lateral-line system 1. A line of sensory receptors along the side of some fishes and amphibians used to detect water movement (phylum Chordata). 2. The external manifestation of a lateral excretory canal of nematodes (phylum Nematoda).

Lateral meristem Meristems that give rise to secondary tissues, typically by the formation of radial files of cells and thereby increasing the diameter and circumference of an axis over time; the vascular cambium and cork cambium.

Lateral root A root that arises from another, older root. Often termed secondary or branch roots.

Laterite Group of hard, red soils from topical areas that show intense weathering and leaching of bases and silica, leaving aluminum hydroxides and iron oxides.

Latewood The small, thick-walled xylem cells produced at the end of the growing season that appear as a dense ring of wood adjacent to the thin-walled earlywood.

Latex Milky juice exuded from some plants.

Latitude the angular distance north or south of the equator of a position on the earth's surface.

Lattice A crystalline-like structure caused by the precise orientation of molecules in a solid or liquid.

Laue pattern The pattern produced on a photographic film when high-frequency electromagnetic waves (such as x-rays) are fired at a crystalline solid.

Laurasia One of the two large continents, the northern one, that formed when the supercontinent Pangaea broke up about 180 million years ago.

Laurer's canal Usually blind canal extending from the base of the seminal receptacle of a digenetic trematode. It probably represents a vestigial vagina.

Law A statement, usually mathematical, which describes some physical phenomena. Compare: hypothesis and theory.

Law of conservation of baryons When a baryon decays or reacts with another particle, the number of baryons is the same on both sides of the equation

Law of conservation of energy The total quantity of energy in a closed system is constant.

Law of conservation of hypercharge Hypercharge is conserved in strong and electromagnetic interactions, but not in weak interactions.

Law of conservation of leptons In a reaction involving leptons, the arithmetic sum of the lepton numbers is the same on each side of the equation

Law of conservation of mechanical energy The sum of the potential and kinetic energies of an ideal energy system is constant.

Law of conservation of momentum When no net external forces are acting on an object, the total vector momentum of the object remains constant

Law of entropy A natural process always takes place in such a direction as to increase the entropy of the universe

Law of heat exchange In any heat transfer system, the heat lost by hot materials equals the heat gained by cold materials.

Law, physical A physical law is a relationship experimentally observed to hold in nature despite what our theories may or may not say about it. Since the goal is to describe nature, we must play by those rules which are imposed by nature. These are equivolent to mathematical postulates: that which is not derived from definitions, but rather assumed to be true so that deducible theorems can be proven. A physical law is generally unsatisfying because it describes without explaining. An example is Newton's Laws. These describe how forces cause an object to change its motion. However, it is possible (in this case) to derive these laws if a different postulate is made instead. One can either say Newton's Third Law implies Conservation of Momentum, or one can say that Conservation of Momentum implies Newton's Third "Law." These are in some sense describing the same phenomenon; it is not clear which (if either) is a "more fundamental" description.

Laws, Newton's This set of physical laws describe how forces cause an object to change its motion. (Newton used the word "action" to describe what we now call "force"; we now have a different meaning for the word action.)

I. An object at rest or in motion will maintain its velocity unless acted upon by a force.

II. An object upon which a force acts will accelerate in proportion to its mass, such that F=ma.

III. For every force acted upon an object, there is a force of equal magnitude which acts upon the object which imposed the force but in the direction opposite to the original force. (This law must be adjusted when relativistic effects, such as the finite speed of information travel, are considered.)

Lawson criterion A condition for the release of energy from a thermonuclear reactor. It is usually stated as the minimum value for the product of the density of the fuel particles and the energy confinement time for energy breakeven. For a half-and-half mixture of deuterium and tritium at ignition temperature, value is between 10^{14} and 10^{15} s/cm^3.

Lay botanist A nonscientist who understands botanical principles and appreciates the value of the plant world.

Layering A method of asexual propagation in which portions of the stem are wounded and covered with a medium, usually soil, to stimulate the production of adventitious roots.

LD50 A chemical dose lethal to 50 percent of a test population.

LDL (low density lipoprotein) Transports fats and cholesterol to the body cells, including the cells lining the bloodstream.

Le Chatelier's principle If a system is in equilibrium, then any change imposed on the system tends to shift the equilibrium to reduce the effect of that applied change.

Leaching The process of removal of ions or molecules by flushing with water.

Leaf margin Refers to the edge of a leaf, the area between the apex and base.

Leaf primordium The first stage of leaf development; a small lateral protuburance formed by an apical shoot meristem that will expand to form a leaf.

Leaf scars A scar left on a stem or twig when the leaf abscises.

Leaf sheath The lower part of a blade or petiole that invests the stem more or less completely.

Leaf-area index A numerical index of the ratio of leaf area to ground area in a plant community.

Leaflet An individual blade unit of a compound leaf.

Learning The relatively permanent change in behavior or potential for behavior that results from experience.

Learning curve A graphical presentation of the response measures in a learning situation. This may be depicted as correct or incorrect responses, or as a proportion of the trials given to the animal on which it gave correct or incorrect responses.

Learning set The acquisition of a learning strategy by the animal; the ability to transfer learning between problems of the same type.

Least nuclear motion, principle of The hypothesis that, for given reactants, the reactions involving the smallest change in nuclear positions will have the lowest energy of activation. (It is also often simply referred to as principle of least motion.)

Leaving group An atom or group (charged or uncharged) that becomes detached from an atom in what is considered to be the residual or main part of the substrate in a specified reaction.

For example, in the heterolytic solvolysis of benzyl bromide in acetic acid

$PhCH_2Br + AcOH \longrightarrow PhCH_2OAc + HBr$

the leaving group is Br^-; in the reaction

$MeS^- + PhCH_2N^+Me_3 \longrightarrow MeSCH_2Ph + NMe_3$

the leaving group is NMe_3; in the electrophilic nitration of benzene, it is H^+. The term has meaning only in relation to a specified reaction. The leaving group is not, in general, the same as the substituent group present in the substrate (e.g. bromo and trimethylammonio in the substrates of the first two examples above.)

A slightly different usage of the term prevails in the (non-mechanistic) naming of transformations, where the actual substituent group present in the substrate (and also in the product) is referred to as the leaving group.

Leeches Segmented worms that are specialized predators and parasites.

Leeward The side away from the direction of a prevailing wind.

Left-hand rule The opposite-chirality version of the right-hand rule.

Left-to-right convention Arrangement of the structural formulae of the reactants so that the bonds to be made or broken form a linear array in which the electrons move from left to right.

Legume Simple, dry, dehiscent fruit that splits along two seams, a pod. Member of the Fabaceae; a type of bean or pea.

Leishman-Donovan (L-D) body Amastigote in the Trypanosomatidae.

Leishmaniasis Infection by a species of Leishmania.

Lek An area where animals assemble for communal courtship display and mating.

Lemma Bract in a grass flower.

Lemniscus One of a pair of internal projections of the epidermis from the neck region of Acanthocephala, which functions in fluid control in the protrusion and invagination of the proboscis.

Lens A transparent object with two refracting surfaces. Usually the surfaces are flat or spherical (spherical lenses). Sometimes, to improve image quality. Lenses are deliberately made with surfaces which depart slightly from spherical (aspheric lenses).

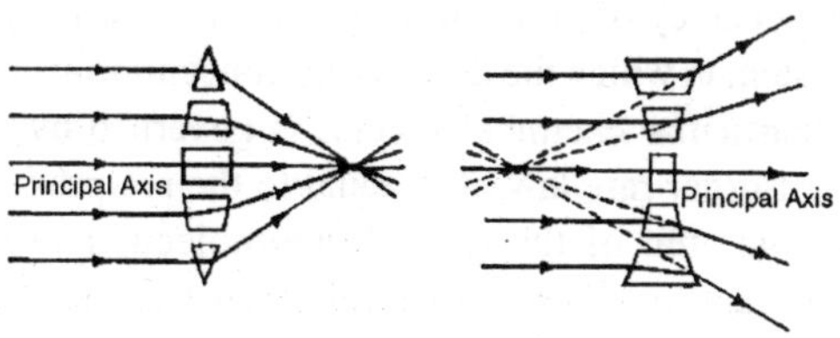

Fig. Lenses

Lentic ecosystem A freshwater ecosystem that has standing water (e.g., lakes or ponds).

Lenticel Isolated areas of loosely arranged cells in the cork surfaces of stems, roots, and fruit that allow interchange of gases between internal tissues and the atmosphere through the periderm.

Lenz's law An induced electric current always flows in such a direction that it opposes the change producing it.

Lepidosaurs A lineage of diapsid reptiles that appeared in the Permian and that includes the modern snakes, lizards, amphisbaenids, and tuataras, and the extinct ichthyosaurs.

Lepospondyls A group of Paleozoic amphibians distinguished by the possession of spool-shaped vertebral centra.

Leptocephalous larva The leaf-shaped larva of freshwater eels and other fishes.

Lepton This type of particle is affected by the weak interaction, but not by the strong interaction. This is to be compared against hadrons. The current theory of the standard model categorizes leptons into families.

Lepton, family of Each family of leptons has four members: a massive particle, a corresponding massive anti-particle, a massless particle, and a corresponding massless anti-particle. The families are distinguished by the mass of the massive particle. (The massive anti-particle has the same mass.) This mass is different for each family. An example of a lepton family is the set of particles: the electron, the positron, the electron neutrino, and the electron anti-neutrino. There is also a muon family of leptons and a tauon family of leptons. There could very well be more families of leptons, but these have not yet been discovered. It would be aestetically pleasing if the number of lepton families were the same as the number of quark families. Currently there are three known families of each.

Lesion An area of tissue that has been destroyed by an agent such as electric current or a chemical.

Leucon The sponge body form that has an extensively branched canal system; the canals lead to chambers lined by choanocytes.

Leucoplast Colorless plastid typically associated with starch formation and storage.

Leukemism Presence of white pelage or plumage in animals with normally pigmented eyes and skin.

Leukocyte Any of several kinds of white blood cells (for example, granulocytes, lymphocytes, monocytes), so-called because they bear no hemoglobin, as do red blood cells.

Leukorrhea White, puslike discharge resulting from infection.

Levelling effect The tendency of a solvent to make all Brønsted acids whose acidity exceeds a certain value appear equally acidic. It is due to the complete transfer to a protophilic solvent of a hydron from a dissolved acid stronger than the conjugate acid of the solvent. The only acid present to any significant extent in all such solutions is the lyonium ion. For example, the solvent water has a levelling effect on the acidities of $HClO_4$, HCl, and HI: aqueous solutions of these acids at the same (moderately low) concentrations have the same acidities. A corresponding levelling effect applies to strong bases in protogenic solvents.

Lewis acid A molecular entity (and the corresponding chemical species) that is an electron-pair acceptor and therefore able to react with a Lewis base to form a Lewis adduct, by sharing the electron pair furnished by the Lewis base. For example:

Me_3B (Lewis acid) + $:NH_3$ (Lewis base)$\longrightarrow Me_3B—N^+H_3$ (Lewis adduct)

Lewis formula (electron dot or lewis structure) Molecular structure in which the valency electrons are shown as dots so placed between the bonded atoms that one pair of dots represents two electrons or one (single) covalent bond, e.g. A double bond is represented by two pairs of dots, etc. Dots representing non-bonded outer-shell electrons are placed adjacent to the atoms with which they are associated, but not between the atoms. Formal charges (e.g. +, -, 2+, etc.) are attached to atoms to indicate the difference between the positive nuclear charge (atomic number) and the total number of electrons (including those in the inner shells), on the formal basis that bonding electrons are shared equally between atoms they join. (Bonding pairs of electrons are usually denoted by lines, representing covalent bonds, as in line formulae.)

Liana A large, woody vine common to the tropical forests; it climbs the tall trees and often trails from the canopy.

Library In molecular biology, a set of clones containing recombinant DNA. Obtained from and representing the genome of the organism.

Lichen An organism composed of a symbiotic association of an ascomycete fungus with algal or cyanobacterial cells.

Lien The spleen.

Life expectancy The average number of years lived by a group of individuals after reaching a given age; the probable number of years of survival for an individual of a given age.

Life range The larger geographic area that an animal utilizes over the course of its life.

Life span The longest period of life reached by a type of organism.

Life-cycle analysis Evaluation of material and energy inputs and outputs at each stage of manufacture, use, and disposal of a product.

Lifetime (mean lifetime), The lifetime of a chemical species which decays in a first-order process is the time needed for a concentration of this species to decrease to 1/e of its original value. Statistically, it represents the mean life expectancy of an excited species. In a reacting system in which the decrease in concentration of a particular chemical species is governed by a first-order rate law, it is equal to the reciprocal of the sum of the (pseudo)unimolecular rate constants of all processes which cause the decay. When the term is used for processes which are not first order, the lifetime depends on the initial concentration of the species, or of a quencher, and should be called apparent lifetime instead.

Ligament A tough, dense band of connective tissue connecting one bone to another.

Ligand If it is possible to indicate a "central atom" in a polyatomic molecular entity, the atoms or groups bound to that atom are called ligands. The term is generally used in connection with metallic "central atoms".

In biochemistry a part of a polyatomic molecular entity may be considered central, and atoms, groups or molecules bound to that part are considered ligands.

Light Hertz was the first to demonstrate experimentally the production and detection of Maxwell's electromagnetic waves, showing visible light to be merely one region of colors which we happen to see. This discovery of course lead directly to radio, television, and radar, which use essentially different "colors" of light that we cannot see (i.e. Which our eyes cannot process).

Light harvesting antenna A complex of several hundred chlorophyll and carotenoid molecules that form a part of each photosystem.

Light intensity The strength of light rays; the degree of brightness dependent on the number of photons striking a given area at a point in time.

Light microscope The type of microscope in which the specimen is viewed under ordinary illumination.

Light reaction that part of the photosynthetic process that, in the presence of light, captures energy to form ATP and NADPH2 to be used to synthesize complex organic molecules in the dark reaction

Light, speed of The speed at which light travels (299,792,458 meters per second in a vacuum — it slows down while traveling through material). Notice that this speed is not infinitely fast. Notice also that the fastest possible speed is numerically the same as the speed of light. This is not because it is the speed of light, but rather because light happens to be traveling at the fastest possible speed, necessarily.

Lignification Impregnated with lignin.

Lignified Impregnated with lignin, such as the secondary cell walls of woody plants.

Lignin A complicated organic molecule found as an important constituent of many secondary cell walls; imparts strength and rigidity to the cellulose microfibrils.

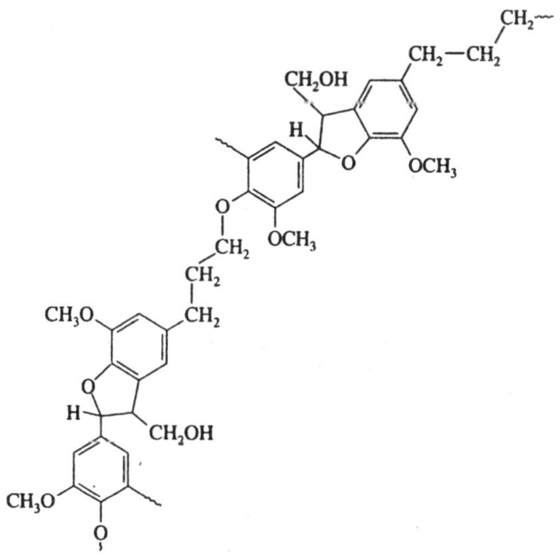

Fig. A Fragment of lignin

Lignite Soft, brown to black coal of low BTU value with original plant components still discernible.

Limax form Form of pseudopodial movement in which entire organism moves without extending a discrete pseudopodium.

Limiting factors Chemical or physical factors that limit the existence, growth, abundance, or distribution of an organism. The principle of limiting factors states that for each physical factor in the environment, there are both minimum and maximum limits (tolerance limits) beyond which a particular species cannot survive. The single factor closest to a tolerance limit for a given species at a given time is the critical factor.

Limiting resource An essential factor whose short supply prevents the growth of a population.

Limnetic zone The open-water zone of a lake or pond; extends to the depth of effective light penetration.

Limnologist A scientist who studies freshwater biology.

Line formula A two-dimensional representation of molecular entities in which atoms are shown joined by lines representing single or multiple bonds, without any indication or implication concerning the spatial direction of bonds. For example, methanol is represented as

Line of force A line so drawn that a tangent to it at any point indicates the direction of the electric field.

Line spectrum A spectrum consisting of monochromatic slit images having wavelengths characteristic of the atoms present in the source

Linear accelerator A device for accelerating particles in a straight line through many stages of small potential difference.

Linear free-energy relation A linear correlation between the logarithm of a rate constant or equilibrium constant for one series of reactions and the logarithm of the rate constant or equilibrium constant for a related series of reactions. Typical examples of such relations (also known as linear Gibbs energy relations) are the Brønsted relation, and the Hammett equation.

The name arises because the logarithm of an equilibrium constant (at constant temperature and pressure) is proportional to a standard free energy (Gibbs energy) change, and the logarithm of a rate constant is a linear function of the free energy (Gibbs energy) of activation. It has been suggested that this name should be replaced by Linear Gibbs Energy Relation, but at present there is little sign of acceptance of this change. The area of physical organic chemistry which deals with such relations is commonly referred to as "Linear Free-Energy Relationships".

Linear solvation energy relationships Equations involving the application of solvent parameters in linear or multiple (linear) regression expressing the solvent effect on the rate or equilibrium constant of a reaction.

Line-shape analysis Determination of rate constants for a chemical exchange from the shapes of spectroscopic lines of dynamic processes. The method is most often used in nuclear magnetic resonance spectroscopy.

Lingual Pertaining to the tongue.

Linkage Tendency of genes located on the same chromosome to be inherited together.

Linkage group Genes linked to the same chromosome that tend to be inherited together.

Linked gene Gene located on the same chromosome as another gene.

Lip cell Thin-walled cells that interrupt the annulus in fern sporangia.

Lipase An enzyme that accelerates the hydrolysis or synthesis of fats.

Lipid A fat, oil, or fatlike compound that usually has fatty acids in its molecular structure. An organic compound consisting mainly of carbon and hydrogen atoms linked by nonpolar covalent bonds. Examples include fats, waxes, phospholipids, and steroids that are insoluble in water.

Lipophilic Literally "fat-loving". Applied to molecular entities (or parts of molecular entities) having a tendency to dissolve in fat-like (e.g. hydrocarbon) solvents.

Liposome Artificial lipoid particle used to deliver antiparasitic drugs directly to macrophages (which eat the particles).

Liquefaction The change to the liquid phase The condensation of a gas to a liquid.

Liquid metal fast breeder A nuclear power plant that converts uranium 238 to plutonium 239; thus, it creates more nuclear fuel than it consumes. Because of the extreme heat and density of its core, the breeder uses liquid sodium as a coolant.

Liter A special name for the cubic decimeter Symbol: L.

Lithogenous sediment A marine sediment that is derived from the breakdown, or weathering, of rocks.

Lithosphere The crust and the top part of the mantle that covers the earth's surface. It is broken into separate lithospheric plates.

Litter Dead organic material such as branches, tree trunks, and dry grass that accumulates on the floor of a forest. Litter acts as additional fuel in a forest fire, producing increased destruction.

Littoral zone The shallow-water zone of a lake or pond; light penetrates to the bottom, and the area is occupied by rooted plants such as water lilies, rushes, and sedges.

Liver A large gland of a dark-red color. It carries out many vital functions such as the formation of urea, manufacture of plasma proteins, synthesis of amino acids, synthesis and storage of glycogen, and many others.

Liverwort A group of the division Bryophyta charcterized by a small, inconspicuous, liver-shaped thallus and that lives in moist environments.

Loam A mixture of sand-, silt-, and clay-sized soil particles.

Lobopodium A blunt, lobelike pseudopodium that is commonly tubular, and is composed of both ectoplasm and endoplasm.

Local chemical messenger A chemical that acts on nearby cells.

Local mate competition Competition for mates that may arise among siblings depending on the population size and operational sex ratio and that may cause the evolution of skewed offspring sex ratios.

Local resource competition model In female philopatric primates, females giving birth to a disproportionate number of males that will disperse and not compete with each other, or high-ranking females giving birth to a disproportionate number of daughters that are able to compete for resources effectively.

Locomotion-fear dichotomy A process during the imprinting process in young birds whereby they initially are more likely to locomote and follow a stimulus object and later develop a fear toward objects; thus, imprinting occurs when the locomotor tendency is high and fear of objects is low.

Loculi Shallow, suckerlike depressions in an adhesive organ of a flatworm.

Locus A section of a chromosome (a gene) that codes for a particular protein.

Logistic equation. A mathematical expression describing an idealized sigmoid curve of population growth.

Logistic growth Growth rates regulated by internal and external factors that establish an equilibrium with environmental resources; species may grow exponentially when resources are unlimited but slowly as the carrying capacity is reached.

London forces Attractive forces between apolar molecules, due to their mutual polarizability. They are also components of the forces between polar molecules. Also called "dispersion forces".

Lone (electron) pair Two paired electrons localized in the valence shell on a single atom. Lone pairs should be designated with two dots.

The term "nonbonding electron pair" is more appropriate, and is found in many modern text books.

Long-day plant A plant that flowers when the length of day exceeds some critical value.

Longevity The length or duration of life; compare to survivorship.

Longitude the angular distance of a position on the earth's surface east or west of the Greenwich Prime Meridian; measured in degrees.

Longitudinal wave A wave in which the vibrations are parallel to the direction of propagation of the wave.

Loop A midpoint of a vibrating segment of a standing wave.

Loop rule The sum of the potential differences encountered in a round trip around any closed loop in a circuit is zero.

Looping movement The type of locomotion exhibited by leeches and some insect larvae whereby they alternate temporary points of attachment to move forward.

Loose connective tissue The type of tissue in which the matrix contains strong, flexible fibers of the protein collagen that are interwoven with fine, elastic, and reticular fibers.

Lophocyte Type of sponge amebocyte that secretes bundles of fibrils.

Lophophore Tentacle-bearing ridge or arm within which is an extension of the coelomic cavity in lophophorate animals (ectoprocts, brachiopods, and phoronids).

Lorica Protective external case found in some protozoa, rotifers, and others.

Loricifera A phylum of aschelminths. The most recent animal phylum to be described; members are commonly called loriciferans.

Loschmidt constant The number of particles per unit volume of an ideal gas at standard temperature and pressure. It has the value $2.68719 \times 10^{25}\ m^{-3}$.

Lotic Of or pertaining to running water, such as a brook or river.

Lotic ecosystems Flowing water ecosystems. They include brooks, streams, and rivers.

Lottery hypothesis The hypothesis that unpredictability plays a key role in the development of communities.

Loudness The sensation that depends principally on the intensity of sound waves reaching the ear.

Low tide the lowest level reached by the falling tide

Low-head hydropower Small-scale hydro technology that can extract energy from small headwater dams; causes much less ecological damage.

Low-quality energy Diffuse, dispersed energy at a low temperature that is difficult to gather and use for productive purposes.

LSD Lysergic acid diethylamide; a synthetic hallucinogenic derivative of lysergic acid.

Lulus Locally unwanted land uses such as toxic waste dumps, incinerators, smelters, airports, freeways, and other sources of environmental, economic, or social degradation.

Lumbar Relating to or near the loins or lower back.

Lumbosacral lumb = the loin; sacrum = sacrum.

Lumen; lm The derived SI unit of luminous flux, defined as the luminous flux emitted by a uniform point source of 1 cd emitting its luminous energy over a solid angle of 1 sr; it thus has units of cd sr.

Lumeniferous aether A substance, which filled all the empty spaces between matter, which was used to explain what medium light was "waving" in. Now it has been discredited, as maxwell's equations imply that electromagnetic radiation can propagate in a vacuum, since they are disturbances in the electromagnetic field rather than traditional waves in some substance, such as water waves.

Luminous flux The part of the total energy radiated per unit of time from a luminous source that is capable of producing the sensation of sight.

Luminous Visible because of the light emitted by its oscillating particles

Lung An organ of the respiratory system in which gas exchange occurs between body fluids (e.g., blood) and air.

Lunules Small, suckerlike discs on the anterior margin of some copepods in the family Caligidae, functioning as organs of adhesion.

Lux The derived SI unit of illuminance equal to the illuminance produced by a luminous flux of 1 lm distributed uniformly over an area of 1 m^2; it thus has units of lm/m^2.

Luxon A particle which travels solely at c (the speed of light in vacuum). All luxons have a rest mass of exactly zero. Though they are massless, luxons do carry momentum. Photons are the prime example of luxons (the name itself is derived from the latin word for light).

Lyate ion The anion produced by hydron removal from a solvent molecule. For example, the hydroxide ion is the lyate ion of water.

Lycophora Ten-hooked larva that hatches from the egg of a cestodarian tapeworm. Also called a decacanth.

Lycophyta A division of "lower" (i.e., non-seed-producing) vascular plants, commonly referred to as club and spike mosses.

Lyman series The series which describes the emission spectrum of hydrogen when electrons are jumping to the ground state. All of the lines are in the ultraviolet.

Lymph The interstitial (intercellular) fluid in the body, also the fluid in the lymphatic system.

Lymph varices Dilated lymph ducts.

Lymphadenitis Inflamed lymph node.

Lymphatic system The one-way system of lymphatic vessels.

Lymphocytes A type of white blood cell; a component of the immune system produced by stem cells in the bone marrow.

Lymphokine A molecule secreted by an activated or stimulated lymphocyte that causes physiological changes in certain other cells.

Lysergic acid diethylamide (LSD) A derivative of lysergic acid alkaloids, first isolated from ergot; strongly hallucinogenic.

Lysosome Cytoplasmic, membrane-bounded organelle that contains digestive and hydrolytic enzymes, which are typically most active at the acid ph found in the lumen of lysosomes.

M

Mach number The ratio of the speed of an object in a given medium to the speed of sound in that medium.

Mach's principle The inertia of any particular particle or particles of matter is attributable to the interaction between that piece of matter and the rest of the universe. Thus, a body in isolation would have no inertia.

Machine A device that multiplies force at the expense of distance or that multiplies distance at the expense of force.

Macro A prefix meaning 'large'.

Macroalgae The multicellular brown, red, and green algae.

Macroelements Essential elements needed for plant growth in relatively large amounts; C, H, O, P, K, N, S, Ca, Fe, Mg.

Macroevolution Evolutionary change on a grand scale, encompassing the origin of novel designs, evolutionary trends, adaptive radiation, and mass extinction.

Macrofauna benthic animals larger than about 0.5 mm

Macrofibril An aggregation of microfibrils in the cell wall, visible with the light microscope.

Macrogamete The larger of the two gamete types in a heterogametic organism, considered the female gamete.

Macromere The largest size class of blastomeres in a cleaving embryo when the blastomeres differ in size from one another.

Macromolecule A very large molecule, generally used in reference to carbohydrates, lipids, proteins, and nucleic acids.

Macronutrient An essential nutrient for which an animal has a large minimal daily requirement (greater than 100 mg) (e.g., calcium, phosphorus, magnesium, potassium, sodium, and chloride along with carbohydrates, lipids, and proteins).

Macroparasite Large parasite that does not multiply in the host of interest. Examples are cestodes, trematodes, and most nematodes in their definitive hosts.

Macrophage A phagocytic cell type in vertebrates that performs crucial functions in the immune response and inflammation, such as presenting antigenic epitopes to T cells and producing several cytokines.

Macrophage migration inhibitory factor (MIF) Cytokine released by sensitized lymphocytes that tends to inhibit migration of macrophages in the immediate vicinity, thus contributing to accumulation of larger numbers of macrophages close to the site of MIF release.

Macroplankton The component of the plankton that consists of large organisms 2 to 20 cm in size.

Macropores Large pore spaces caused by invertebrates and larger animals, including reptiles and mammals, that permeate the soil. They drain water not held by capillarity.

Macroscopic A physical entity or process of large scale, the scale of ordinary human experience. Specifically, any phenomena in which the individual molecules and atoms are neither measured, nor explicitly considered in the description of the phenomena.

Madreporite Sievelike structure that is the intake for the water-vascular system of echinoderms.

Magma Molten rock from deep in the earth's interior; called lava when it spews from volcanic vents.

Magnetite A magnetic, iron-containing material.

Magnetic anomalies Magnetic bands in the sea floor that run parallel to the mid-ocean ridge.

Magnetic confinement A technique for enclosing a nuclear fusion reaction in a powerful magnetic field inside a vacuum chamber.

Magnetic equivalence Nuclei having the same resonance frequency in nuclear magnetic resonance spectroscopy and also identical spin-spin interactions with the nuclei of a neighbouring group are magnetically equivalent. The spin-spin interaction between magnetically equivalent nuclei does not appear, and thus has no effect on the multiplicity of the respective NMR signals. Magnetically equivalent nuclei are necessarily also chemically equivalent, but the reverse is not necessarily true.

Magnetic field A region in which a magnetic force can be detected.

Magnetic field intensity The force exerted by a magnetic field on a unit N pole situated in the field.

Magnetic flux Lines of flux through a region of a magnetic field, considered collectively.

Magnetic flux density The magnetic flux through a unit area normal to the magnetic field; also called magnetic induction.

Magnetic force A force associated with motion of electric charges.

Magnetic monopole A hypothetical particle which constitutes sources and sinks of the magnetic field. Magnetic monopoles have never been found, but would only cause fairly minor modifications to maxwell's equations. They also seem to be predicted by some grand-unified theories. If magnetic monopoles do exist, they do not seem to be very common in our universe.

Magnetization transfer NMR method for determining kinetics of chemical exchange by perturbing the magnetization of nuclei in a particular site or sites and following the rate at which magnetic equilibrium is restored. The most common perturbations are saturation and inversion, and the corresponding techniques are often called "saturation transfer" and "selective inversion-recovery".

Magnetosphere A region of the upper atmosphere in which the motion of charged particles is governed primarily by the magnetic field of the earth

Magnification Two kinds of magnification are useful to describe optical systems and they must not be confused, since they aren't synonymous. Any optical system which produces a real image from a real object is described by its linear magnification. Any system which one looks through to view a virtual image is described by its angular magnification. These have different definitions, and are based on fundamentally different concepts.

Magnus effect A rotating cylinder in a moving fluid drags some of the fluid around with it, in its direction of rotation. This increases the speed in that region, and thus the pressure is lower. Consequently, there is a net force on the cylinder in that direction, perpendicular to the flow of the fluid. This is called the magnus effect.

Major histocompatibility complex (MHC). Complex of genes coding for proteins inserted in the cell membrane; the proteins are the basis of self-nonself recognition by the immune system.

Major mineral Mineral requirement needed in relatively large amounts.

Malacostraca The class of crustaceans whose members are characterized by having appendages modified for crawling along the substrate, as in lobsters, crayfish, and crabs. Alternatively, the abdomen and body appendages may be used in swimming, as in shrimp.

Malacostracan Any member of the crustacean subclass Malacostraca, which includes both aquatic and terrestrial forms of crabs, lobsters, shrimps, pillbugs, sand fleas, and others.

Malate Four-carbon organic acid found with aspartate to be the first products of the C4 pathway of photosynthesis.

Male dominance polygyny A mating system in which males compete and acquire dominance ranks that influence their access to females, with higher-ranking males often having greater mating activity.

Male parasitism The permanent attachment of a male to a female in some deep-sea fishes.

Malignant tumor A mass of cancerous cells that have left their site of origin, migrated through the body, invaded normal tissues, and are growing out of control.

Malleability The property of a metal that enables it to be hammered or rolled into sheets.

Malleus The ossicle attached to the tympanum in middle ears of mammals.

Malnourishment A nutritional imbalance caused by lack of specific dietary components or inability to absorb or utilize essential nutrients.

Malpighian tubules Blind tubules opening into the hindgut of nearly all insects and some myriapods and arachnids, and functioning primarily as excretory organs.

Malthusian growth A population explosion followed by a population crash; also called irruptive growth.

Malus' law The light intensity I of a ray with initial intensity I_0 travelling through a polarizer at an angle theta between the polarization of the light ray and the polarization axis of the polarizer is given by

$I = I_0 \cos 2ø$.

Mamelon Ventral, serrated projection on the ventral surface of a male nematode of the family Syphaciidae. Its function is unknown.

Mammalia The class of vertebrates whose members are at least partially covered by hair, have specialized teeth, and are endothermic. Young are nursed from mammary glands. The mammals.

Mammals Vertebrates that have hair and mammary glands.

Mammary gland The breast. In female mammals, the mammary glands produce and secrete milk to nourish developing young.

Man and Biosphere (MAB) program A design for nature preserves that divides protected areas into zones with different purposes. Critical ecosystem functions and endangered wildlife are protected in a central core region where limited scientific study is the only human access allowed. Ecotourism and research facilities are located in a relatively pristine buffer zone around the core, while sustainable resource harvesting and permanent habitation are allowed in multiple-use peripheral regions.

Mandibles Third pair of appendages from the anterior in Crustacea; second pair in Insecta; primarily function in feeding; derived from appendages on primitive fourth (first postoral) somite.

Mangal a tropical community of mangrove plants and associated organisms

Manganese nodules Lumps of minerals (including manganese and other valuable minerals) that are found on the sea floor beyond the continental shelf.

Mange Dermatitis caused by species of mites, often designated with the causative organism. For example, Sarcoptes causes sarcoptic mange.

Mangroves Shrubs and trees that live along the seashore in tropical and subtropical regions and tolerate inundation by seawater.

Mantle 1. The semi-liquid region between the crust and core of the earth.

2. The outer layer of tissue that secretes the shell of molluscs.

Mantle cavity The space between the mantle and the visceral mass of molluscs.

Manubrium The portion projecting from the oral side of a jellyfish medusa, bearing the mouth; oral cone; presternum or anterior part of sternum; handle-like part of malleus of ear.

Marasmus Malnutrition, especially of infants, caused by a diet deficient in both calories and protein.

Marcus equation A general expression which correlates the Gibbs energy of activation (ΔG) with the driving force (rG) of the reaction:

$$\Delta G = (\lambda/4)(1 + \Delta rG^{o}/\lambda)2$$

where λ is the reorganization energy and $\Delta_r G^o$ is the standard free energy of the reaction corrected for the electrostatic work required to bring the reactants together. $\lambda/4$ is the intrinsic barrier of the reaction. Originally developed for outer-sphere electron transfer reactions, the Marcus equation has later been applied also to atom and group transfer reactions.

Margin The edge of a flattened structure; in leaves, the lateral edge of the blade.

Marginal bodies Sensory pits or short tentacles between the marginal loculi of the opisthaptor of an aspidogastrean trematode.

Marginal meristem The meristem along the margin of a leaf primordium responsible with the formation and shape of the leaf.

Marginal value theorem A model that predicts when an organism should cease foraging in one patch and travel to another.

Mariculture The culture of marine organisms. In open mariculture (or semi-culture), organisms are cultured in natural environments; in closed mariculture (or intensive mariculture), organisms are cultured in a controlled environment.

Marine archaeology The discovery, salvage, and interpretation of material remains of humankind's past that have been preserved in the sea.

Marine Living in or pertaining to the sea.

Marine natural products Chemical compounds that are obtained from marine organisms.

Marine snow Detritus and other particulate organic matter that is found in the water column.

Maritime culture A human culture with a very close relationship with the sea.

Market equilibrium The dynamic balance between supply and demand under a given set of conditions in a "free" market (one with no monopolies or government interventions).

Markownikoff rule "In the addition of hydrogen halides to unsymmetrically constituted [unsaturated] hydrocarbons, the halogen atom becomes attached to the carbon bearing the lesser number of hydrogen atoms." Originally formulated by Markownikoff (Markovnikov) to generalize the orientation in additions of hydrogen halides to simple alkenes, this rule has been extended to polar addition reactions as follows. " In the heterolytic addition of a polar molecule to an alkene or alkyne, the more electronegative (nucleophilic) atom (or part) of the polar molecule becomes attached to the carbon atom bearing the smaller number of hydrogen atoms."

This is an indirect statement of the common mechanistic observation, that the more electropositive (electrophilic) atom (or part) of the polar molecule becomes attached to the end of the multiple bond that would result in the more stable carbenium ion (whether or not a carbenium ion is actually formed as a reaction intermediate in the addition reaction). Addition in the opposite sense is commonly called "anti-Markovnikov addition". Markownikoff.

Marrara Nasopharyngeal blockage by a parasite. Also called halzoun.

Marsh Wetland without trees; in North America, this type of land is characterized by cattails and rushes.

Marsupial One of the pouched mammals of the subclass Metatheria.

Mass The fundamental unit of measurement equivalent to the weight of a substance when compared with the weight of hydrogen.

Mass burn Incineration of unsorted solid waste.

Mass density Mass per unit volume of a substance.

Mass number Sum of the number of protons and neutrons in the nucleus of an atom; designated by a superscript to the upper left of the elemental symbol.

Mass spectrograph Instrument used to determine the mass of ionized particles.

Masseter A chewer.

Mast cell Type of cell in various tissues that releases pharmacologically active substances with a role in inflammation.

Mastax The pharyngeal apparatus of rotifers used for grinding ingested food.

Mastication Chewing.

Mastigont Axoneme of a cilium or flagellum together with its basal fibrils and organelles.

Mastigophora The protozoan subphylum where members possess one or more flagella that are used for locomotion; autotrophic, heterotrophic, or saprozoic.

Mastitis Infection of the udder of cattle.

Mastoid A bony process of the skull.

Maternal Pertaining to the mother.

Maternal condition model In polygynous mammals, females that are high ranking, well nourished, or otherwise in good condition giving birth to a greater proportion of sons since their sons will outreproduce their daughters and vice versa for females that are in poor condition.

Mating system The species-typical pattern of mate finding, reproduction, and parenting of offspring.

Matric potential The water potential component caused by the attraction of water molecules to a hydrophilic matrix.

Matrix isolation A term which refers to the isolation of a reactive or unstable species by dilution in an inert matrix (argon, nitrogen, etc.), usually condensed on a window or in an optical cell at low temperature, to preserve its structure for identification by spectroscopic or other means.

Matrix The intercellular substance of a tissue, or that part of a tissue into which an organ or process is set; site of the Krebs cycle.

Matter Anything that has the properties of mass and inertia.

Matter wave A property of matter that is directly proportional to Planck's constant and inversely proportional to mass and velocity.

Maturation To complete the natural development of an animal system. Improvement in the performance of behavior occurs as parts of the nervous system and other structures complete development.

Maurer's clefts Blotches on the surface of an erythrocyte infected with Plasmodium falciparum.

Maxilla One member of a pair of mouthparts located just posterior to the mandibles of many arthropods.

Maxillae Fifth pair of appendages in Crustacea, primarily feeding in function, derived from appendages on primitive sixth (third postoral) somite; homologous to labium in insects. The maxillae of insects are the third pair of head appendages, homologous to maxillules of Crustacea.

Maxillary gland In malacostracan crustaceans, the excretory organs are located near the maxillary segments and are termed maxillary glands.

Maxilliped One of the pairs of head appendages located just posterior to the maxilla in crustaceans, a thoracic appendage that has become incorporated into the feeding mouthparts.

Maxillopodan eye Naupliar eye of crustacean class Maxillopoda; has a tapetum (crystalline reflective layer).

Maxillules Fourth pair of appendages in Crustacea, primarily feeding in function; derived from appendages on primitive fifth (second postoral) somite; homologous to maxillae in insects.

Maximum sustainable yield The maximum level of fishing effort that a fish stock can withstand without causing major upsets in the abundance of its stock.

Maxwell's demon A thought experiment illustrating the concepts of entropy. We have a container of gas which is partitioned into two equal sides; each side is in thermal equilibrium with the other. The walls and the partition of the container are perfect insulators.

Now imagine there is a very small demon who is waiting at the partition next to a small trap door. He can open and close the door with negligible work. Let's say he opens the door to allow a fast-moving molecule to travel from the left side to the right, or for a slow-moving molecule to travel from the right side to the left, and keeps it closed for all other molecules. The net effect would be a flow of heat — from the left side to the right — even though the container was in thermal equilibrium. This is clearly a violation of the second law of thermodynamics.

So what is wrong. It turns out that information has to do with entropy as well. In order to sort out the molecules according to speeds, the demon would be having to keep a memory of them — and it turns out that increase in entropy of the maintenance of this simple memory would more than make up for the decrease in entropy due to the heat flow.

Meadow An open clearing in a landscape in which low-growing, herbaceous plants dominate.

Mean The central tendency of a series of measurements or observations, calculated by summing all values in a distribution and dividing the total by the number of values.

Meatus A passageway.

Mechanical equivalent of heat The conversion factor that relates heat units to work units; 4.19 / cal.

Mechanical vector Vector that transmits disease organism by mechanical means only. Contrast with biological vector.

Mechanical wave A wave that originates in the displacement of a portion of an elastic medium from its normal position, causing it to oscillate about an equilibrium position.

Mechanics, statistical Describes the mechanics (motions and interactions) of many particles via statistical analysis. In other words, instead of considering the motions of every particle individually, statistics allows the consideration of how particles act on average and give meaningful properties of the collection of particles rather than information about what is happening to any individual particle.

Mechanism A detailed description of the process leading from the reactants to the products of a reaction, including a characterization as complete as possible of the composition, structure, energy and other properties of reaction intermediates, products, and transition states. An acceptable mechanism of a specified reaction (and there may be a number of such alternative mechanisms not excluded by the evidence) must be consistent with the reaction stoichiometry, the rate law, and with all other available experimental data, such as the stereochemical course of the reaction. Inferences concerning the electronic motions which dynamically interconvert successive species along the reaction path (as represented by curved arrows, for example) are often included in the description of a mechanism.

It should be noted that for many reactions all this information is not available and the suggested mechanism is based on incomplete experimental data. It is not appropriate to use the term mechanism to describe a statement of the probable sequence in a set of stepwise reactions. That should be referred to as a reaction sequence, and not a mechanism.

Mechanism-based inhibition Irreversible inhibition of an enzyme due to its catalysis of the reaction of an artificial substrate. Also called "suicide inhibition".

Mechanoreceptor A sensory receptor that is sensitive to mechanical stimulation, such as changes in pressure or tension.

Meconium A dark green mucilaginous material in the intestine of the full-term fetus, being a mixture of the secretions of the intestinal glands and some amniotic fluid.

Medial On or near the plane that divides a bilateral animal into mirror images. Also median.

Median (parietal) eye A photoreceptor located middorsally on the head of some vertebrates; it is associated with the vertebrate epithalamus.

Mediastinum The membranes enclosing the tissues and organs between the two lungs.

Mediated transport. Transport of a substance across a cell membrane mediated by a carrier molecule in the membrane.

Mediterranean climate areas Specialized landscapes with warm, dry summers; cool, wet winters; many unique plant and animal adaptations; and many levels of endemism.

Medium Any region through which a wave disturbance propagates Mechanical waves require a matter medium Electromagnetic waves propagate through a vacuum and various matter media.

Medulla oblongata Portion of the brain stem located between the pons and the spinal cord.

Medusa A jellyfish, or the free-swimming stage in the life cycle of Cnidarians.

Megacolon Flabby distended colon caused by chronic Chagas' disease.

Megagametogenesis The production of megagametes (large gametes) from megaspores in the ovules of angiosperms.

Megagametophyte The gametophyte stage containing eight haploid nuclei within the embryo sac.

Megalopolis Also known as a megacity or supercity; megalopolis indicates an urban area with more than 10 million inhabitants.

Megaplankton The component of the plankton that consists of very large organisms over 20 cm in size.

Megasporangium Sporangium that contains megaspores.

Megaspore Haploid cell produced by meiosis in the ovules of angiosperms; a single megasporocyte produces four megaspores, only one of which remains functional.

Megaspore mother cell Diploid cell in megasporangium that, upon undergoing meiosis, yields megaspores.

Megasporocyte Also known as the megaspore mother cell; this diploid cell undergoes meiosis to produce four haploid megaspores in the ovules of angiosperms.

Megasporogenesis The process of megaspores (large spores) being produced via meiosis in the ovules of angiosperms.

Megawatt (MW) Unit of electrical power equal to one thousand kilowatts or one million watts.

Mehlis' glands Unicellular mucous and serous glands surrounding the ootype of a flatworm.

Meiofauna Small invertebrates found in the interstices between sand grains.

Meiosis Process of cell division by which egg and sperm cells are formed, involving a diminution in the amount of genetic material. Comprises two successive nuclear divisions with only one round of DNA replication, which produces four haploid daughter cells from an initial diploid cell.

Meiotic Cell division that reduces the number of chromosomes from diploid to haploid.

Meisenheimer adduct A cyclohexadienyl derivative formed as Lewis adduct from a nucleophile (Lewis base) and an aromatic or heteroaromatic compound, also called Jackson-Meisenheimer adduct. In earlier usage the term "Meisenheimer complex" was restricted to the typical Meisenheimer alkoxide adducts of nitro-substituted aromatic ethers, e.g.

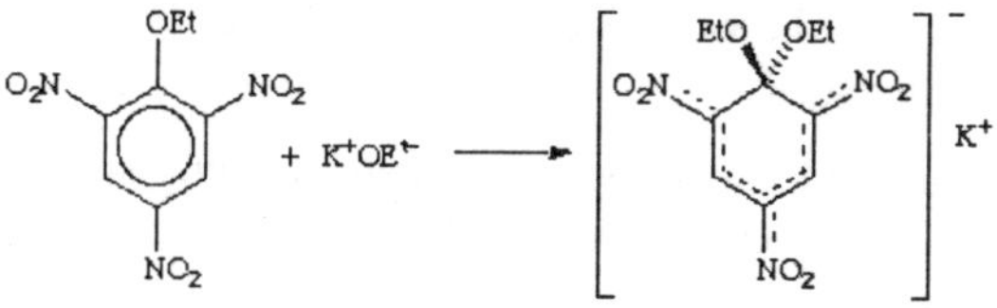

Analogous cationic adducts, such as

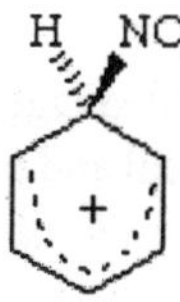

considered to be reaction intermediates in electrophilic aromatic substitution reactions, are called "Wheland intermediates", and sometimes, inappropriately, -complexes.

Meissner effect The decrease of the magnetic flux within a superconducting metal when it is cooled below the transition temperature. That is, superconducting materials reflect magnetic fields

Meissner's corpuscles Tactile sensory receptors in the superficial dermis.

Melanin Black or dark-brown pigment found in plant or animal structures.

Melanism Dark coloration of the skin, fur, or feathers because of the presence of the pigment melanin.

Melanophore Black or brown chromatophore containing melanin.

Melatonin A hormone secreted by the pineal gland. Functions in regulating photoperiodicity.

Melon A fatty structure on the forehead of some cetaceans that is used to direct sound waves emitted during echolocation.

Melting point The temperature at which a substance changes its state from solid to liquid.

Membrane In living organisms, a phospholipid bilayer impregnated with protein and certain other compounds; functions in partitioning of cellulose activities.

Membranelle Short, transverse rows of cilia, fused at their bases, serving to move food particles toward the oral groove of a protozoan.

Memory cell A lymphocyte capable of initiating the antibody- mediated immune response on detection of a specific antigen molecule for which it is genetically programmed. It circulates freely in the blood and lymph, and may live for years.

Memory The capacity of an organism to form lasting connections based on past experiences; the ability to store and use information.

Meninges Any of three membranes (arachnoid, dura mater, pia mater) that envelop the vertebrate brain and spinal cord. Also, solid connective tissue sheath enclosing the central nervous system of some vertebrates.

Menisci A crescent-shaped fibrocartilage.

Meniscus The crescent-shaped surface at the edge of a liquid column

Menopause In the human female, that time of life when ovulation ceases; cessation of the menstrual cycle.

Menses The period of shedding of the lining (endometrium) of the uterus and associated fluids if an ovum is not fertilized, most notably in primates.

Menstrual cycle The period of the regularly recurring physiologic changes in the endometrium that culminates in its shedding (menstruation).

Menstruation Loss of blood and tissue from the uterus at the end of a female primate's reproductive cycle.

Meristem The tissue or zone from which new cells are produced by cell division.

Meristematic tissue Within some seaweeds, specific tissue sites where most cell division for growth occurs.

Meroblastic Partial cleavage occurring in zygotes having a large amount of yolk at the vegetal pole; cleavage restricted to a small area on the surface of the egg.

Merogony Multiple fission to produce merozoites; schizogony.

Meroplankton Planktonic organisms that spend only part of their life in the plankton. Compare holoplankton.

Merostomata The class of arthropods whose members are aquatic and possess book gills on the opisthosoma. Eurypterids (extinct) and horseshoe crabs.

Merozoite A very small trophozoite at the stage just after cytokinesis has been completed in multiple fission of a protozoan.

Mesenchyme Undifferentiated mesoderm. It will eventually develop into muscle, blood vessels, skeletal elements, and (other) connective tissues.

Mesenterial filament Any of the long, thin tubes attached to the gut of corals and other Cnidarians that are involved in digestion and absorption.

Mesenteries Infoldings of gastroderm and mesoglea extending into the gastrovascular cavity of Cnidarians; sheets of peritoneum from which the digestive tract is suspended in coelomates.

Mesentery Peritoneal fold serving to hold the viscera in position.

Mesic Referring to a region that receives adequate precipitation to maintain biological productivity.

Mesocarp The middle layer of the fruit wall, located between the exocarp and endocarp.

Mesocercaria Juvenile stage of the digenetic trematode Alaria. It is an unencysted form between the cercaria and the metacercaria.

Mesocoel Middle body coelomic compartment in some deuterostomes, anterior in lophophorates, corresponds to hydrocoel in echinoderms.

Mesoderm The third germ layer, formed in the gastrula between the ectoderm and endoderm; gives rise to connective tissues, muscle, urogenital and vascular systems, and the peritoneum.

Mesoglea The layer of jellylike or cement material between the epidermis and gastrodermis in Cnidarians and ctenophores; also may refer to jellylike matrix between epithelial layers in sponges.

Mesohyl Gelatinous matrix surrounding sponge cells; mesoglea, mesenchyme.

Mesolecithal Pertaining to a zygote (or ovum) having a moderate amount of yolk concentrated in the vegetal pole.

Mesolytic cleavage Cleavage of a bond in a radical ion whereby a radical and an ion are formed. The term reflects the mechanistic duality of the process, which can be viewed as homolytic or heterolytic depending on how the electrons are attributed to the fragments.

Mesomeric effect The effect (on reaction rates, ionization equilibria, etc.) attributed to a substituent due to overlap of its p or pi orbitals with the p or pi orbitals of the rest of the molecular entity. Delocalization is thereby introduced or extended, and electronic charge may flow to or from the substituent. The effect is symbolized by M.

Strictly understood, the mesomeric effect operates in the ground electronic state of the molecule. When the molecule undergoes electronic excitation or its energy is increased on the way to the transition state of a chemical reaction, the mesomeric effect may be enhanced by the electromeric effect, but this term is not much used, and the mesomeric and electromeric effects tend to be subsumed in the term resonance effect of a substituent.

Mesomerism Essentially synonymous with resonance. The term is particularly associated with the picture of pi electrons as less localized in an actual molecule than in a Lewis formula.

The term is intended to imply that the correct representation of a structure is intermediate between two or more Lewis formulae.

Meson These particles are not fundamental. These are made up of quarks and anti-quarks. Mesons are a sub-category of hadrons; that is to say, all mesons are hadrons, but not all hadrons are mesons. A meson is made up of a constituent quark and anti-quark pair. These need not be the same flavor. However, they must have opposite color charge, leaving the meson uncharged with respect to the strong interaction. It could still be electrically or otherwise charged.

Mesonephros The middle of three pairs of embryonic renal organs in vertebrates. Functional kidney of fishes and amphibians; its collecting duct is a Wolffian duct.

Mesopelagic zone The pelagic environment from a depth of approximately 100 to 200 m (350 to 650 ft) to 1,000 m (3,000 ft).

Mesophase The phase of a liquid crystalline compound between the crystalline and the isotropic liquid phase.

Mesophyll Photosynthetic middle layer in the blade of a leaf; typically composed of palisade and spongy parenchyma.

Mesophyte A plant that grows in soils that contain moderate or intermediate amounts of moisture.

Mesoplankton The component of the plankton that consists of organisms 0.2 to 2 mm in size.

Mesosome The portion of the body in lophophorates and some deuterostomes that contains the mesocoel.

Mesosphere The atmospheric layer above the stratosphere and below the thermosphere; the middle layer; temperatures are usually very low.

Mesothorax The middle of the three thoracic segments of an insect; usually contains the second pair of legs and the first pair of wings.

Mesozoa A phylum of animals whose members are parasites of marine invertebrates. With a two-layered body organization. Dioecious, complex life histories. Orthonectids and dicyemids.

Mesozoic The geological era beginning about 250 million years ago. During this time gymnosperm forests, dinosaurs, mammals, birds, and flowering plants appeared. By the end of the Mesozoic, about 70 million years ago, dinosaurs became extinct.

Messenger RNA A single-stranded polyribonucleotide; formed in the nucleus from a DNA template and carries the transcribed genetic code to the ribosome where the genetic code is translated into protein.

Metabolic energy That energy obtained from ATP produced in metabolism.

Metabolism A group of processes that includes digestion, production of energy (respiration), and synthesis of molecules and structures by organisms; the sum of the constructive (anabolic) and destructive (catabolic) processes.

Metabolites Chemical substances required in metabolism.

Metacentric Chromosome with centromere at or near the middle.

Metacercaria Stage between the cercaria and adult in the life cycle of most digenetic trematodes; usually encysted and quiescent.

Metacestode Developmental stage of a cestode after metamorphosis of the oncosphere; a juvenile cestode.

Metacoel Posterior coelomic compartment in some deuterostomes and lophophorates; corresponds to somatocoel in echinoderms.

Metacommunication Communication about communication, where one signal changes the meaning of signals that follow.

Metacryptozoite Merozoite developed from a cryptozoite.

Metacyclic Stage in the life cycle of a parasite that is infective to its definitive host.

Metacyst Cystic stage of a parasite that is infective to a host.

Metamere A repeated body unit along the longitudinal axis of an animal; a somite, or segment.

Metamerism Division of the body along the anteroposterior axis into a serial succession of segments, each of which contains identical or similar representatives of all the organ systems of the body; primitively in arthropods, including externally a pair of appendages and internally a pair of nerve ganglia, a pair of nephridia, a pair of gonads, paired blood vessels and nerves, and a portion of the digestive and muscular systems.

Metamorphic rock A type of rock, either granitic or sedimentary in origin, and structurally changed by high temperature and pressure.

Metamorphosis Type of development in which one or more juvenile types differ markedly in body form from the adult; occurs in numerous animal phyla. Also applies to the actual process of changing from larval to adult form.

Metanauplius Later naupliar larvae of some crustaceans; that is, occurring after several naupliar stages but before another larval type or preadult in the developmental sequence.

Metanephridium An excretory organ found in many invertebrates; it consists of a tubule that has one end opening at the body wall and the opposite end in the form of a funnel-like structure that opens to the body cavity.

Metanephros Embryonic renal organs of vertebrates arising behind the mesonephros; the functional kidney of reptiles, birds, and mammals. It is drained from a ureter.

Metaphase Stage in mitosis when chromosomes become aligned in the middle of the cell and firmly attached to the mitotic spindle but have not yet segregated toward opposite poles.

Metapodosoma Portion of the podosoma that bears the third and fourth pairs of legs of a tick or mite.

Metapolar cells Posterior tier of cells in the calotte of a dicyemid mesozoan.

Metasome The portion of the body in lophophorates and some deuterostomes that contains the metacoel.

Metathesis A bimolecular process formally involving the exchange of a bond (or bonds) between similar interacting chemical species so that the bonding affiliations in the products are identical (or closely similar) to those in the reactants. For example:

Metathorax The posterior of the three segments of an insect thorax; it usually contains the third pair of walking legs and the second pair of wings.

Meter A unit of length in the metric system equivalent to 1,650,763.73 wave-lengths of the orange-red light emitted by krypton-86 One of the seven fundamental units of measure.

Methane CH_4, natural gas.

Methanogen A bacterium that obtains energy from CO_2 and H_2 and forms methane.

Metraterm Muscular, distended termination of the uterus of a digenetic trematode.

Metric system A system of measurement that is based on decimal multiples and subdivisions.

Micellar catalysis The acceleration of a chemical reaction in solution by the addition of a surfactant at a concentration higher than its critical micelle concentration so that the reaction can proceed in the environment of surfactant aggregates (micelles). (Rate enhancements may be due, for example, to higher concentration of the reactants in that environment, more favourable orientation and solvation of the species, or enhanced rate constants in the micellar pseudophase of the surfactant aggregate.) Micelle formation can also lead to a decreased reaction rate.

Micelle Surfactants in solution are often association colloids, that is, they tend to form aggregates of colloidal dimensions, which exist in equilibrium with the molecules or ions from which they are formed. Such aggregates are termed micelles.

Michelson-Morley experiment Possibly the most famous null-experiment of all time, designed to verify the existence of the proposed "lumeniferous aether" through which light waves were thought to propagate. Since the earth moves through this aether, a lightbeam fired in the earth's direction of motion would lag behind one fired sideways, where no aether effect would be present. This difference could be detected with the use of an interferometer.

Micro A prefix meaning 'small', as in 'microscope', 'micrometer', 'micrograph'. Also, a metric prefix meaning 10^{-6}.

Microatoll A small, flat, atoll-shaped coral structure generally found in protected coral lagoons.

Microbial loop A component of epipelagic food webs in which dissolved organic matter is cycled through the picoplankton and nanoplankton back into the main part of the food web.

Microbody Membrane-bound organelles that are the site of certain enzymatic conversions; example: peroxisomes and glyoxisomes.

Microevolution A change in the gene pool of a population across generations.

Microfauna benthic animals smaller than about 0.1 mm

Microfibril An elongated strand of cellulose molecules.

Microfilament Component of the cytoskeleton; involved in cell shape, motion, and growth. Helical protein filament formed by the polymerization of globular actin molecules.

Microfilaria The prelarval stage of filarial worms. Found in the blood of humans and the tissues of the vector.

Microfossils The microscopic shells and other remains of marine organisms that make up biogenous sediments.

Microgamete The smaller of the two gamete types in a heterogametic organism, considered the male gamete.

Microgametocyte Cell that gives rise to microgametes.

Microgametogenesis The production of microgametes (small gametes) from microspores in the anthers of flowering plants.

Micro-hydro generators Small power generators that can be used in low-level rivers to provide economical power for four to six homes, freeing them from dependence on large utilities and foreign energy supplies.

Micromere The smallest size class of blastomeres in a cleaving embryo when the blastomeres differ in size from one another.

Microneme One of the types of structures composing the apical complex in the phylum Apicomplexa, slender and elongate, leading to the anterior and thought to function in host cell penetration.

Microniscus Intermediate larval stages of the isopod suborder Epicaridea, parasitic on free-living copepods.

Micronucleus A small body of DNA that contains the hereditary information of ciliates (Protista); exchanged between protists during conjugation. It undergoes meiosis before functioning in sexual reproduction.

Micronutrient A dietary element essential in only small quantities (e.g., iron, chlorine, copper, and vitamins).

Microorganisms Single-celled living organisms that can be seen only with the aid of a microscope.

Microparasite Small (or very small) parasite that multiplies within the host of interest. Examples are protistan and prokaryotic parasites.

Microplankton The component of the plankton that consists of organisms 20 to 200 mm (0.02 to 0.2 mm) in size.

Micropores Spaces between soil particles that hold water by means of capillary forces.

Micropredator Temporary parasite.

Micropyle The small opening through which the cells emerge from a gemmule (phylum Porifera).

Microscopic A physical entity or process of small scale, too small to directly experience with our senses. Specifically, any phenomena on the molecular and atomic scale, or smaller.

Microscopic diffusion control (encounter control) The observable consequence of the limitation that the rate of a bimolecular chemical reaction in a homogeneous medium cannot exceed the rate of encounter of the reacting molecular entities.

If (hypothetically) a bimolecular reaction in a homogeneous medium occurred instantaneously when two reactant molecular entities made an encounter, the rate of reaction would be an encounter-controlled rate, determined solely by rates of diffusion of reactants. Such a hypothetical "fully diffusion controlled rate" is also said to correspond to "total microscopic diffusion control", and represents the asymptotic limit of the rate of reaction as the rate constant for the chemical conversion of the encounter pair into product (or products) becomes large relative to the rate constant for separation (or dissociation) of the encounter pair.

"Partial microscopic diffusion control" is said to operate in a homogeneous reaction when the rates of chemical conversion and of separation are comparable. (The degree of microscopic diffusion control cannot usually be determined with any precision.)

Microscopic reversibility, principle of In a reversible reaction, the mechanism in one direction is exactly the reverse of the mechanism in the other direction. This does not apply to reactions that begin with a photochemical excitation.

Microscopy Examination with a microscope.

Microspora The protozoan phylum characterized by members having unicellular spores; intracellular parasites in nearly all major animal groups.

Microsporangium The structure (sporangium) in which microspores are produced.

Microspore Haploid cells produced by meiosis in the anthers of angiosperms; four microspores are produced from a single microsporocyte.

Microspore mother cell Diploid cell in microsporangium that undergoes meiosis to produce microspores.

Microsporocyte Also known as the microspore mother cell, this diploid cell undergoes meiosis to produce four haploid microspores in the anthers of angiosperms.

Microsporogenesis The process of microspores (small spores) being produced via meiosis in the anthers of angiosperms.

Microsporophyll A leaflike structure giving rise to one or more microsporangia.

Microtubule A hollow cylinder of tubulin subunits; involved in cell shape, motion, and growth; functional unit of cilia and flagella. It is one of three major classes of filaments of the cytoskeleton.

Microvillus Narrow, cylindrical cytoplasmic projection from epithelial cells; microvilli form the brush border of several types of epithelial cells. Also, microvilli with unusual structure cover the surface of cestode tegument.

Mictic eggs Pertaining to the haploid eggs of rotifers. If it isn't fertilized, the egg develops parthenogenetically into a male; if fertilized, mictic eggs secrete a heavy shell and become dormant, hatching in the spring into amictic females.

Midbrain The portion of the brain between the pons and forebrain.

Middle lamella Layer of adhesive material (primarily pectins) found between adjacent cell walls.

Mid-ocean ridge The continuous chain of volcanic submarine mountains that extends around the earth. It includes the Mid-Atlantic Ridge and East Pacific Rise.

Midrib The central large vein of a leaf.

Miescher's tubules Sarcocysts; tissue cysts of Sarcocystis.

Migration 1. The (usually intramolecular) transfer of an atom or group during the course of a molecular rearrangement.

2. The movement of a bond to a new position, within the same molecular entity, is known as "bond migration".

Allylic rearrangements, e.g.,

$RCH{=}CHCH_2X \longrightarrow RCH(X)CH{=}CH_2$

exemplify both types of migration.

Migratory aptitude The term is applied to characterize the relative tendency of a group to participate in a rearrangement. In nucleophilic rearrangements (migration to an electron-deficient centre), the migratory aptitude of a group is loosely related to its capacity to stabilize a partial positive charge, but exceptions are known, and the position of hydrogen in the series is often unpredictable.

Migratory insertion A combination of migration and insertion. The term is mainly used in organometallic chemistry.

M = metal

Milankovitch cycles Periodic variations in tilt, eccentricity, and wobble in the earth's orbit; Milutin Milankovitch suggested that it is responsible for cyclic weather changes.

Millikan oil drop experiment A famous experiment designed to measure the electronic charge. Drops of oil were carried past a uniform electric field between charged plates. After charging the drop with x-rays, he adjusted the electric field between the plates so that the oil drop was exactly balanced against the force of gravity. Then the charge on the drop would be known. Millikan did this repeatedly and found that all the charges he measured came in integer multiples only of a certain smallest value, which is the charge on the electron.

Milpa agriculture An ancient farming system in which small patches of tropical forests are cleared and perennial polyculture agriculture practiced and is then followed by many years of fallow to restore the soil; also called swidden agriculture.

Mimicry When one species resembles one or more other species; often protection is afforded the mimic species.

Mineral A naturally occurring, inorganic, crystalline solid with definite chemical composition and characteristic physical properties.

Mineralocorticoids Hormones of the adrenal cortex, especially aldosterone, that regulate salt balance.

Miracidium The ciliated, free-swimming first stage larva of a digenean trematode that undergoes further development in the body of a snail.

Misuse alert Radioactivity is a process, not a thing, and not a substance. It is just as incorrect to say "U-235 emits radioactivity" as it is to say "current flows." A malfunctioning nuclear reactor does not release radioactivity, though it may release radioactive materials into the surrounding environment. A patient being treated by radiation therapy does not absorb radioactivity, but does absorb some of the radiation (alpha, beta, gamma) given off by the radioactive materials being used.

This misuse of the word radioactivity causes many people to incorrectly think of radioactivity as something one can get by being near radioactive materials. There is only one process which behaves anything like that, and it is called artificially induced radioactivity, a process mainly carried out in research laboratories. When some materials are bombarded with protons, neutrons, or other nuclear particles of appropriate energy, their nuclei may be transmuted, creating unstable isotopes which are radioactive.

Mitigation Repairing or rehabilitating a damaged ecosystem or compensating for damage by providing a substitute or replacement area.

Mitochondrion Membrane-bounded organelle that specializes in aerobic respiration (oxidative phosphorylation) and produces most of the ATP in eukaryotic cells.

Mitosis Nuclear division in which there is an equal qualitative and quantitative division of the chromosomal material between the two resulting nuclei; ordinary cell division.

Mitotic apparatus Collectively, the asters, spindle, centrioles, and microtubules of a dividing cell.

Mixed perennial polyculture Growing a mixture of different perennial crop species (where the same plant persists for more than one year) together in the same plot; imitates the diversity of a natural system and is often more stable and more suitable for sustainable agriculture than monoculture of annual plants.

Mixed semidiurnal tide A tidal pattern with two successive high tides of different heights each day.

Mixing control The experimental limitation of the rate of reaction in solution by the rate of mixing of solutions of the two reactants. It can occur even when the reaction rate constant is several powers of 10 less than that for an encounter-controlled rate. Analogous effects of the limitation of reaction rates by the speed of mixing are encountered in heterogeneous (solid/liquid, solid/gas, liquid/gas) systems.

MKSA The system of physical units based on the fundamental metric units: meter kilogram, second and ampere.

Möbius aromaticity A monocyclic array of orbitals in which there is a single out-of-phase overlap (or, more generally, an odd number of out-of-phase overlaps) reveals the opposite pattern of aromatic character to Hückel systems; with 4n electrons it is stabilized (aromatic), whereas with 4n + 2 it is destabilized (antiaromatic). In the excited state 4n + 2 Möbius pi-electron systems are stabilized, and 4n systems are destabilized. No examples of ground-state Möbius pi systems are known, but the concept has been applied to transition states of pericyclic reactions. The name is derived from the topological analogy of such an arrangement of orbitals to a Möbius strip.

Modal action pattern (MAP) A spatiotemporal behavior pattern that is common to members of a species; different individuals perform the pattern in a recognizably similar fashion.

Moderator A material that slows down neutrons.

Modern physics The physics developed since about 1900, which includes relativity and quantum mechanics.

Modern synthesis The combination of principles of population genetics and Darwinian evolutionary theory.

Modification A change, as in morphology, usually associated with a functional advantage.

Moiety In physical organic chemistry moiety is generally used to signify part of a molecule, e.g. in an ester R^1COOR^2 the alcohol moiety is R^2O. The term should not be used for a small fragment of a molecule.

Molar L. Molaris = a grinder.

Mole The term mole is short for the name gram-molar-weight; it is not a shortened form of the word molecule. (However, the word molecule does also derive from the word molar.)

Molecular biology The study of the biochemical structure and function of organisms.

Molecular entity Any constitutionally or isotopically distinct atom, molecule, ion, ion pair, radical, radical ion, complex, conformer etc., identifiable as a separately distinguishable entity. Molecular entity is used here as a general term for singular entities, irrespective of their nature, while chemical species stands for sets or ensembles of molecular entities. Note that the name of a compound may refer to the respective molecular entity or to the chemical species, e.g. methane, may mean a single molecule of CH_4 (molecular entity) or a molar amount, specified or not (chemical species), participating in a reaction. The degree of precision necessary to describe a molecular entity depends on the context. For example "hydrogen molecule" is an adequate definition of a certain molecular entity for some purposes, whereas for others it is necessary to distinguish the electronic state and/

or vibrational state and/or nuclear spin, etc. of the hydrogen molecule.

Molecular genetics The study of the biochemical structure and function of DNA.

Molecular mass The molecular mass of something is the mass of one mole of it (in cgs units), or one kilomole of it (in MKS units). The units of molecular mass are gram and kilogram, respectively. The cgs and MKS values of molecular mass are numerically equal. The molecular mass is not the mass of one molecule. Some books still call this the molecular weight. One definition of molar is "Pertaining to a body of matter as a whole: contrasted with molecular and atomic." The mole is a measure appropriate for a macroscopic amount of material, as contrasted with a microscopic amount (a few atoms or molecules).

Molecular mechanics calculation An empirical calculational method intended to give estimates of structures and energies for conformations of molecules. The method is based on the assumption of "natural" bond lengths and angles, deviation from which leads to strain, and the existence of torsional interactions and attractive and/or repulsive van der Waals and dipolar forces between non-bonded atoms. The method is also called "(empirical) force-field calculations".

Molecular metal A non-metallic material whose properties resemble those of metals, usually following oxidative doping; e.g. polyacetylene following oxidative doping with iodine.

Molecular orbital A one-electron wavefunction describing an electron moving in the effective field provided by the nuclei and all other electrons of a molecular entity of more than one atom. Such molecular orbitals can be transformed in prescribed ways into component functions to give "localized molecular orbitals". Molecular orbitals can also be described, in terms of the number of nuclei (or "centres") encompassed, as two-centre, multi-centre, etc. molecular orbitals, and are often expressed as a linear combination of atomic orbitals.

An orbital is usually depicted by sketching contours on which the wavefunction has a constant value (contour map) or by indicating schematically the envelope of the region of space in which there is an arbitrarily fixed high (say 96%) probability of finding the electron occupying the orbital, giving also the algebraic sign (+ or -) of the wavefunction in each part of that region.

Molecular rearrangement The term is traditionally applied to any reaction that involves a change of connectivity (sometimes including hydrogen), and violates the so-called "principle of minimum structural change". According to this oversimplified principle, chemical species do not isomerize in the course of a transformation, e.g. substitution, or the change of a functional group of a chemical species into a different functional group is not expected to involve the making or breaking of more than the minimum number of bonds required to effect that transformation. For example, any new substituents are expected to enter the precise positions previously occupied by displaced groups.

The simplest type of rearrangement is an intramolecular reaction in which the product is isomeric with the reactant (one type of "intramolecular isomerization"). An example is the first step of the Claisen rearrangement:

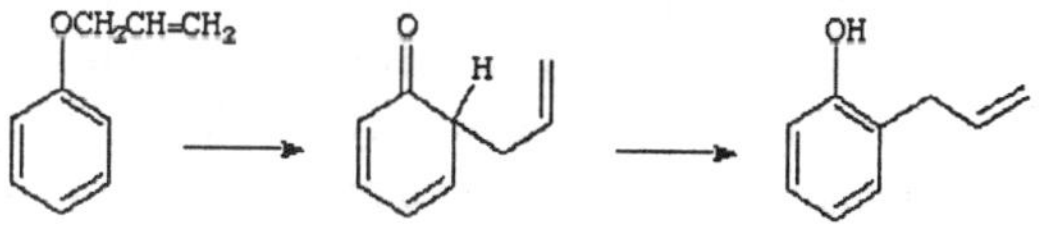

The definition of molecular rearrangement includes changes in which there is a migration of an atom or bond (unexpected on the basis of the principle of minimum structural change), as in the reaction

$CH_3CH_2CH_2Br + AgOAc \longrightarrow (CH_3)_2CHOAc + AgBr$

where the rearrangement stage can formally be represented as the "1,2-shift" of hydride between adjacent carbon atoms in the carbocation

$CH_3CH_2CH_2^+ \longrightarrow (CH_3)_2CH^+$

Such migrations occur also in radicals, e.g.:

The definition also includes reactions in which an entering group takes up a different position from the leaving group, with accompanying bond migration. An example of the latter type is the "allylic rearrangement":

$(CH_3)_2C{=}CHCH_2Br + OH^- \rightarrow (CH_3)_2C(OH)CH{=}CH_2 + Br^-$

A distinction is made between "intramolecular rearrangements" (or "true molecular rearrangements") and "intermolecular rearrangements" (or "apparent rearrangements"). In the former case the atoms and groups that are common to a reactant and a product never separate into independent fragments during the rearrangement stage (i.e. the change is intramolecular), whereas in an "intermolecular rearrangement" a migrating group is completely free from the parent molecule and is re-attached to a different position in a subsequent step.

Molecular weight The sum of all the atomic weights of a molecule.

Molecularity The number of reactant molecular entities that are involved in the "microscopic chemical event" constituting an elementary reaction. (For reactions in solution this number is always taken to exclude molecular entities that form part of the medium and which are involved solely by virtue of their solvation of solutes.) A reaction with a molecularity of one is called "unimolecular", one with a molecularity of two "bimolecular" and of three "termolecular".

Molecule An electrically neutral entity consisting of more than one atom ($n > 1$). Rigorously, a molecule, in which $n > 1$ must correspond to a depression on the potential energy surface that is deep enough to confine at least one vibrational state.

Mollusca The phylum of coelomate animals whose members possess a head-foot, visceral mass, mantle, and mantle cavity. Most molluscs also possess a radula and a shell. The molluscs. Bivalves, snails, octopuses, and related animals.

Molt A process in arthropods, such as crustaceans and insects, as well as in some vertebrates, such as snakes, whereby the organism sheds its outer exoskeleton or skin periodically as it grows. Also refers to changing the pelage or feathers in a mammal or bird.

Momentum The product of the mass and velocity of a moving body

Monera The kingdom of life whose members are characterized by having cells that lack a membrane-bound nucleus, as well as other internal, membrane-bound organelles (they are prokaryotic); bacteria.

Monerans Members of the kingdom Monera, which consists of prokaryotic organisms.

Monestrous A species in which the female is receptive for only a few days once each year.

Monitored, retrievable storage Holding wastes in underground mines or secure surface facilities such as dry casks where they can be watched and repackaged, if necessary.

Monkey wrenching Environmental sabotage such as driving large spikes in trees to protect them from loggers, vandalizing construction equipment, pulling up survey stakes for unwanted developments, and destroying billboards.

Monocarpic Denoting a plant that flowers only once.

Monochromatic light Light composed of a single color.

Monocot One of the two primary groups of angiosperms characterized by a single cotyledon, parallel venation of leaves, and floral parts in threes.

Monocotyledon A class of angiosperms in which the seedlings typically possess one cotyledon. Commonly abbreviated to monocot.

Monoculture An agricultural system in which only one crop species is cultivated.

Monoculture agroforestry Intensive planting of a single species; an efficient wood production approach, but one that encourages pests and disease infestations and conflicts with wildlife habitat or recreation uses.

Monocyte A type of leukocyte that becomes a phagocytic cell (macrophage) after moving into tissues.

Monoecious Hermaphroditic; an individual that contains reproductive systems of both sexes.

Monogamous Having one mate at a time.

Monogamy The condition of having a single mate at any one time.

Monogenea The class of Platyhelminthes that has members that are called monogenetic flukes; most ectoparasites on vertebrates (usually on fishes, occasionally on turtles, frogs, copepods, squids); one life-cycle form in only one host; bear an opisthaptor. Examples: Disocotyle, Gyrodactylus, Polystoma.

Monogononta A class of rotifers containing members that possess one ovary; mastax not designed for grinding; produce mictic and amictic eggs. Example: Notommata.

Monohybrid A hybrid offspring of parents different in one specified character.

Monohybrid cross A mating between two individuals heterozygous for one particular trait.

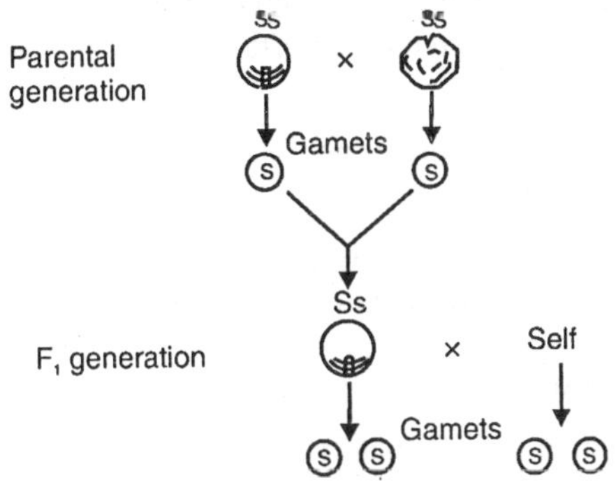

Fig. Monohybrid cross

Monomer A molecule of simple structure, but capable of linking with others to form polymers.

Monophyletic groups Groups that contain an ancestor and all the descendants of that ancestor.

Monophyletic origin A group of organisms that evolved from a single ancestral type.

Monophyly The condition that a taxon or other group of organisms contains the most recent common ancestor of the group and all of its descendants; contrasts with polyphyly and paraphyly.

Monoplacophora The class of molluscs whose members have a single, arched shell; a broad, flat foot; and certain serially repeated structures. Neopilina.

Monosaccharide A simple sugar that cannot be decomposed into smaller sugar molecules; the most common are pentoses (such as ribose) and hexoses (such as glucose).

Monostome Fluke that lacks a ventral sucker.

Monounsaturated fat Composed of fatty acid chains in which there is only a single C-C double bond; examples: canola oil and olive oil.

Monoxenous Living within a single host during a parasite's life cycle.

Monozoic Tapeworms with a single proglottid, do not undergo strobilation to form chain of proglottids.

Monozygotic Twins that arise from a single zygote, hence, two genetically identical individuals.

Monsoon A seasonal reversal of wind patterns caused by the different heating and cooling rates of the oceans and continents.

Montane coniferous forests Coniferous forests of the mountains consisting of belts of different forest communities along an altitudinal gradient.

Moral agents Beings capable of making distinctions between right and wrong and acting accordingly. Those whom we hold responsible for their actions.

Moral subjects Beings that are not capable of distinguishing between right or wrong or that are not able to act on moral principles and yet are susceptible of being wronged by others. This category assumes some rights or inherent values in moral subjects that gives us duties or obligations towards them.

Morbidity Illness or disease.

More O'ferrall-jencks diagram Visualization of the potential energy surfaces for a reacting system, as a function of two chosen coordinates. It is particularly useful to discuss structural effects on the transition state geometry for processes occurring either by stepwise or concerted routes. The use of such diagrams, first suggested for elimination reactions, was later extended to acid-base catalysis and to certain other reactions.

Structural changes influencing vibrational modes of the transition states cause changes in transition state geometry. Changes in the direction of the reaction coordinate (reactant or product stabilizing- or destabilizing factors) cause changes according to the Hammond principle. Structural changes perpendicular to the reaction coordinate (anti-Hammond effects, perpendicular effects) cause changes opposite to the Hammond behaviour, i.e., the easier the process related to the structural change, the more advanced it will be at the transition state.

Morphine A pain-relieving and addictive compound derived from the opium poppy (Papaver somniferum).

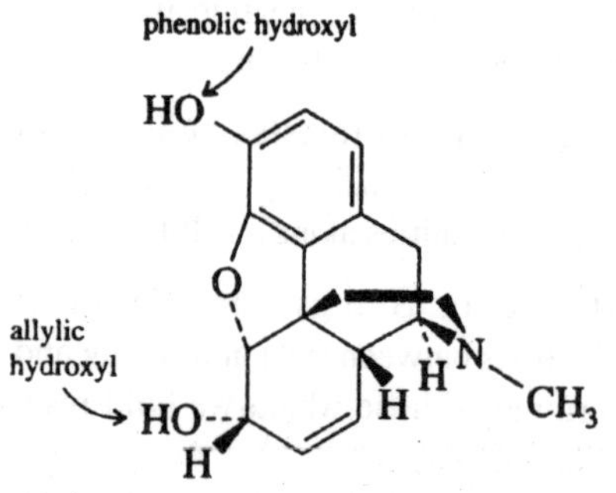

Fig. Morphine

Morphogenesis Development of the architectural features of organisms; formation and differentiation of tissues and organs.

Morphology The science of structure. Includes cytology, the study of cell structure; histology, the study of tissue structure; and anatomy, the study of gross structure.

Mortality Death rate in a population; the probability of dying.

Morula A stage in the embryonic development of some animals that consists of a solid ball of cells.

Mosaic An organism whose tissues are made up of two or more genetically different types.

Mosaic cleavage. Embryonic development characterized by independent differentiation of each part of the embryo; determinate cleavage.

Mosaic evolution A change in a portion of an organism (e.g., a bird wing) while the basic form of the organism is retained.

Moss A group of terrestrial, nonvascular plants; the dominant plant body is the gametophyte with the sporophyte embedded in it.

Motility Ability to move.

Motivation Internal processes that arouse and direct behavior. Today this refers more to examination of the consequences of behavior and feedback to the animal concerning the consequences of those actions.

Motor (efferent) neuron or nerve A neuron or nerve that transmits impulses from the central nervous system to an effector such as a muscle or gland.

Motor unit A motor neuron and the muscle fibers associated with it.

Mountain tundra That portion of tundra vegetation confined to alpine meadows. Low-growing grasses, sedges, and forbs with a very short growing season; permafrost is typical.

Mucigel A slimy material secreted by and covering the root cap and root hairs.

Mucilaginous Containing a mucilage, usually composed of mucopolysaccharides.

Mucin Any of a group of glycoproteins secreted by certain cells, especially those of salivary glands.

Mucous cell A glandular cell that secretes mucus.

Mucron Apical anchoring device on an acephaline gregarine protozoan.

Mucus Viscid, slippery secretion rich in mucins produced by secretory cells such as those in mucous membranes.

Mud flat A muddy bottom that is exposed at low tide.

Muellerian mimicry Evolution by one species to resemble the coloration, body shape, or behavior of an unrelated species that is protected from predators by a venomous stinger, bad taste, or some other adaptation.

Mulch Protective ground cover, including manure, wood chips, straw, seaweed, leaves, and other natural products, or synthetic materials, such as heavy paper or plastic, that protect the soil, save water, and prevent weed growth.

Muller's larva A free-swimming ciliated larva that resembles a modified ctenophore, characteristic of many marine polyclad turbellarians.

Muller's ratchet The steady accumulation of mutations in a population of asexual organisms over time.

Mullerian mimicry Occurs when two similar species are both distasteful to predators.

Multi-centre bond Representation of some molecular entities solely by localized two-electron two-centre bonds appears to be unsatisfactory. Instead, multi-centre bonds have to be considered in which electron pairs occupy orbitals encompassing three or more atomic centres. Examples include the three-centre bonds in diborane, the delocalized pi bonding of benzene, and bridged carbocations.

Multi-centre reaction A synonym for pericyclic reaction. The number of "centres" is the number of atoms not bonded initially, between which single bonds are breaking or new bonds are formed in the transition state. This number does not necessarily correspond to the ring size of the transition state for the pericyclic reaction. Thus, a Diels-Alder reaction is a "four-centre reaction". This terminology has largely been superseded by the more detailed one developed for the various pericyclic reactions.

Multigenic (polygenic) inheritance Inheritance in which the genetic control for a trait results in the phenotypic expression varying continuously.

Multiple alleles A condition in which more than two alleles exist for a given trait.

Multiple fission Asexual reproduction by the splitting of a cell or organism into many cells or organisms.

Multiple fruit A fruit derived from the fusion of the ovaries of several flowers in an inflorescence; example: pineapple.

Multiple use Many uses that occur simultaneously; used in forest management; limited to mutually compatible uses.

Muon experiment An experiment which demonstrates verifies the prediction of time dilation by special relativity. Muons, which are short-lived subatomic particles, are created with enormous energy in the upper atmosphere by the interaction of energetic cosmic rays. Muons have a very short halflife in their own reference frame, about 2.2 us. Since they are travelling very close to c, however, time dilation effects should become important. A naive calculation would indicate that, without special relativistic effects, the muons would travel on the average only about 700 m before decaying, never reaching the surface of the earth. Observations reveal, however, that significant numbers of muons do reach the earth. The explanation is that muon is in a moving frame of reference, and thus time is slowed down for the muons relative to the earth, effectively extending the halflife of the muons relative to the earth, allowing some of them to reach the surface.

Muscle fiber The contractile unit of a muscle.

Muscle tissue The type of tissue that allows movement. The three kinds are skeletal, smooth, and cardiac. Tissue made of bundles of long cells called muscle fibers.

Musclel Musculus = a contractile organ.

Muscularis mucosae Smooth muscle fibers around the mucosa of the gut wall, surrounding the lamina propria and surrounded by the submucosa.

Mushroom Common name for a group of fungi (Basidiomycetes) that produce an aboveground reproductive structure.

Mutagens Agents, such as chemicals or radiation, that damage or alter genetic material (DNA) in cells.

Mutation A change, either spontaneous or by external factors, in the genetic material of a cell; mutations in the gametes (sex cells) can be inherited by future generations of organisms.

Mutation pressure The constant resupplying of mutations to a gene pool due to a base mutation rate.

Mutational analysis An analysis that examines the effects of specific mutants for the effects on particular phenomena with respect to morphology, physiology, or behavior.

Mutual inductance The ratio of the induced emf in one circuit to the rate of change of current in the coil of another circuit.

Mutualism A type of interaction in which two different species derive benefit from their association and in which the association is necessary to both; often symbiotic.

Mycelium A network of fungal hyphae.

Mycetome Specialized organ in some insects that bears mutualistic bacteria.

Mycobiont The fungal partner in a mutualistic relationship such as mycorrhizae or lichens.

Mycologist A person who studies fungi.

Mycorrhiza Symbiotic association between a fungus and a plant root.

Mycotoxin A toxic compound formed by the hyphae of common molds growing under a variety of conditions, especially in contaminated foods.

Myelin A fatty material forming the medullary sheath of nerve fibers.

Myiasis Infection by fly maggots.

Myocyte Contractile cell (pinacocyte) in sponges.

Myofibril A contractile filament within muscle or muscle fiber.

Myogenic Originating in muscle, such as heartbeat arising in vertebrate cardiac muscle because of inherent rhythmical properties of muscle rather than because of neural stimuli.

Myoglobin The oxygen-transporting pigment of muscle tissue.

Myomere The muscle plate or portion of a somite that develops into voluntary muscle.

Myosin A large protein of contractile tissue that forms the thick myofilaments of striated muscle. During contraction it combines with actin to form actomyosin.

Myotome That part of a somite destined to form muscles; the muscle group innervated by a single spinal nerve.

Myriapods Members of the four noninsect classes of the subphylum Uniramia. Includes centipedes, millipedes, pauropods, and symphylans.

Myxini The class of vertebrates whose members are fishlike, jawless, without paired appendages, and possess four pairs of tentacles around the mouth. Hagfishes.

Myxozoa The protozoan phylum characterized by members having spores of multicellular origin; the myxozoans.

Myzorhynchus Apical stalked, suckerlike organ on the scolex of some tetraphyllidean cestodes

N

NAD A coenzyme that acts as an electron acceptor; particularly important in respiration.

NADP A molecule that acts as an electron intermediate during photosynthesis. It is reduced during the light reactions and oxidized during the Calvin Cycle.

Nagana Disease of ruminants caused by Trypanosoma brucei brucei or T. Congolense.

Naiad The aquatic immature stage of any hemimetabolous insect.

Nail The horny cutaneous plate of the dorsal surface of the distal end of a finger or toe.

Naked grain A grain that separates easily from the surrounding bracts.

Nanoplankton The component of the plankton that consists of very small organisms 2 to 20 mm (0.002 to 0.2 mm) in size; they are too small to catch in a standard plankton net.

Narcissistic reaction A chemical reaction that can be described as the conversion of a reactant into its mirror image, without rotation or translation of the product, so that the product enantiomer actually coincides with the mirror image of the reactant molecule. Examples of such reactions are cited under the entries fluxional and degenerate rearrangement.

Narcotic Any psychoactive compound that is dangerously addictive; a compound that induces central nervous system depression resulting in numbness, lethargy, and/orsleep.

Nares Openings into the nasal cavity, both internally and externally, in the head of a vertebrate.

Nasal gland Salt-excreting gland in the nostrils of seabirds and reptiles

Natal Pertaining to birth.

Natality The production of new individuals by birth, hatching, germination, or cloning.

Native species A local species that has not been introduced.

Natural history The study of where and how organisms carry out their life cycles.

Natural increase Crude death rate subtracted from crude birth rate.

Natural killer (NK) cell A lymphocyte present in nonimmunized individuals that exhibits independent cytolytic activity against tumor cells; also called cytotoxic T cells.

Natural landscaping A landscaping design that uses carefully contrived meadows, forests, and natural-looking ponds and marshes to create an illusion of wild nature.

Natural law A theory that has been tested many thousands of times and found always to be true, e.g., the law of gravity.

Natural resources Goods and services supplied by the environment.

Natural selection The mechanism for evolutionary change in which environmental pressures cause certain genetic combinations in a population to become more abundant; genetic combinations best adapted for present environmental conditions tend to become predominant.

Naturally acquired active immunity The type of immunity that develops when an individual's immunologic system comes into contact with an appropriate antigenic stimulus during the course of normal activities; it usually arises as the result of recovering from an infection.

Nauplius Typically the earliest larval stage(s) of crustaceans; has only three pairs of appendages: antennules, antennae, and mandibles_all primarily of locomotive function.

Navicular navicul = a little ship. A carpal bone shaped like a boat.

Navigation The process by which an animal uses various cues to determine its position in reference to a goal.

Neap tides The tides with a small tidal range. They occur around the times when the moon is in quarter. Compare spring tides.

Neascus Strigeoid metacercaria with a spoon-shaped forebody.

Necrosis Cell or tissue death.

Nectar A sweet, syrupy exudate produced by some flowers as an attractant for pollinators.

Nectar guide Color patterns present on petals that direct insects toward the nectar; often not visible to human eye.

Nectary A gland that secretes nectar.

Negative feedback principle The idea that in a system where there are self-propagating circumstances, those new circumstances tend to act against previously existing circumstances. Such a principle is really a restatement of a conservation law.

Neighbouring group participation The direct interaction of the reaction centre (usually, but not necessarily, an incipient carbenium centre) with a lone pair of electrons of an atom or with the electrons of a sigma or pi bond contained within the parent molecule but not conjugated with the reaction centre. A distinction is sometimes made between n, sigma, and pi participation.

A rate increase due to neighbouring group participation is known as "anchimeric assistance". "Synartetic acceleration" is the special case of anchimeric assistance ascribed to participation by electrons binding a substituent to a carbon atom in a -position relative to the leaving group attached to the -carbon atom. According to the underlying model, these electrons then provide a three-centre bond (or "bridge") "fastening together" (as the word "synartetic" is intended to suggest) the - and -carbon atoms between which the charge is divided in the intermediate bridged ion formed (and in the transition state preceding its formation). The term synartetic acceleration is not widely used.

Nekton Term for actively swimming organisms, essentially independent of wave and current action. Compare with plankton.

Nematocyst An organelle characteristic of the Cnidaria that is used in defense, food gathering, and attachment.The stinging structure of Cnidarians.

Nematoda The phylum of aschelminths that contains members commonly called either roundworms or nematodes. Tripoblastic, bilateral, vermiform, unsegmented, and pseudocoelomate.

Nematogen State in the life cycle of a dicyemid mesozoan.

Nematomorpha The phylum of aschelminths commonly called horsehair worms.

Nemertea The phylum that has members commonly called the proboscis worms; elongate, flattened worms found in marine mud and sand; triploblastic; complete digestive tract with anus; closed circulatory system.

Neo-Darwinism A modified version of Darwin's evolutionary theory that eliminates elements of the Lamarckian inheritance of acquired characteristics and pangenesis that were present in Darwin's formulation; this theory originated with August Weissmann in the late nineteenth century and, after incorporating Mendelian genetic principles, has become the currently favored version of Darwinian evolutionary theory.

Neolithic The Stone Age period following the advent of agriculture.

Neo-Malthusian A belief that the world is characterized by scarcity and competition in which too many people fight for too few resources. Named for Thomas Malthus, who predicted a dismal cycle of misery, vice, and starvation as a result of human overpopulation.

Neopterygian Any of a large group of bony fishes that includes most modern species.

Neoteny An evolutionary process by which organismal development is retarded relative to sexual maturation; produces a descendant that reaches sexual maturity while retaining a morphology characteristic of the preadult or larval stage of an ancestor.

Nephric nephros = kidney.

Nephridiopore An external excretory opening in invertebrates.

Nephridium One of the segmentally arranged, paired excretory tubules of many invertebrates, notably the annelids. In a broad sense, any tubule specialized for excretion and/or osmoregulation; with an external opening and with or without an internal opening.

Nephron Functional unit of kidney structure of vertebrates, consisting of a Bowman's capsule, an enclosed glomerulus, and the attached uriniferous tubule.

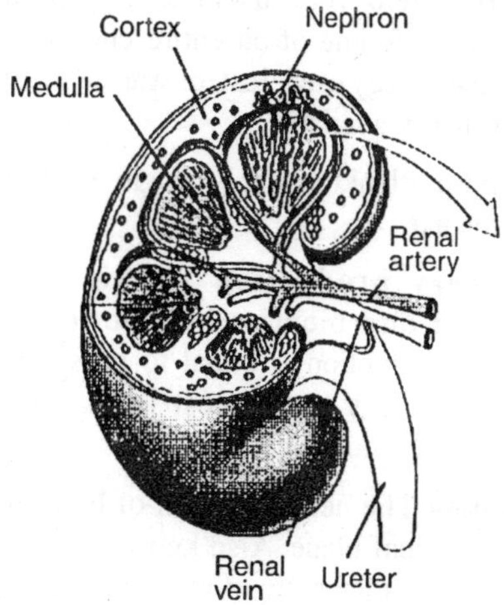

Nephrostome Ciliated, funnel-shaped opening of a nephridium.

Neritic Portion of the sea overlying the continental shelf, specifically from the subtidal zone to a depth of 200 m.

Neritic ecosystems The marine ecosystem that consists of relatively shallow water extending from the littoral zone to the edge of the continental shelves.

Nerve A bundle of neurons or nerve cells outside the central nervous system.

Nerve cell A cell specialized to originate or transmit nerve impulses.

Nerve cord A long, compact bundle of nerve cells that is part of the central nervous system.

Nerve net A diffuse, two-dimensional plexus of bi- or multipolar neurons; found in Cnidarians.

Nerve tract A bundle of neurons, particularly axons, held together by connective tissue.

Nervous tissue The type of tissue composed of individual cells called neurons and supporting neuroglial cells.

Nested hierarchy A pattern in which species are ordered into a series of increasingly more inclusive clades according to the taxonomic distribution of synapomorphies.

Net energy yield Total useful energy produced during the lifetime of an entire energy system minus the energy used, lost, or wasted in making useful energy available.

Net plankton Plankton that is caught in a standard plankton net.

Net primary production The total energy converted into biomass by autotrophs. Net primary production is equal to gross primary production less energy lost in maintenance functions of autotrophs.

Net venation The netlike pattern of branching of veins on a leaf blade. Also known as reticulate venation; characteristic of most dicot leaves.

Neuroendocrine system The combination of the nervous and endocrine systems.

Neurofibril node Regular gaps in a myelin sheath around a nerve fiber; formerly called node of Ranvier.

Neurogenic Originating in nervous tissue, as does the rhythmical beat of some arthropod hearts.

Neuroglia Tissue supporting and filling the spaces between the nerve cells of the central nervous system.

Neurohormone A chemical transmitter produced by nervous tissue. Uses the bloodstream or other body fluids for distribution to its target site.

Neurolemma Delicate nucleated outer sheath of a nerve cell; sheath of Schwann.

Neurolemmocyte The cell that surrounds a fiber of a peripheral nerve and forms the neurolemmal sheath and myelin; formerly called Schwann cell.

Neuromast Cluster of sense cells on or near the surface of a fish or amphibian that is sensitive to vibratory stimuli and water.

Neuromuscular junction The junction between nerve and muscle; myoneural junction or neuromuscular cleft.

Neuron A nerve cell that consists of a cell body and its processes.

Neuropeptide A hormone produced by secretory nervous tissue.

Neuropodium Lobe of parapodium nearer the ventral side in polychaete annelids.

Neurosecretions Hormonelike chemical substances that are produced by specialized neurons that affect various physiological processes in the body.

Neurosecretory cell Any cell (neuron) of the nervous system that produces a hormone.

Neurotoxins Toxic substances, such as lead or mercury, that specifically poison nerve cells.

Neurotransmitter Chemical substance secreted by the terminal end of an axon that stimulates a muscle fiber contraction or an impulse in another neuron.

Neurulation External changes along the upper surface of a chordate embryo that result in the formation of the neural tube.

Neuston Organisms that live on the surface of the sea.

Neutral Soil ph of 7, neither acid nor alkaline.

Neutral weak current A subatomic reaction in which leptons collide without change in the charges of the colliding particles.

Neutron Uncharged atomic particles of essentially the same mass as protons. In the most common stable atoms the number of neutrons is equivalent to the number of protons.

Neutrophil Most abundant of polymorphonuclear leukocytes; an important phagocyte; so called because it stains with both acidic and basic stains.

New towns Experimental urban environments that seek to combine the best features of the rural village and the modern city.

Newton The unit of force; a derived unit having the dimensions kg m/s2 The force required to accelerate a one-kilogram mass at a rate of one meter per second each second.

Newton's first and second laws of motion F = d(mv)/dt F is the net (total) force acting on the body of mass m. The individual forces acting on m must be summed vectorially. In the special case where the mass is constant, this becomes F = ma.

Newton's law of universal gravitation The force of attraction between any two particles of matter in the universe is directly proportional to the product of their masses and inversely proportional to the square of the distance between their centers of mass

Newton's second law of motion The acceleration of a body is directly proportional to the net force exerted on the body, is inversely proportional to the mass of the body, and has the same direction as the net force; also called the law of acceleration.

Newton's third law of motion If one body exerts a force on a second body, then the second body exerts a force equal in magnitude and opposite in direction on the first body; also called the law of interaction.

Nicotine adenine dinucleotide A local electron carrier that transfers hydrogen atoms and electrons within metabolic pathways; a free-moving carrier, not membrane bound in a transport system; NAD^+.

Nictitating membrane Third eyelid, a transparent membrane of birds and many reptiles and mammals, that can be pulled across the eye.

Nidus Specific locality of a given disease; result of a unique combination of ecological factors that favors the maintenance and transmission of the disease organism.

Nihilists Those who believe the world has no meaning or purpose other than a dark, cruel, unceasing struggle for power and existence.

Nitrate (NO_3^{2-})An important nutrient in the ocean.

Nitrate-forming bacteria Bacteria that convert nitrites into compounds that can be used by green plants to build proteins.

Nitrene Generic name for HN: and substitution derivatives thereof, containing an electrically neutral univalent nitrogen atom with four non-bonding electrons. Two of these are paired; the other two may have parallel spins (triplet state) or antiparallel spins (singlet state). The name is the strict analogue of carbene and, as a generic name, it is preferred to a number of alternatives proposed ("imene", "imine radical", "azene", "azylene", "azacarbene", "imin", "imidogen").

Nitrenium ion The cation H_2N^+ and its N-hydrocarbyl derivatives R_2N^+, in which the nitrogen has a positive charge, and two unshared electrons. A synonymous term is aminylium ion.

Nitrogen (N_2)A colorless and tasteless gas that is an essential constituent of proteins.

Nitrogen base The basic component of nucleic acids; composed of a nitrogen-containing molecule of purine or pyrimidine.

Nitrogen cycle The circulation and reutilization of nitrogen in both inorganic and organic phases.

Nitrogen fixation Reduction of molecular nitrogen to ammonia by some bacteria and cyanobacteria, often followed by nitrification, the oxidation of ammonia to nitrites and nitrates by other bacteria.

Nitrogen oxides Highly reactive gases formed when nitrogen in fuel or combustion air is heated to over 65°C (120° F) in the presence of oxygen or when bacteria in soil or water oxidize nitrogen-containing compounds.

Nitrogenase The enzyme involved in the conversion of atmospheric nitrogen gas into ammonia.

Nitrogen-fixing bacteria Bacteria that convert nitrogen from the atmosphere or soil solution into ammonia that can then be converted to plant nutrients by nitrite- and nitrate-forming bacteria.

Nitrous oxide NO_2, an atmospheric gas that traps infrared radiation and also can form nitric acid in rain droplets, resulting in acid rain.

Nociceptor A sensory receptor responding to potentially harmful stimuli; produces a sensation of pain.

Nocturnal Animals whose primary activity occurs during the dark portion of the daily cycle.

Node The point on a stem at which leaves and buds are attached.

Node of Ranvier A constriction of mylelinated nerve fibers at regular intervals at which the myelin sheath is absent and the axon is enclosed only by sheath cell processes; also known as a neurofibril node.

Nodule A tumorlike growth on the roots of certain higher plants that encloses a population of nitrogen-fixing bacteria.

Noether theorem A theorem which demonstrates that symmetries are what gives rise to conserved quantities. For instance, translational symmetry (the fact that the laws of physics work the same in all places) gives rise to conservation of momentum, since position and momentum are complementary. Additionally, conservation of energy is indicated by time symmetry, and conservation of angular momentum is indicated by isotropy.

No-hair conjecture The conjecture (proved in the 1970s and 1980s) within general relativity that a black hole has only three salient external characteristics: mass, angular momentum, and electric charge. All other properties (including baryon number, lepton number, strangeness, etc.) Are destroyed as matter falls into the horizon.

There is some indication that quantum mechanical considerations in quantum gravity will result in a "quantum hair" coming into play. However, that 1. Would constitute a prediction of a theory which does not yet formally exist, and 2. Is utterly insignificant for solar-massed black holes, the only types that can be formed today.

Noise Sound produced by irregular vibrations in matter which is unpleasant to the listener.

Nomenclature The study of the naming of organisms in the fashion that reflects their evolutionary relationships.

Non Abelian This describes an algebraic operation (multiplication, addition, division, subtraction, etc.). If the operation is non-Abelian, then the elements do not commute. Although regular multiplication commutes (3x4 = 4x3 = 12) and addition also commutes (3+4 = 4+3 = 7), there are mathematical elements (matrices) which do not commute under multiplication. (If A and B are matrices, then A times B does not equal B times A.) Operations which are invariant (unchanging) when the elements are exchanged (operations which do commute) are called Abelian. These ideas are named for the developer, Niels Abel.

Nonamniote lineage The vertebrate lineage leading to modern amphibians.

Nonbiodegradable A chemical that cannot be broken down by bacteria or other organisms. It is said to be persistent.

Nonclassical carbocation A carbocation the ground state of which has delocalized (bridged) bonding pi- or sigma-electrons.

Noncyclic photophosphorylation The formation of ATP utilizing the Z-scheme of photosynthesis.

Nondisjunction. Failure of a pair of homologous chromosomes to separate during meiosis, leading to one gamete with n + 1 chromosomes and another gamete with n-1 chromosomes.

Noninterference A principle that requires us to refrain from interfering with the lives or destinies of other moral agents or moral subjects except to satisfy essential needs.

Nonpoint sources Scattered, diffuse sources of pollutants, such as runoff from farm fields, golf courses, construction sites, etc.

Nonpolar covalent bond The type of bond that is formed when electrons spend as much time orbiting one nucleus as the other; thus, the distribution of charges is symmetrical.

Nonrenewable resources Materials or services from the environment that are not replaced or replenished by natural processes at a rate comparable to our use of the resource; a resource depleted or exhausted by use.

Nonseptate Lacking cross walls; commonly used in reference to certain fungal hyphae. Also termed aseptate.

Nonseptate hypha Fungal hyphae that lack septa or cross walls.

Nonshattering Seeds and/or fruits do not split off and scatter from the fruiting head; a trait associated with domesticated plants.

Nonshivering thermogenesis The hormonal triggering of heat production.

Norepinephrine A neurotransmitter released from the axon ends of some nerve fibers; noradrenaline.

Normal A line drawn perpendicular to a line or surface.

North/South division A description of the fact that most of the world's wealthier countries tend to be in North America, Europe, and Japan while the poorer countries tend to be located closer to the equator.

Northern coniferous forest The biome characterized by cool summers, cold winters, and short growing seasons. Characteristic animals include insects, snowshoe hares, lynx, wolves, caribou, and moose.

Notochord An elongated cellular cord, enclosed in a sheath, which forms the primitive axial skeleton of chordate embryos and adult cephalochordates.

Notopodium Lobe of parapodium nearer the dorsal side in polychaete annelids.

N-type germanium "Electron-rich" germanium consisting of equal numbers of free electrons and bound positive charges so that the net charge is zero.

Nucellus Tissue in the ovule within which the embryo sac develops; integuments surround the nucellus.

Nuchal ligament L. The back of the head or neck; ligare = to bind.

Nuclear binding force The force that acts within the small distances between nucleons.

Nuclear change A change in the identity of atomic nuclei.

Nuclear envelope Double membrane forming the surface boundary of a eukaryotic nucleus; consists of outer and inner membranes perforated by nuclear pores.

Nuclear fusion A process in which two smaller atomic nuclei fuse into one larger nucleus and release energy; the source of power in a hydrogen bomb.

Nuclear mass defect The arithmetic difference between the mass of a nucleus and the larger sum of its uncombined constituent particles.

Nuclear membrane The membrane surrounding the nucleus of eucaryotic cells.

Nuclear pores Pitted regions in the nuclear envelope through which processed mRNA migrates to the ribosomes.

Nuclear reactor A device in which the controlled fission of certain substances is used to produce new substances and energy.

Nucleic acid One of a class of molecules composed of joined nucleotides; chief types are deoxyribonucleic acid (DNA), found in cell nuclei (chromosomes) and mitochondria, and ribonucleic acid (RNA), found both in cell nuclei (chromosomes and nucleoli) and in cytoplasmic ribosomes.

Nucleofuge A leaving group that carries away the bonding electron pair. For example, in the hydrolysis of an alkyl chloride, Cl- is the nucleofuge. The tendency of atoms or groups to depart with the bonding electron pair is called nucleofugality.

Nucleoid The region in a prokaryotic cell where the chromosome is found.

Nucleolar organizer An area on certain chromosomes associated with the formation of the nucleolus.

Nucleolus A deeply staining body within the nucleus of a cell and containing RNA; nucleoli are specialized portions of certain chromosomes that carry multiple copies of the information to synthesize ribosomal RNA.

Nucleon Either a proton or a neutron; the term neucleon is used when one is not concerned with distinguishing between protons and neutrons, and rather is refering to either usually as a resident of the nucleus.

Nucleophile, nucleophilic A nucleophile (or nucleophilic reagent) is a reagent that forms a bond to its reaction partner (the electrophile) by donating both bonding electrons.

A "nucleophilic substitution reaction" is a heterolytic reaction in which the reagent supplying the entering group acts as a nucleophile. For example

MeO^- (nucleophile) + Et-Cl $\longrightarrow$ MeOEt + Cl^- (nucleofuge)

The term "nucleophilic" is also used to designate the apparent polar character of certain radicals, as inferred from their higher relative reactivity with reaction sites of lower electron density.

Nucleophilic catalysis Catalysis by a Lewis base, involving formation of a Lewis adduct as a reaction intermediate. For example, the hydrolysis of acetic anhydride in aqueous solution catalysed by pyridine:

$C_5H_5N + (CH_3CO)_2O \longrightarrow [C_5H_5NCOCH_3]^+ + CH_3CO_2^-$

$[C_5H_5NCOCH_3]^+ + H_2O\ C_5H_5N + \longrightarrow CH_3CO_2H + H^+aq$

Nucleophilicity 1. The property of being nucleophilic.

2. The relative reactivity of a nucleophilic reagent. (It is also sometimes referred to as "nucleophilic power".) Qualitatively, the concept is related to Lewis basicity. However, whereas Lewis basicity is measured by relative equilibrium constants,

$B: + A \overset{k}{\rightleftharpoons} B^+A^-$ (equilibrium constant K)

nucleophilicity of a Lewis base is measured by relative rate constants of different nucleophilic reagents towards a common substrate, most commonly involving formation of a bond to carbon.

Nucleoplasm Protoplasm of nucleus, as distinguished from cytoplasm.

Nucleoprotein. A molecule composed of nucleic acid and protein; occurs in the nucleus and cytoplasm of all cells.

Nucleosome A repeating subunit of chromatin in which one and three-quarter turns of the double-helical DNA are wound around eight molecules of histones.

Nucleotide A molecule consisting of phosphate, 5-carbon sugar (ribose or deoxyribose), and a purine or a pyrimidine; the purines are adenine and guanine, and the pyrimidines are cytosine, thymine, and uracil.

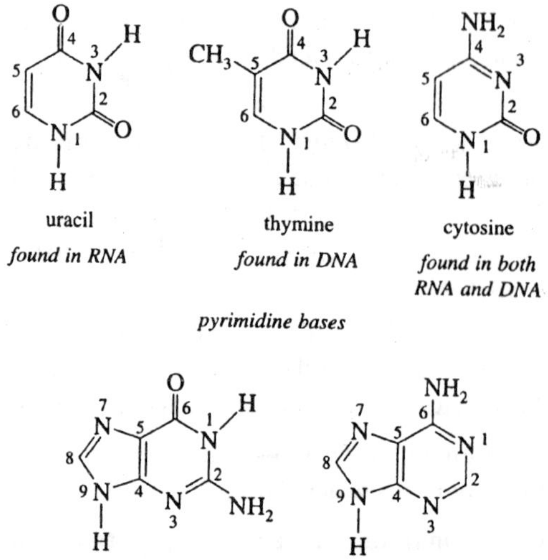

Fig. **Nucleotide**

Nucleus The center — usually of the central piece of an atom. The nucleus of an atom is composed of protons and neutrons, which are themselves composed of quarks. The nucleus of an atom is about 1000 to 10,000 times smaller than the size of an atom, where the size of the atom is defined by the orbit of the electrons.

Nuclide An atom of a particular mass and of a particular element.

Nuda The class of ctenophorans whose members lack tentacles and have a flattened body with a highly branched gastrovascular cavity.

Nuées ardentes Deadly, denser-than-air mixtures of hot gases and ash ejected from volcanoes.

Null experiment An experiment which, after being executed, yields no result. Null experiments are just as meaningful as non-null experiments; if current theory predicts an observable effect (or predicts there should be no observable effect), and experimentation (within the required accuracy) does not yield said effect, then the null experiment has told us something about our theory.

Null hypothesis In a statistical sense, to the statement that there will not be any difference between effects of experimental treatments.

Numbers pyramid A diagram showing the relative population sizes at each trophic level in an ecosystem; usually corresponds to the biomass pyramid.

Numerical calculations Calculations done using numbers, as opposed to symbolically; typically on a computer.

Numerical taxonomy A system of classification in which there is no attempt to distinguish true and false similarities.

Nuptial flight The mating flight of insects, especially that of the queen with male or males.

Nurse cells Single cells or layers of cells surrounding or adjacent to other cells or structures for which the nurse cells provide nutrient or other molecules (for example, for insect oocytes or Trichinella spp. Juveniles).

Nut A dry, indehiscent one-seeded fruit with a hard pericarp (the shell).

Nutrient A raw material other than carbon dioxide and water that is needed by an autotroph to produce organic matter. Examples are nitrate and phosphate.

Nutrient regeneration The release of nutrients from organic matter by decomposers.

Nutrition The study of the sources, actions, and interactions of nutrients.

Nyctinasty The "sleep movements" of leaves in response to change in turgor pressure of cells at the base of the petiole.

Nymph An immature stage (following hatching) of a hemimetabolous insect that lacks a pupal stage.

Nymphochrysalis Nonfeeding, prenymph stage in the life cycle of a chigger mite.

O

Obligate anaerobe A microorganism that can survive only under anaerobic conditions.

Obligate symbiont Organism that is physiologically dependent on establishing a symbiotic relationship with another.

Observability Not all of the members of a group are always seen the same proportion of time. Animals of different ages, sexes, or of different dominance status may be seen more or less than other animals.

Observational learning The tendency to perform an appropriate action or response as the result of having observed another animal's performance of similar actions in the same situation.

Observational method In an historical context, the notion, developed by C. Lloyd Morgan, that only data gathered by direct experiment and watching the animals under study could provide the basis for generalizations. This was in contrast to Romanes's notion that much about animal behavior could be gained by inference from what was seen. In the modern sense, this refers to any of a number of techniques used to watch and record the actions and interactions of animals.

Observations Objectively measuring or recording the results of testing a theory; watching and evaluating.

Obtect pupa Pupa with wings and legs tightly appressed to its body and covered by an external cuticle.

Obturator L. A disk covering an opening.

Ocean desert The concept of a region in the ocean in which lack of nutrients is reflected in a lack of plants and animals; a region of low biological productivity.

Ocean shorelines Rocky coasts and sandy beaches along the oceans; support rich, stratified communities.

Ocean thermal electric conversion (OTEC) Energy derived from temperature differentials between warm ocean surface waters and cold deep waters. This differential can be used to drive turbines attached to electric generators.

Oceanic Pertaining to the portion of the marine environment that overlies the deep ocean basins.

Oceanic ecosystems The marine ecosystem that extends from the continental shelves into the unproductive open ocean.

Oceanic islands Islands in the ocean; formed by breaking away from a continental landmass, volcanic action, coral formation, or a combination of sources; support distinctive communities.

Oceanic zone The pelagic environment beyond the shelf break.

Ocellus A simple eye or eyespot in many types of invertebrates.

Octave The interval between a given musical tone and one with double or half the frequency.

Octomerous Eight parts, specifically, symmetry based on eight.

Ocytes A group of prokaryotes currently classified among the Archaebacteria but possibly a sister group of eukaryotes.

Odontophore Tooth-bearing organ in molluscs, including the radula, radular sac, muscles, and cartilages.

Off-center, on-surround cell A ganglion cell that generates an action potential when light shines on the periphery of its receptive field but is inhibited when light shines on the center of its receptive field.

Offset allowances A controversial component of air quality regulations that allows a polluter to avoid installation of control equipment on one source with an "offsetting" pollution reduction at another source.

Off-surround The photoreceptor cells in the outer portion of a receptive field that have an inhibitory effect on ganglion cells if stimulated by light.

Ohm The derived SI unit of electric resistance, defined as the resistance between two points on a conductor when a constant potential difference of 1 v produces a current of 1 a in the conductor; it thus has units of v/a.

Ohm's law V = IR, where V is the potential across a circuit element, I is the current through it, and R is its resistance. This is not a generally applicable definition of resistance. It is only applicable to ohmic resistors, those whose resistance R is constant over the range of interest and V obeys a strictly linear relation to I.

Materials are said to be ohmic when V depends linearly on R. Metals are ohmic so long as one holds their temperature constant. But changing the temperature of a metal changes R slightly. Therefore such a device as an electric light bulb increases its temperature as it warms up, which is why it glows slightly brighter for a very brief time just after it is turned on.

For non-ohmic resistors, R is current-dependent and the definition R = dV/dI is far more useful. This is sometimes called the dynamic resistance. Solid state devices such as thermistors are non-ohmic, and non-linear. A thermistor's resistance decreases as it warms up, so its dynamic resistance is negative. Tunnel diodes and some electrochemical processes have a complicated I-V curve with a negative resistance region of operation.

The dependence of resistance on current is partly due to the change in the device's temperature with increasing current, but other subtle processes also contribute to change in resistance in solid state devices.

Oil A triglyceride that is liquid at room temperature.

Olbers' paradox If the universe is infinite, uniform, and unchanging then the entire sky at night would be bright — about as bright as the sun. The further you looked out into space, the more stars there would be, and thus in any direction in which you looked your line-of-sight would eventually impinge upon a star. The paradox is resolved by the big bang theory, which puts forth that the universe is non-uniform, dynamic, and (probably) finite.

Olfaction The act of smelling; the sense of smell.

Olfactory sacs Structures on both sides of the head of fishes that are sensitive to chemical stimuli.

Oligochaeta The class of annelids whose members are characterized by having few setae and no parapodia. Monoecious with direct development. The earthworm (Lumbricus) and Tubifex.

Oligopod larva Usual larva in Coleoptera and Neuroptera, with a well-developed head and thoracic legs.

Oligotrophic Condition of rivers and lakes that have clear water and low biological productivity (oligo 5 little; trophic 5 nutrition); are usually clear, cold, infertile headwater lakes and streams.

Omasum The third compartment of the stomach of a ruminant mammal.

Omentum omentum = fat skin. A folded peritoneal membrane from the stomach to the dorsal body wall.

Ommatidia The functional units of the invertebrate compound eye, each one of which has a lens, focusing cone, photoreceptive cells, and forms an image.

Ommatidium One of the optical units of the compound eye of arthropods.

Omnivore An animal that uses a variety of animal and plant material in its diet.

On-center, off-surround cell A ganglion cell that generates an action potential when light shines on the center of its receptive field and is inhibited when light shines on the periphery of its receptive field.

Onchocercoma Subcutaneous nodule containing masses of the nematode Onchocerca volvulus.

Onchosphere The larva of the tapeworm contained within the external embryonic envelope and armed with six hooks and cilia. Typically referred to as a coracidium when released into the water.

Oncogene Any of a number of genes that are associated with neoplastic growth (cancer). The gene in its benign state, either inactivated or carrying on its normal role, is a proto-oncogene.

Oncomiracidium A ciliated larva of a monogenetic trematode.

Oncosphere Rounded larva common to all cestodes, bears hooks.

One-gene—one-polypeptide theory The concept that one gene in DNA codes for a sequence of amino acids in a specific polypeptide.

Onium ion 1. A cation (with its counterion) derived by addition of a hydron to a mononuclear parent hydride of the nitrogen, chalcogen and halogen family, e.g. H_4N^+ ammonium ion.

2. Derivatives formed by substitution of the above parent ions by univalent groups, e.g. $(CH_3)_2S^+H$ dimethylsulfonium, $(CH_3CH_2)_4N^+$ tetraethylammonium.

3. Derivatives formed by substitution of the above parent ions by groups having two or three free valencies on the same atom. Such derivatives are, whenever possible, designated by a specific class name. E.g. $R_2C{=}NH_2^+$ iminium ion.

Ontogenetic niche The multitude of inherited ecological and social traits that are, in effect, passed on from generation to generation and that play integral roles, in concert with genetic inheritance, influencing behavior development.

Ontogeny The course of development of an individual from egg to senescence.

Onychophora A phylum of terrestrial animals with 14 to 43 pairs of unjointed legs, oral papillae, and two large antennae. Onycophorans live in humid tropical areas of the world. Their ancestors may have been an evolutionary transition between annelids and arthropods. Velvet worms or walking worms.

Oocyst Cystic form in the Apicomplexa, resulting from sporogony; the oocyst may be covered by a hard, resistant membrane (as in Eimeria), or it may not (as in Plasmodium).

Oocyst residuum Cytoplasmic material not incorporated into the sporocyst within an oocyst; seen as an amorphous mass within an oocyst.

Oocyte Stage in formation of ovum, just preceding first meiotic division (primary oocyte) or just following first meiotic division (secondary oocyte).

Ooecium Brood pouch; compartment for developing embryos in ectoprocts.

Oogamy The union of a large nonmotile egg with a small motile or nonmotile male sperm.

Oogenesis The process by which an egg cell forms from an oocyte.

Oogenotop Female genital complex of a flatworm, including oviduct, ootype, Mehlis' glands, common vitelline duct, and upper uterus.

Oogonium In algae and fungi, a specialized egg-containing cell; a single-celled female gametangium, containing one or more eggs.

Ookinete The motile zygote of malarial parasites.

Oomycetes An assemblage of organisms, previously grouped with the fungi, but currently assigned to the kingdom Protista based in part on the presence of cellulose cell walls, rather than chitinous walls as found in the fungi.

Oostegites Modified thoracic epipods in females of the crustacean superorder Peracarida. They form a pouch for brooding embryos.

Ootheca Egg packet secreted by some insects; may be covered with sclerotin.

Ootid Stage of formation of ovum after second meiotic division following expulsion of second polar body.

Ootype Part of oviduct in flatworms that receives ducts from vitelline glands and Mehlis' gland.

Opalinata The protozoan subphylum where members are cylindrical; covered with cilia. Examples: Opalina, Zelleriella.

Open access The concept of international law that permits free access by any nation to marine resources existing outside national jurisdictions.

Open canopy A forest where tree crowns cover less than 20 percent of the ground; also called woodland.

Open circulatory system A circulatory system in which blood is not confined to vessels in a part of its circuit within an animal; blood bathes tissues in blood sinuses.

Open range Unfenced, natural grazing lands; includes woodland as well as grassland.

Open system A system that exchanges energy and matter with its environment.

Operant (instrumental) conditioning A type of learning that involves the animal's forming a variety of associations during which it learns to associate its behavior with the consequences of that behavior. That is, the sequence of events is dependent upon the behavior of the animal.

Operational definition. A definition which describes an experimental procedure by which a numeric value of the quantity may be determined.

Example: Length is operationally defined by specifying a procedure for subdividing a standard of length into smaller units to make a measuring stick, then laying that stick on the object to be measured, etc.

Very few quantities in physics need to be operationally defined. They are the fundamental quantities, which include length, mass and time. Other quantities are defined from these through mathematical relations.

Operational sex ratio The number of sexually mature males and females in a population.

Operator A gene responsible for the activation and deactivation of the structural genes.

Operculum A cover. 1. The cover of a gill chamber of a bony fish (Chordata).

2. The cover of the genital pores of a horseshoe crab (Meristomata, Arthropoda).

3. The cover of the aperature of a snail shell (Gastropoda, Mollusca).

Operon A genetic unit consisting of a cluster of genes under the control of other genes, found in prokaryotes.

Ophiuroidea The class of echinoderms whose members have arms sharply set off from the central disk. Tube feet without suction disks. Brittle stars.

Ophthalmic Pertaining to the eye.

Opisthaptor Posterior attachment organ of a monogenetic trematode.

Opisthomastigote A form of Trypanosomatidae with the kinetoplast at the posterior end. The flagellum runs through a long reservoir to emerge at the anterior. There is no undulating membrane. An example is Herpetomonas.

Opisthosoma The portion of the body of a chelicerate arthropod that contains digestive, reproductive, excretory, and respiratory organs.

Opposite arrangement Two leaves borne per node and arranged across the stem from each other.

Opposite Leaf arrangement in which they are attached at a node directly across from one another on the stem.

Opsonization Modification of the surface characteristics of an invading particle or organism by binding with antibody or a nonspecific molecule in such a manner as to facilitate phagocytosis by host cells.

Optical center The point in a thin lens through which the secondary axes pass.

Optical density A property of a transparent material that is a measure of the speed of light through it.

Optical sign conventions. In introductory courses in physics a sign convention is used for objects and images in which the lens equation must be written $1/p + 1/q = 1/f$. Often the rules for this sign convention are presented in a convoluted manner. A simple and easy to remember rule is this: p is the object-to-lens distance. q is the lens to image distance. The coordinate axis along the optic axis is in the direction of passage of light through the lens, this defining the positive direction. Example: If the axis and the light direction is left-to-right (as is usually done) and the object is to the left of the lens, the object-to-lens distance is positive. if the object is to the right of the lens (virtual object), the object-to-lens distance is negative. It works the same for images.

For refractive surfaces, define the surface radius to be the directed distance from a surface to its center of curvature. Thus a surface convex to the incident light is positive, one concave to the incident light is negative. The surface equation is then $n/s + n'/s' = (n'-n)/R$ where s and s' are the object and image distances, and n and n' the refractive index of the incident and emergent media, respectively.

For mirrors, the equation is usually written $1/s + 1/s' = 2/R = 1/f$. A diverging mirror is convex to the incoming light, with negative f. From this fact we conclude that R is also negative. This form of the equation is consistent with that of the lens equation, and the interpretation of sign of focal length is the same also. But violence is done to the definition of R we used above, for refraction. One can say that the mirror folds the length axis at the mirror, so that emergent rays to a real image at the left represent a positive value of s'. We are forced also to declare that the mirror also flips the sign of the surface radius. For reflective surfaces, the radius of curvature is defined to be the directed distance from a surface to its center of curvature, measured with respect to the axis used for the emergent light. With this qualification the convention for the signs of s' and R is the same for mirrors as for refractive surfaces.

In advanced optics courses, a cartesian sign convention is used in which all things to the left of the lens are negative, all those to the right are positive. When this is used, the lens equation must be written $1/p + 1/f = 1/q$. (The sign of the $1/p$ term is opposite that in the other sign convention). This is a particularly meaningful version, for $1/p$ is the measure of vergence (convergence or divergence) of the rays as they enter the lens, $1/f$ is the amount the lens changes the vergence, and $1/q$ is the vergence of the emergent rays.

Optical yield In a chemical reaction involving chiral reactants and products, the ratio of the optical purity of the product to that of the precursor, reactant or catalyst. This should not be confused with "enantiomeric excess". The optical yield is in no way related to the chemical yield of the reaction.

Optimality model A model that attempts to predict the particular combination of costs and benefits of a behavior that will maximize an individual's inclusive fitness.

Oral Having to do with the mouth. The end of an animal containing the mouth.

Oral cavity The cavity within the mouth.

Oral sucker The sucker on the anterior end of a tapeworm, fluke, or leech.

Orbital Discrete pathways or bands surrounding an atom through which electrons move.

Orbital steering A concept expressing that the stereochemistry of approach of two reacting species is governed by the most favourable overlap of their appropriate orbitals.

Orbital symmetry The behaviour of an atomic or localized molecular orbital under molecular symmetry operations characterizes its orbital symmetry. For example, under a reflection in an appropriate symmetry plane, the phase of the orbital may be unchanged (symmetric), or it may change sign (antisymmetric), i.e. the positive and negative lobes are interchanged.

A principal context for the use of orbital symmetry is the discussion of chemical changes that involve "conservation of orbital symmetry". If a certain symmetry element (e.g. the reflection plane) is retained along a reaction pathway, that pathway is "allowed" by orbital symmetry conservation if each of the occupied orbitals of the reactant(s) is of the same symmetry type as a similarly (e.g. singly or doubly) occupied orbital of the product(s). This principle permits the qualitative construction of correlation diagrams to show how molecular orbitals transform (and how their energies change) during idealized chemical changes (e.g. cycloadditions).

An idealized single bond is a sigma bond, i.e., it has cylindrical symmetry, whereas a p-orbital or pi-bond orbital has pi symmetry, i.e. it is antisymmetric with respect to reflection in a plane passing through the atomic centres with which it is associated. In ethene, the pi-bonding orbital is symmetric with respect to reflection in a plane perpendicular to and bisecting the C-C bond, whereas. the pi-antibonding orbital is antisymmetric with respect to this operation.

Considerations of orbital symmetry are frequently grossly simplified in that, for example, the pi orbitals of a carbonyl group would be treated as having the same symmetry as those of ethene, and the fact that the carbonyl group in, for example, camphor, unlike that in formaldehyde, has no mirror planes would be ignored. These simplified considerations nevertheless afford the basis of one approach to the understanding of the rules which indicate whether pericyclic reactions are likely to occur under thermal or photochemical conditions.

Order Taxonomic rank consisting of a group of related families.

Order of magnitude Compares sizes relative to some unit of measure.

Ordinate The value corresponding to the vertical distance of a point on a graph from the X axis The Y coordinate.

Ordovician A geological period beginning around 500 million years ago, believed to be the time when plants may have begun to migrate onto land.

Organ A structure composed of different tissues, such as root, stem, leaf, or reproductive (e.g., flower) parts.

Organ of Ruffini Sensory receptor in the skin believed to be a sensor for touch-pressure, position sense of a body part, and movement. Also known as corpuscle of Ruffini.

Organ system A set of interconnected or interdependent parts that function together in a common purpose or produce results that cannot be achieved by one of them acting alone.

Organelle A body within the cytoplasm of eukaryotic cells; several different types of organelles occur, each with a specialized function such as the chloroplast, which functions in photosynthesis.

Organic Living or once living material; compounds containing carbon formed by living organisms.

Organic compound A molecule that contains carbon, hydrogen, and usually oxygen.

Organic evolution The change in an organism over time; a change in the sum of all genes in a population.

Organic molecule A molecule that contains one or more carbon atoms.

Organization, level of The extent of specialization and organization of the cells of an organism. Organization may be at the cellular, tissue, or organ level.

Organizational effects of hormones Changes resulting from the presence of hormones at critical time periods such that specific developmental pathways for specific brain regions and developing gonadal tissues are influenced to become either female or malelike.

Organizer Area of an embryo that directs subsequent development of other parts.

Oriental sore Disease caused by Leishmania tropica. Also called Jericho boil, Delhi boil, Aleppo boil, or cutaneous leishmaniasis.

Orientation The way an organism positions itself in relation to external cues.

Origin The origin is the point from which everything is referenced. In a coordinate system, the origin is the location of "zero" - the (artifically) special place which separates the positive direction from the negative direction.

Oroya fever Clinical form of Carrion's disease, caused by the bacterium Bartonella bacilliformis and transmitted by sand flies.

Orthogenesis A unidirectional trend in the evolutionary history of a lineage as revealed by the fossil record; also, a now discredited, anti-Darwinian evolutionary theory, popular around 1900, postulating that genetic momentum forced lineages to evolve in a predestined linear direction that was independent of external factors and often led to decline and extinction.

Oscillator The internal mechanism that is the clock in a biological rhythm.

Oscilloscope A cathode-ray tube with associated electronic circuits that enable external voltages to deflect the electron beam of the cathode-ray tube simultaneously along both horizontal and vertical axes.

Osculum Excurrent opening in a sponge.

Osmoconformer An organism whose body fluids have the same or very similar osmotic pressure as that of its aquatic environment; a marine organism that does not utilize energy in osmoregulation.

Osmole. Molecular weight of a solute, in grams, divided by the number of ions or particles into which it dissociates in solution.

Osmoregulator Any substance, either organic or inorganic, that functions to change the solute potential of a solution, and thereby controls the water relations of that solution by osmosis.

Osmosis The movement of water through a selectively permeable membrane due to the existence of a concentration gradient.

Osmotic conformerso rganisms that tolerate large variations of internal ionic concentrations without serious damage by remaining isosmotic with the water around them.

Osmotic potential. Osmotic pressure.

Osmotic pressure In hypoosmotic conditions, the internal fluid pressure that develops from the osmotic inflow of water.

Osmotic shock Alternating seawater and dilute freshwater in intertidal zones causes erratic osmotic balance in plants growing in these areas.

Osmotroph A heterotrophic organism that absorbs dissolved nutrients.

Osphradium A sense organ in aquatic snails and bivalves that tests incoming water.

Ossicles Small separate pieces of echinoderm endoskeleton. Also, tiny bones of the middle ear of vertebrates.

Osteichthyes The class of fishes whose members are characterized by the presence of a bony skeleton, a swim bladder, and an operculum. Bony fishes.

Osteoblast A bone-forming cell.

Osteoclast A large, multinucleate cell that functions in bone dissolution.

Osteocyte A bone cell that is characteristic of adult bone, has developed from an osteoblast, and is isolated in a lacuna of the bone substance.

Osteon Unit of bone structure; Haversian system.

Otolith Calcareous concretions in the membranous labyrinth of the inner ear of lower vertebrates, or in the auditory organ of certain invertebrates.

Outbreeding depression A reduction in the fitness of offspring resulting from matings among individuals from different populations or demes, possibly caused by the breaking up of coadapted gene complexes.

Outcrosser A flower that must be cross-pollinated to successfully complete the reproductive process.

Outer-sphere (electron transfer) An outer-sphere electron transfer is a reaction in which the electron transfer takes place with no or very weak (4 -16 kJ mol-1) electronic interaction between the reactants in the transition state. If instead the donor and the acceptor exhibit a strong electronic coupling, the reaction is described as inner-sphere electron transfer. The two terms derive from studies concerning metal complexes and it has been suggested that for organic reactions the term "nonbonded" and "bonded" electron transfer should be used.

Outgroup. In phylogenetic systematic studies, a species or group of species closely related to but not included within a taxon whose phylogeny is being studied, and used to polarize variation of characters and to root the phylogenetic tree.

Outwelling The export of detritus and other organic matter from estuaries to other ecosystems.

Ovarian cycle The cycle in the ovary during which the oocyte matures and ovulation occurs.

Ovary The primary reproductive organ of a female; where eggs (ova) are produced.

Overburden Overlying layers of noncommercial sediments that must be removed to reach a mineral or coal deposit.

Overdispersion Nonrandom dispersion of individuals in a habitat, such as when a minority of host individuals bears a majority of parasites.

Overnutrition Receiving too many calories.

Overpopulation Too many organisms (especially humans) for the available resources to support in a sustainable way.

Overshoot The extent to which a population exceeds the carrying capacity of its environment.

Overturn The sinking of surface water that has become more dense than the water below.

Ovicapt Sphincter on the oviduct of a flatworm.

Oviger Leg that carries eggs in pycnogonids.

Oviparity Reproduction in which eggs are released by the female; development of offspring occurs outside the maternal body.

Oviparous An animal that releases eggs.

Ovipositor A modification of the abdominal appendages of some female insects that is used for depositing eggs in or on some substrate (Arthropoda, Hexapoda).

Ovisac External sac attached to the somite that bears openings of gonoducts in females of many Copepoda. Fertilized eggs pass into the ovisacs for embryonation.

Ovovitellarium Mixed mass of ova and vitelline cells; found in the monogenean genus Gyrodactylus and in a few tapeworms.

Ovoviviparity Reproduction in which eggs develop within the maternal body without additional nourishment from the parent and hatch within the parent, or immediately after laying.

Ovoviviparous Describes reproduction in which embryos develop within the maternal body without additional nourishment from the parent and hatch within the parent or immediately after emerging.

Ovulation The release of an egg from the ovary.

Ovule Structure that will become a seed following fertilization; literally an integumented megasporangium that contains the embryo sac before fertilization.

Ovuliferous scale In conifers, an axillary, scalelike shoot that bears one or more ovules. Each strobilus may contain one or many ovuliferous scales, depending on the species.

Ovum Mature female germ cell (egg).

Oxidation 1. The complete, net removal of one or more electrons from a molecular entity (also called "de-electronation").

2. an increase in the oxidation number of any atom within any substrate.

3. Gain of oxygen and/or loss of hydrogen of an organic substrate.

All oxidations meet criteria 1. and

2., and many meet criterion

3., but this is not always easy to demonstrate.Alternatively, an oxidation can be described as a transformation of an organic substrate that can be rationally dissected into steps or primitive changes. The latter consist in removal of one or several electrons from the substrate followed or preceded by gain or loss of water and/or hydrons or hydroxide ions, or by nucleophilic substitution by water or its reverse and/or by an intramolecular molecular rearrangement.

Oxidative coupling The coupling of two molecular entities through an oxidative process, usually catalysed by a transition metal compound and involving dioxygen as the oxidant; e.g.,

$2\ CO + 2\ MeOH + 1/2\ O_2\ (+\ catalyst) \longrightarrow MeOOCCOOMe + H_2O$

Oxidative phosphorylation The electron transport system associated with aerobic respiration and mitochondria. In the release of energy through a series of cytochromes, three molecules of ATP are made.

Oxidization The loss of electrons or hydrogen from an atom or molecule.

Oxygen An element that is one of the constituents of water, organic matter, and many other chemicals. Oxygen gas (O_2), composed of two oxygen atoms, is needed for respiration and is produced by photosynthesis.

Oxygen cycle The circulation and reutilization of oxygen in the biosphere.

Oxygen debt The amount of oxygen that must be supplied following physical exercise to convert accumulated lactic acid (lactate) to glucose.

Oxygen minimum layer A layer of water at a depth of approximately 500 m (1,600 ft) where oxygen is depleted.

Oxygen minimum zone the ocean layer below the photic zone where dissolved oxygen concentration is lowest

Oxygen sag Oxygen decline downstream from a pollution source that introduces materials with high biological oxygen demands.

Oxyhemoglobin Compound formed when oxygen combines with hemoglobin.

Oyster reef A dense oyster bed present in some estuaries and other marine environments.

Ozone A highly reactive molecule containing three oxygen atoms; a dangerous pollutant in ambient air. In the stratosphere, however, ozone forms an ultraviolet absorbing shield that protects us from mutagenic radiation.

P

P53 protein A tumor suppressor protein with critical functions in normal cells. A mutation in the gene that encodes it, p53, can result in loss of control over cell division and thus cancer.

P680 The reaction center for Photosystem II; a chlorophyll A molecule that is bound to a membrane protein and has a peak absorbance at 680 nm.

P700 The reaction center for Photosystem I; a chlorophyll A molecule that is bound to a membrane protein and has a peak absorbance at 700 nm.

Pacinian corpuscle A sensory receptor in skin, muscles, body joints, body organs, and tendons that is involved with the vibratory sense and firm pressure on the skin; also called a lamellated corpuscle.

Paedogenesis Reproduction by immature or larval animals caused by acceleration of maturation. Progenesis.

Paedomorphosis Displacement of ancestral juvenile features to later stages of the ontogeny of descendants.

Pain receptor A modified nerve ending that, when stimulated, gives rise to the sense of pain.

Pair bond An affiliation between an adult male and an adult female for reproduction. Characteristic of monogamous species.

Paleolithic Old Stone Age; a cultural period during which early humans obtained food solely by foraging; ending in some areas approximately 10,000 years ago.

Paleontologist A person who studies fossils and all aspects of extinct life.

Paleozoic The geological era beginning 590 million years ago and ending about 250 million years ago. During this time plants invaded, established, and diversified on the land

Palisade parenchyma Parenchyma cells in the leaf mesophyll characterized by uniform rows of tightly packed cells with many chloroplasts beneath the upper epidermis.

Palisade The vertical photosynthetic cells below the upper epidermis in leaf tissue; these cells are a specialized parenchyma.

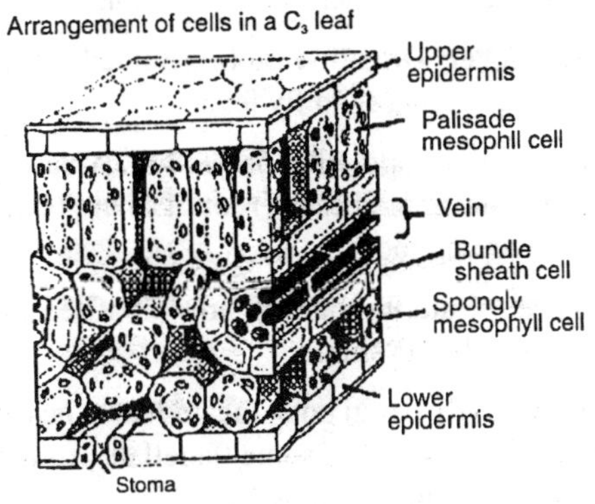

Fig. Palisade

Palmate An arrangement of leaflets (or lobes on a simple leaf), each originating from a common point, usually the axial end

Palmately compound leaf Leaflets radiating from a common point.

Pampiniform pampin = a tendril. A threadlike plexus.

Pancreas A gland located in the abdomen that produces both digestive enzymes (exocrine pancreas) and hormones (endocrine pancreas). Key hormones produced by the pancreas are insulin and glucagon, which play roles in regulating blood glucose levels.

Pancreatic islet An island of special tissue in the pancreas.

Pangaea The single large landmass, or supercontinent, that broke up to form today's continents

Pangenesis Darwin's hypothesis that hereditary characteristics are carried by individual body cells that produce particles that collect in the germ cells.

Panicle A branched inflorescence with the branches bearing loose flower clusters.

Pansporoblast Myxosporidean sporoblast that gives rise to more than one spore. Also called a sporoblast mother cell.

Panthalassa The large ocean that surrounded the supercontinent Pangaea and which was the ancestor of the modern Pacific Ocean.

Papatasi fever Virus disease transmitted by sand flies. Also called sand fly fever.

Papilla A small nipplelike projection. A vascular process that nourishes the root of a hair, feather, or developing tooth.

Papula Respiratory processes on skin of sea stars; also, pustules on skin.

Parabasal body Golgi body located near the basal body (kinetosome) of some flagellate protozoa, from which the parabasal filament runs to the basal body.

Parabasal filament Fibril, with periodicity visible in electron micrographs, that courses between the parabasal body and a kinetosome.

Parabiosis The fusion of two individuals, resulting in mutual physiological intimacy.

Parabolic mirrors Curved mirrors that focus light from a large area onto a single, central point, thereby concentrating solar energy and producing high temperatures.

Parabronchi The tiny air tubes within the lung of a bird across which gas exchange occurs.

Parallel circuit An electric circuit in which two or more components connected across two common points in the circuit so as to provide separate conducting paths for the current

Parallel venation A type of venation in which the main veins of a leaf are parallel or nearly so, but converge at the apex and base.

Parallelogram method The graphic method of finding the resultant of two vectors that do not act along a straight line.

Paralytic shellfish poisoning A condition caused when humans eat shellfish that have become contaminated with the

Paramagnetic Substances having a magnetic susceptibility greater than 0 are paramagnetic. They are drawn into a magnetic field.

Paramagnetism The property of a substance by which it is feebly attracted by a strong magnet.

Paramastigote Form of trypanosomatid in which the kinetosome and kinetoplast are beside the nucleus.

Paramere Copulatory appendage in male cimicid bugs.

Paramylon bodies Organelles containing the starch-like substance paramylon; in some algae and flagellates.

Parapatric speciation Speciation that occurs in small, local populations, called demes.

Paraphyletic groups Groups that do not include all descendants of a common ancestor.

Paraphyly The condition that a taxon or other group of organisms contains the most recent common ancestor of all members of the group but excludes some descendants of that ancestor; contrasts with monophyly and polyphyly.

Parapodia Paired lateral extensions on each segment of polychaetes (Annelida); may be used in swimming, crawling, and burrowing.

Parapodium One of the paired lateral processes on each side of most segments in polychaete annelids; variously modified for locomotion, respiration, or feeding.

Parapolar cells Cells making up the ciliated somatoderm immediately behind the calotte of a mesozoan.

Parasite An organism that lives in or on another organism, deriving nourishment at the expense of its host, usually without killing it.

Parasitic Denoting an association where one living organism benefits at the expense of another.

Parasitic castration Condition in which a parasite causes retardation in development or atrophy of host gonads, often accompanied by failure of secondary sexual characteristics to develop.

Parasitism The condition of an organism living in or on another organism (host) at whose expense the parasite is maintained; destructive symbiosis.

Parasitoid Organism that is a typical parasite early in its development but that finally kills the host during or at the completion of development; often used in reference to many insect parasites of other insects.

Parasitologist Quaint person who seeks truth in strange places; a person who sits on one stool, staring at another.

Parasitophorous vacuole Vacuole within a host cell that contains a parasite.

Parasympathetic nervous system Portion of the autonomic nervous system that arises from the brain and sacral region of the spinal cord.

Parasympathetic One of the subdivisions of the autonomic nervous system, whose fibers originate in the brain and in anterior and posterior parts of the spinal cord.

Paratenic host Host in which a parasite survives without undergoing further development. Also known as a transport host.

Parathyroid gland One of the small glands located within a lobe of the thyroid gland.

Paraxial (crystalline) rod Rod that runs alongside the axoneme in the flagellum of a kinetoplastid flagellate.

Parenchyma A spongy mass of mesenchyme cells filling spaces around viscera, muscles, or epithelia in acoelomate animals. Depending on the species, parenchyma may function in providing skeletal support, nutrient storage, motility, reserves of regenerative cells, transport of materials, structural interactions with other tissues, modifiable tissue for morphogenesis, oxygen storage, and perhaps other functions that have yet to be determined.

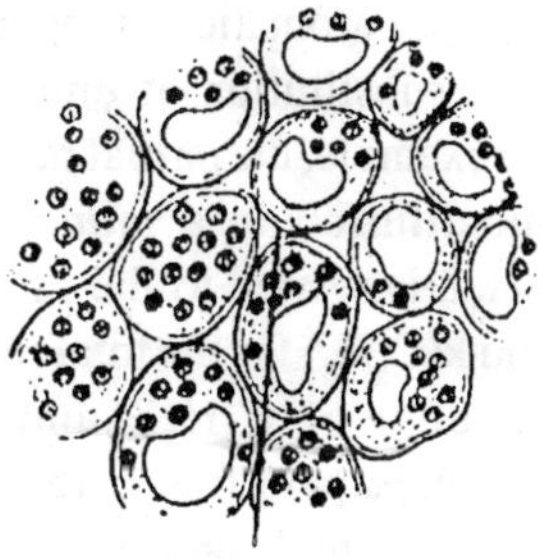

Fig. Parenchyma

Parent material Undecomposed mineral particles and unweathered rock fragments beneath the subsoil; weathering of this layer produces new soil particles for the layers above.

Parental investment Any investment in an offspring that increases its chances of survival and reproduction at the expense of the parents' ability to invest in other offspring.

Parental manipulation The selective providing of care to some offspring at the expense of other offspring so as to maximize the parents' reproductive success.

Parietal Bone of the cranium or coelomic membrane, as parietal peritoneum.

Parietal cell Cell of a gastric gland that secretes hydrochloric acid and intrinsic factor.

Parotid The parotid gland at the base of the ear.

Pars prostatica Dilation of the ejaculatory duct of a flatworm, surrounded by unicellular prostate cells.

Parsec The unit of distance defined as the distance indicated by an earth-orbit parallax of 1 arcsec. It equals about 3.086×10^{16} m.

Parthenocarpy Fruit development without fertilization.

Parthenogenesis Unisexual reproduction involving the production of young by females not fertilized by males; common in rotifers, cladocerans, aphids, bees, ants, and wasps. A parthenogenetic egg may be diploid or haploid.

Partial rate factor The rate of substitution at one specific site in an aromatic compound relative to the rate of substitution at one position in benzene. For example, the partial rate factor fpZ for para-substitution in a monosubstituted benzene C_6H_5Z is related to the rate constants $k(C_6H_5Z)$ and $k(C_6H_6)$ for the total reaction (i.e. at all positions) of C_6H_5Z and benzene, respectively, and% para (the percentage para-substitution in the total product formed from C_6H_5Z) by the relation

$$f_p^z = \frac{6k(C_6H_5Z)}{k(C_6H_6)} \frac{\%\text{para}}{100}$$

Similarly for meta-substitution:

$$f_m^z = \frac{6k(C_6H_5Z)}{2k(C_6H_6)} \frac{\%\text{meta}}{100}$$

Particle In classical mechanics and quantum mechanics there are objects called particles which are sometimes conceptualized as infinitesimal balls. Some particles are deemed fundamental particles and are considered particles which are not made up of other particles. Particles typically come in two types: fermions and bosons. Fermions are further categorized as leptons and hadrons. Hadrons are further categorized as mesons and baryons. In a field theory one does not consider particles, but instead, oscillations of the field.

Particle, fundamental Those particles which are not made up of other particles. Just as molecules are combinations of atoms, and atoms are combinations of particles, some particles are also combinations of other particles. Currently, leptons, quarks, and gauge bosons are considered fundamental.

Particulate material Atmospheric aerosols, such as dust, ash, soot, lint, smoke, pollen, spores, algal cells, and other suspended materials; originally applied only to solid particles but now extended to droplets of liquid.

Parts per billion (ppb) Number of parts of a chemical found in one billion parts of a particular gas, liquid, or solid mixture.

Parturition The process of childbirth.

Paruterine organ Fibromuscular organ in some cestodes that replaces the uterus.

Pascal Unit

Pascal's Principle of Hydrostatics Pascal actually has three separate principles of hydrostatics. When a textbook refers to Pascal's Principle it should specify which is meant.

Paschen series The series which describes the emission spectrum of hydrogen when the electron is jumping to the third orbital. All of the lines are in the infrared portion of the spectrum.

Passive avoidance learning A form of operant conditioning, in which the animal does not make an overt response, but rather is trained to avoid making a particular response by being exposed to some type of noxious stimulus each time the response occurs.

Passive continental margin A continental margin that is located at the "trailing edge" of a continent and as a result shows little geological activity. Compare active continental margin.

Passive heat absorption The use of natural materials or absorptive structures without moving parts to gather and hold heat; the simplest and oldest use of solar energy.

Passive immunization Immune state in an animal created by inoculation with serum (containing antibodies) or lymphocytes from an immune animal, rather than by exposure to the antigen.

Pasture Enclosed domestic meadows or managed grazing lands.

Patch model A type of optimal foraging model that assumes that prey occur in discrete clumps and that seeks to predict where and when organisms will forage for these prey.

Patches Regions of localized concentrations of resources.

Patchiness Within a larger ecosystem, the presence of smaller areas that differ in some physical conditions and thus support somewhat different communities; a diversity-promoting phenomenon.

Patent Stage in an infection at which infectious agents produce evidence of their presence, such as eggs or cysts. Contrast with prepatent.

Pathogen An organism that produces disease in a host organism, disease being an alteration of one or more metabolic functions in response to the presence of the organism.

Pathogenesis Production and development of disease.

Pathogenicity Capability of an agent to produce disease.

Pauli exclusion principle No two identical fermions in a system, such as electrons in an atom, can have an identical set of quantum numbers.

Paurometabolous metamorphosis A form of insect development in which immatures resemble parents, and molting is restricted to the immature stages.

Pauropoda A class of arthropods whose bodies are small, soft, 11 segmented, and have 9 pairs of legs.

Peanut worms Burrowing invertebrates with an unsegmented body and an anterior end that can be pulled into the body.

Peat Deposits of moist, acidic, semidecayed organic matter.

Peat moss A relatively sterile, inert medium composed of partially decomposed plants of the genus Sphagnum. Exceptionally high water-holding capacity.

Pebrine An infectious disease of silkworms caused by the protozoan Nosema bombicis.

Peck order. A hierarchy of social privilege in a flock of birds.

Pecten Any of several types of comblike structures on various organisms, for example, a pigmented, vascular, and comblike process that projects into the vitreous humor from the retina at a point of entrance of the optic nerve in the eyes of all birds and many reptiles.

Pectin A complex polysaccharide in the middle lamella and primary walls of plant cells.

Pectoral Of or pertaining to the breast or chest; to the pectoral girdle; or to a pair of horny shields of the plastron of certain turtles.

Pectoral fin Each of the pair of fins just behind the head of fishes.

Pedal locomotion The type of locomotion exhibited by flatworms, some Cnidarians, and the gastropod molluscs. This locomotion involves waves of activity in the muscular system, which is applied to the substratum.

Pedalium Flattened blade at the base of the tentacles in cubozoan medusae (Cnidaria).

Pedicel An individual stalk of a flower that is part of an inflorescence.

Pedicellariae Pincerlike structures found on the body wall of many echinoderms. They are used in cleaning and defense.

Pedipalps Second pair of appendages in chelicerate arthropods, modified variously in different groups.

Peduncle A stalk. Also, a band of white matter joining different parts of the brain.

Pelage Hairy covering of mammals.

Pelagic Pertaining to the open ocean.

Pelagic organism An organism that lives above the bottom in the open sea.

Pellagra Lassitude, torpor, dermatitis, diarrhea, dementia, and death brought about by a diet deficient in tryptophan and niacin.

Pellicle A thin, frequently noncellular covering of an animal (e.g., the protective and supportive pellicle of protists occurs just below the plasma membrane); may be composed of a cell membrane, cytoskeleton, and other organelles.

Peltier effect The change in temperature produced at a junction between two dissimilar metals or semiconductors when an electric current passes through the junction.

Pelvic Situated at or near the pelvis, as applied to girdle, cavity, fins, and limbs.

Pelvic fin Each of the second pair of ventral fins of fishes.

Pelycosaur Any of a group of carnivorous Permian synapsids distinguished by powerful jaws, stabbing teeth, and a large skin-covered sail on the back.

Pen in squids, a thin, chitinous structure extending the length of the mantle tissue

Pendulum A body suspended so that it can swing back and forth about an axis.

Penicillin An antibiotic produced by certain species of Penicillium. An ascomycete fungus.

Pennsylvanian A geological period of the Carboniferous beginning about 320 million years ago.

Pentadactyl With five digits, or five fingerlike parts, to the hand or foot.

Pentaradial symmetry A form of radial symmetry found in the echinoderms in which body parts are arranged in fives around an oral-aboral axis.

Pentastomida A phylum of worms that are all endoparasites in the lungs or nasal passageways of carnivorous vertebrates. Tongue worms.

Pentose A 5-carbon sugar.

Penumbra The partially illuminated part of a shadow.

PEP carboxylase The enzyme responsible for CO2 fixation in the primary fixation of C4 metabolism.

Pepo A berry with the outer wall formed from receptacle tissue fused to the exocarp; the fleshy interior is mesocarp and endocarp.

Peptidase An enzyme that breaks down simple peptides, releasing amino acids.

Peptide bond. A bond that binds amino acids together into a polypeptide chain, formed by removing an OH from the carboxyl group of one amino acid and an H from the amino group of another to form an amide group CO-NH.

Peptides Two or more amino acids linked by a peptide bond.

Percolation The movement of water by gravitational pull down through soil particles.

Pereiopod Thoracic appendage of a crustacean.

Perennial A plant that overwinters and continues to grow for many years. It may reproduce every year, or only on rare occasions.

Perianth The petals and sepals together.

Pericardium Area around heart; membrane around heart.

Pericarp The fruit wall that develops from the ovary wall.

Pericycle Root tissue sandwiched between the endodermis and the phloem; it is the outermost layer of the stele; meristematic region that gives rise to branch roots.

Pericyclic reaction A chemical reaction in which concerted reorganization of bonding takes place throughout a cyclic array of continuously bonded atoms. It may be viewed as a reaction proceeding through a fully conjugated cyclic transition state. The number of atoms in the cyclic array is usually six, but other numbers are also possible. The term embraces a variety of processes, including cycloadditions, cheletropic reactions, electrocyclic reactions and sigmatropic rearrangements, etc. (provided they are concerted).

Periderm Protective tissue that replaces the epidermis after secondary growth begins; it includes the cork, the cork cambium, and sometimes other cells.

Perigynous A flower in which the base of the sepals, petals, and stamens form a cup around the ovary.

Period 1. The time for one complete cycle, vibration, revolution, or oscillation

2. The time required for a single wavelength to pass a given point.

Periodic motion Motion repeated in each of a succession of equal time intervals.

Periostracum Outer horny layer of a mollusc shell.

Peripheral nervous system The nerves and ganglia of the nervous system that lie outside of the brain and spinal cord.

Peripheral protein A protein on the surface of a biological membrane.

Peripheral Structure or location distant from center, near outer boundaries.

Periproct Region of aboral plates around the anus of echinoids.

Perisarc Sheath covering the stalk and branches of a hydroid.

Periselectivity The differentiation between two symmetry-allowed processes, for example the [2+4] vs. [4+6] cycloaddition of cyclopentadiene to tropone.

Perissodactyl Pertaining to an order of ungulate mammals with an odd number of digits.

Peristalsis Movement provided by the smooth muscles to the contents of the digestive tract, ureters, urethra, uterine duct, and vas deferens.

Peristomium The segment of the body of an annelid that surrounds the mouth.

Perithecium A flask-shaped ascocarp.

Peritoneum The membrane that lines the coelom and covers the coelomic viscera.

Peritreme Elongated sclerite extending forward from the stigma of certain mites, mainly in the suborder Mesostigmata.

Peritrophic membrane Noncellular, delicate membrane lining an insect's midgut.

Permafrost The permanently frozen soil in polar regions.

Permanent parasite Parasite that lives its entire adult life within or on a host.

Permanent retrievable storage Placing waste storage containers in a secure building, salt mine, or bedrock cavern where they can be inspected periodically and retrieved, if necessary, for repacking or for transfer if a better means of disposal or reuse is developed.

Permanent wilting point The soil moisture content at the point when a given plant's root system can no longer absorb water.

Permeability A property of membranes allowing all substances to pass freely.

Permeability of free space The ratio of the magnetic flux density in a substance to the external field strength for vacuum. It is equal to 4 pi x 10^{-7} h/m.

Permease. A transporter molecule; a molecule in the cell membrane that makes it possible for another molecule (to which the membrane is not otherwise permeable) to be transported across the membrane, that is, mediated transport.

Permian The last geological period of the Paleozoic era beginning about 286 million years ago and ending 251 million years ago. The end of the Permian was a time of abrupt extinction in the diversity of flora and fauna, especially of sea life.

Permittivity of free space The ratio of the electric displacement to the intensity of the electric field producing it in vacuum. It is 8.854 x 10^{-12} f/m.

Peroxisomes Small organelles containing enzymes of the glyoxylate cycle, catalase, and peroxidases.

Persistent The term persistent is used to characterize radicals which have lifetimes of several minutes or greater in dilute solution in inert solvents. Persistence is a kinetic or reactivity property. In contrast, radical stability, which is a thermodynamic property, is expressed in terms of the C-H bond strength of the appropriate hydrocarbon. The lifetime of a radical is profoundly influenced by steric shielding of the radical centre by bulky substituents.

Pest Any organism that reduces the availability, quality, or value of a useful resource.

Pest resurgence Rebound of pest populations due to acquired resistance to chemicals and nonspecific destruction of natural predators and competitors by broadscale pesticides.

Pesticide Any chemical that kills, controls, drives away, or modifies the behavior of a pest.

Pesticide rain Long-range transport of pesticides by air currents and deposition through precipitation in sites far from its origin; analogous to acid rain.

Pesticide treadmill A need for constantly increasing doses or new pesticides to prevent pest resurgence.

Petal The often showy flower component attached just inside the sepals; petals are usually colorful to attract pollinators.

Petaloids Describes flowerlike arrangement of respiratory podia in irregular sea urchins.

Petiole A stalklike portion of a leaf connecting the blade to the stem or branch.

Petrifaction A fossil formed when plant parts are infiltrated or replaced by mineral substances such that the structure is preserved but the fossil is actually rock.

Petroleum products Oil and the materials made from crude natural petroleum.

Peyer's patches Lymphoid tissue in the wall of the intestine; not circumscribed by a tissue capsule.

Pfund series The series which describes the emission spectrum of hydrogen when the electron is jumping to the fifth orbital. All of the lines are in the infrared portion of the spectrum.

PGA 3-phosphoglyceric acid, the first compound produced in the Calvin cycle.

Ph (potential of hydrogen). A symbol referring to the relative concentration of hydrogen ions in a solution; ph values are from 0 to 14, and the lower the value, the more acid or hydrogen ions in the solution. Equal to the negative logarithm of the hydrogen ion concentration.

Ph scale The numerical scale that measures acidity and alkalinity, ranging from 0 (most acidic) to 14 (most basic); ph stands for potential hydrogen and refers to the concentration of hydrogen ions (H^+).

Phaeophyta The group of organisms commonly termed the brown algae, consisting of about 1500 species, including the kelps.

Phagocyte Any cell that engulfs and devours microorganisms or other particles.

Phagocytosis The engulfment of a particle by a phagocyte or a protozoan.

Phagolysosome The organelle that is formed when a lysosome combines with a vesicle.

Phagotroph A heterotrophic organism that ingests solid particles for food.

Pharmaceuticals Medicinal drugs.

Pharynx The part of the digestive tract between the mouth cavity and the esophagus that, in vertebrates, is common to both digestive and respiratory tracts. In cephalochordates the gill slits open from it.

Phase 1. A condition of matter

2. In any periodic phenomenon, a number that describes a specific stage within each oscillation

3. The angular relationship between current and voltage in an A-C circuit

4. The number of separate voltage waves in a commercial A-C supply.

Phase angle 1. Of any periodic function, the angle obtained by multiplying the phase by 360 if the angle is to be expressed in degrees, or by 2pi if in radians

2. The angle between the voltage and current vectors

Phase-transfer catalysis The phenomenon of rate enhancement of a reaction between chemical species located in different phases (immiscible liquids or solid and liquid) by addition of a small quantity of an agent (called the "phase-transfer catalyst") that extracts one of the reactants, most commonly an anion, across the interface into the other phase so that reaction can proceed. These catalysts are salts of "onium ions" (e.g. tetraalkylammonium salts) or agents that complex inorganic cations (e.g. crown ethers). The catalyst cation is not consumed in the reaction although an anion exchange does occur.

Phasmid One of a pair of glands or sensory structures found in the posterior end of certain nematodes.

Phasor A representation of the concepts of magnitude and direction in a reference plane; a rotating vector

Phellem Also termed cork; the outer, protective tissue of stems and roots composed of nonliving cells with suberized cell walls, formed centrifugally by the cork cambium or phellogen.

Phelloderm A tissue laid down to the inside of the phellogen; the inner part of the periderm.

Phellogen The cork cambium.

Phenetic Refers to the use of a criterion of overall similarity to classify organisms into taxa; contrasts with classifications based explicitly on a reconstruction of phylogeny.

Phenotype The expression that results from an interaction of one or more gene pairs and the environment.

Phenotype matching A mechanism by which kin may recognize one another; individuals use kin as a referent whose phenotypes are learned by association.

Phenotypic gradualism. The hypothesis that new traits, even those that are strikingly different from ancestral ones, evolve by a long series of small, incremental steps.

Phenotypic ratio The relative numbers of progeny in each phenotypic category produced by a genetic cross.

Pheromone Chemical substance released by one organism that influences the behavior or physiological processes of another organism.

Philopatric Remaining near the place of birth after sexual maturation.

Phloem The food-conducting tissue of vascular plants, composed of sieve elements, various kinds of parenchyma cells, fibers, and sclereids.

Phloem rays The part of the vascular ray which is located in the secondary phloem.

Phonoreceptor A specialized nerve ending that responds to sound.

Phonotaxis Orientation with respect to sound.

Phoresis Form of symbiosis when the symbiont, the phoront, is mechanically carried about by its host. Neither is physiologically dependent on the other.

Phoronida A phylum of marine animals whose members live in permanent chitinous tubes in muddy, sandy, or solid substrates. Feed via an anterior lophophore with two parallel rings of long tentacles.

Phoronids Tube-dwelling, unsegmented invertebrates that possess a horseshoe-shaped or circular lophophore.

Phosphagen A term for creatine phosphate and arginine phosphate, which store and may be sources of high-energy phosphate bonds.

Phosphate An important nutrient in the ocean.

Phosphatide A lipid with phosphorus, such as lecithin. A complex phosphoric ester lipid, such as lecithin, found in all cells.

Phospholipid A type of lipid molecule occurring in a bilayer in biological membranes; a lipid with two fatty acids and a phosphate group attached to glycerol.

Phosphorus cycle The movement of phosphorus atoms from rocks through the biosphere and hydrosphere and back to rocks.

Phosphorylation The addition of a phosphate group to a molecule.

Photic zone The surface layer where there is enough light for photosynthesis to occur.

Photoautotroph Organism that synthesizes organic matter using the energy of light.

Photochemical oxidants Products of secondary atmospheric reactions.

Photodegradable plastics Plastics that break down when exposed to sunlight or to a specific wavelength of light.

Photoelastic Pertaining to certain materials that become double refracting when strained.

Photoelectric effect An effect explained by a. Einstein that demonstrate that light seems to be made up of particles, or photons. Light can excite electrons (called photoelectrons in this context) to be ejected from a metal. Light with a frequency below a certain threshold, at any intensity, will not cause any photoelectrons to be emitted from the metal. Above that frequency, photoelectrons are emitted in proportion to the intensity of incident light.

The reason is that a photon has energy in proportion to its wavelength, and the constant of proportionality is the planck constant. Below a certain frequency — and thus below a certain energy — the incident photons do not have enough energy to knock the photoelectrons out of the metal. Above that threshold energy, called the workfunction, photons will knock the photoelectrons out of the metal, in proportion to the number of photons (the intensity of the light). At higher frequencies and energies, the photoelectrons ejected obtain a kinetic energy corresponding to the difference between the photon's energy and the workfunction.

Photoelectrons Electrons emitted from a light-sensitive material when it is illuminated with light of sufficiently short wavelength.

Photoinhibition reduction of photsynthetic rates due to too much light

Photolysis The cleavage of one or more covalent bonds in a molecular entity resulting from absorption of light, or a photochemical process in which such cleavage is an essential part. The term is used incorrectly to describe irradiation of a sample, although in the combination flash photolysis this usage is accepted.

Photometry The quantitative measurement of visible radiation from light sources.

Photon A quantum of light energy; the carrier of the electromagnetic interaction.

Photooxidation The change in the structure of a molecule due to exposure to light; sometimes referred to as bleaching.

Photoperiodism The system within organisms that causes certain events, including the onset of reproduction, to be related to the length of day.

Photophore An organ that produces bioluminescence.

Photophosphorylation The formation of ATP utilizing light energy in photosynthesis.

Photopigment A molecule in visual receptor cells (e.g., rhodopsin) that responds to light energy.

Photoreceptors Sensory cells that contain photopigments and respond specifically to light energy.

Photorespiration In C_3 plants, respiration rates increase

Photosynthesis The biochemical process by which green plants and some bacteria capture light energy and use it to produce chemical bonds. Carbon dioxide and water are consumed while oxygen and simple sugars are produced.

Photosynthetic efficiency The percentage of available light captured by plants and used to make useful products.

Photosynthetic pigment A molecule such as chlorophyll that is responsible for capturing solar energy in photosynthesis.

Photosynthetic unit A group of associated chlorophyll molecules, including antenna molecules and a central chlorophyll a collector molecule.

Photosystem I The second part of the Z-scheme in which a chlorophyll a molecule absorbs most effectively at 700 nm.

Photosystem II The first part of the Z-scheme in which the chlorophyll a molecule absorbs most effectively at 680 nm.

Phototaxis A taxis in which light is the orienting stimulus. An involuntary tendency for an organism to turn toward (positive) or away from (negative) light.

Phototrophs Organisms capable of using CO_2 in the presence of light as a source of metabolic energy.

Phototropism The bending of a plant toward a unidirectional light source.

Photovoltaic cell An energy-conversion device that captures solar energy and directly converts it to electrical current.

Photovoltaic effect The generation of a potential difference across a P-N junction as a consequence of the absorption of incident light of appropriate frequency.

Phragmoplast A system of microtubules and vesicles that arises between two daughter nuclei at telophase and forms the cell plate.

pH-rate profile A plot of observed rate coefficient, or more usually its decadic logarithm, against pH of solution, other variables being kept constant.

Phycobilins A group of photosynthetic pigments that includes phycocyanin, a bluish pigment in cyanobacteria (blue-green algae), and phycoerythrin, a red pigment in red algae.

Phycobilosomes Groups of pigmented cells containing phycocyanin and phycoerythrin.

Phycocolloid One of several starch-like chemicals found in some seaweeds. They are of significant commercial importance.

Phycocyanin A blue photosynthetic pigment found in cyanobacteria and red algae.

Phycoerythrin A red photosynthetic pigment found in cyanobacteria and red algae.

Phyletic gradualism. A model of evolution in which morphological evolutionary change is continuous and incremental and occurs mainly within unbranched species or lineages over long periods of geological time; contrasts with punctuated equilibrium.

Phyletics Phylogenetic systematics, cladistics.

Phyllopodium Leaflike swimming appendage of branchiopod crustaceans.

Phylogenetic species concept. An irreducible (basal) cluster of organisms, diagnosably distinct from other such clusters, and within which there is a parental pattern of ancestry and descent.

Phylogenetic systematics The study of the phylogenetic relationships among organisms in which true and false similarities are differentiated; cladistics.

Phylogenetic tree A branching diagram depicting the evolutionary relationships between major groups of organisms

Phylogeny The origin and diversification of any taxon, or the evolutionary history of its origin and diversification, usually presented in the form of a dendrogram.

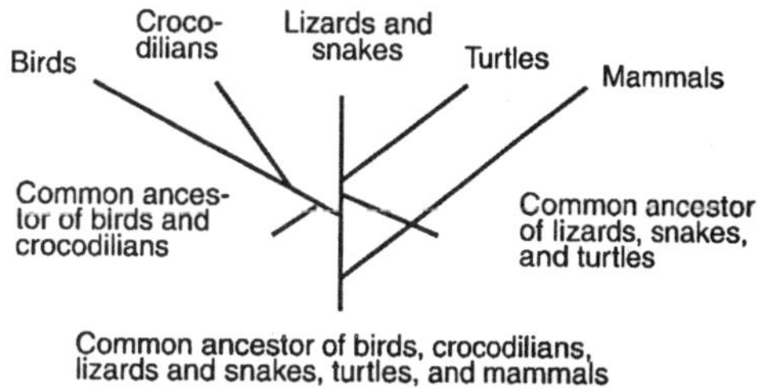

Fig. Phylogeny

Phylum A chief category, between kingdom and class, of taxonomic classifications into which are grouped organisms of common descent that share a fundamental pattern of organization.

Physical change A change in which the composition and identifying properties of a substance remain unchanged Physical quantity A measurable aspect of the universe, such as length.

Physical or abiotic factors Nonliving factors, such as temperature, light, water, minerals, and climate, that influence an organism.

Physics The science that deals with the relationships between matter and energy The second half of life.

Physiological dependence The condition in which there is a physical need for a drug to avoid withdrawal symptoms.

Physiological limits The tolerance range of given species for temperature induced rates of metabolic activity.

Physiology A branch of biology dealing with the organic processes and phenomena of an organism or any of its parts or of a particular bodily process.

Physostomous swim bladder In fish, a swim bladder with an air passage or duct to the esophagus.

Phytochrome The protein pigment responsible for the phenomenon of photoperiodism.

Phytoflagellates Members of the class Phytomastigophorea, plantlike flagellates.

Phytomastigophorea The protozoan class where members usually have chloroplasts; mainly autotrophic, some heterotrophic. Examples: Euglena, Volvox, Chlamydomonas.

Phytophagous Organisms that feed on plants.

Phytoplankton Microscopic, free-floating, autotrophic organisms that function as producers in aquatic ecosystems.

Pia mater The innermost meninx that is in direct contact with the brain and spinal cord.

Picoplankton The component of the plankton that consists of extremely small organisms, mostly bacterioplankton, 0.2 to 2 mm (0.0002 to 0.002 mm) in size; they are too small to catch in a standard plankton net.

Piezoelectric effect The property of certain natural and synthetic crystals to develop a potential difference between opposite surfaces when subjected to a mechanical stress, and conversely

Pigments Molecules that reflect and absorb light at particular wavelengths.

Pilidium Free-swimming, hat-shaped larva of nemertine worms.

Pilidium larva Free-swimming, hat-shaped larva of nemertean worms characterized by an apical tuft of cilia.

Piloting The use of familiar landmarks to find a direction or area.

Pinacocyte Thin, flat cell covering the outer surface, and some of the inner surface, of poriferans.

Pinacoderm The layer of pinacocytes in sponges.

Pineal gland An endocrine gland located near the midline of the brain that produces melatonin, a hormone involved in biological rhythms, particularly in annual cycles.

Pinkeye Bacterial conjunctivitis, sometimes transmitted by flies of the genus Hippolates.

Pinna The external ear. Also a feather, wing, or fin or similar part.

Pinnate Denoting an arrangement of leaflets (or lobes on a simple leaf) along a main central unit.

Pinnately compound leaf Leaflets attached on both sides of a common axis.

Pinnipeds Mammals with paddle-shaped flippers: seals, eared seals (sea lions and fur seals), and the walrus.

Pinocytosis Taking up of fluid by endocytosis; cell drinking.

Pioneer community The first community to become established in an area.

Pioneer organism An organism that first colonizes soil or pioneer species.In primary succession on a terrestrial site, the plants, lichens, and microbes that first colonize the site.

Pipestem fibrosis Thickening of the walls of a bile duct as the result of the irritating presence of a parasite.

Piroplasm Any of the class Piroplasmea, while in a circulating erythrocyte.

Pistil The female reproductive part of a flower, consisting of stigma, style, and ovary. Also termed the carpel.

Pistillate Denoting a unisexual flower having a pistil but no stamens (=carpellate).

Pistillate flower A flower having carpels but no stamens.

Pit A pore in a secondary cell wall.

Pit organ Receptor of infrared radiation (heat) on the head of some snakes (pit vipers).

Pitch The identification of a certain sound with a definite tone; depends on the frequency which the ear receives

Pith ray The region or tissue located between vascular bundles in a stem and connecting the pith and cortex; also termed interfascicular region or medullary ray.

Pituitary pituitar = mucous secretion. An endocrine gland at the base of the brain.

Pituitary gland The master gland of the endocrine system of vertebrate animals. Located directly below the hypothalamus of the brain, the pituitary produces or releases a variety of hormones that target other endocrine glands of the body.

Pivot point The point from which the lengths of all torque arms are measured.

Placenta Structure by which an unborn child or animal is attached to its mother's uterine wall and through which it is nourished.

Placode Localized, plate-like thickening of vertebrate head ectoderm from which a specialized structure develops; such structures include eye lens, special sense organs, and certain neurons.

Placoderms A group of heavily armored jawed fishes of the Lower Devonian to Lower Carboniferous.

Placoid scale Type of scale found in cartilaginous fishes, with basal plate of dentin embedded in the skin and a backward-pointing spine tipped with enamel.

Placozoa A phylum of small, flattened, marine animals that feed by forming a temporary digestive cavity. Tricoplax adherans.

Planck constant; h The fundamental constant equal to the ratio of the energy of a quantum of energy to its frequency. It is the quantum of action. It has the value 6.626196×10^{-34} j s.

Planck radiation law A law which described blackbody radiation better than its predecessor, thus resolving the ultraviolet catastrophe. It is based on the assumption that electromagnetic radiation is quantized.

Planck's constant A fundamental constant in nature that determines what values are allowed for physical quantities in quantum mechanics; $h = 6.63 \times 10^{-34}$ j s.

Planidium First instar of hypermetamorphic, parasitic Diptera and Hymenoptera, which is apodous but moves actively by means of thoracic and caudal setae.

Planktivore An animal that feeds on plankton.

Plankton Animals (zooplankton) and unicellular algae (phytoplankton) that have only limited locomotory capabilities and are therefore distributed by water movements.

Planktonic Small organisms that passively float or drift in a body of water.

Plant growth regulator Any molecule that exhibits hormone-like effects in a plant, whether synthesized or naturally occurring.

Plantae One of the five kingdoms of life; characterized by being eukaryotic and multicellular, and having rigid cell walls and chloroplasts.

Plantigrade Pertaining to animals that walk on the whole surface of the foot (for example, humans and bears); compares with digitigrade.

Plants Members of the kingdom Plantae, which consists of photosynthetic, eukaryotic, and mostly multicellular organisms.

Planula Free-swimming, ciliated larval type of Cnidarians; usually flattened and ovoid, with an outer layer of ectodermal cells and an inner mass of endodermal cells.

Planuloid ancestor Hypothetical form representing ancestor of Cnidaria and Platyhelminthes.

Plasma plasma = form. An undifferentiated tissue from which another tissue or tissues may form.

Plasma cell A mature, differentiated B lymphocyte chiefly occupied with antibody synthesis and secretion; a plasma cell lives for only five to seven days.

Plasma membrane Outermost membrane of a cell; its surface has molecular regions that detect changes in external conditions and act as a selective barrier to the passage of ions and molecules between the cell and its environment. The external covering of a protozoan.

Plasmalemma The cell membrane.

Plasmid A small circle of DNA that may be carried by a bacterium in addition to its genomic DNA.

Plasmodesma A cytoplasmic strand that connects adjacent plant cells through pores in the cell wall.

Plasmodium Multinucleate ameboid mass, syncytial.

Plasmolysis The osmotic removal of water from the cytoplasm and vacuole, causing the cytoplasm to pull away from the cell wall and clump in the center.

Plasmotomy Division of a multinucleate cell into multinucleate daughter cells, without accompanying mitosis.

Plastic song One of the stages of song development in birds during which the bird has begun to sing using the species-typical pattern, but with a certain degree of variation.

Plasticity Cell wall "stretchability" induced by auxin.

Plastid A membranous organelle in plant cells functioning in photosynthesis and/or nutrient storage, for example, chloroplast.

Plastron Ventral body shield of turtles; structure in corresponding position in certain arthropods; thin film of gas retained by epicuticle hairs of aquatic insects.

Plate tectonics the collective geologic processes that move the crustal plates of the earth and cause continental drifting and seafloor spreading

Plate The anode of an electronic tube

Platelet A tiny, incomplete cell in the blood that releases substances initiating blood clotting.

Platyhelminthes The phylum of flatworms; bilateral acoelomates.

Play behavior Certain locomotor, social, and manipulative behavior patterns exhibited by young and some adult mammals and birds.

Pleiotropic Pertaining to a gene producing more than one effect; affecting multiple phenotypic characteristics.

Pleistocene A geological period, which began about 2 million years ago, characterized by a series of ice ages.

Pleopod One of the swimming appendages on the abdomen of a crustacean.

Plerocercoid larva Metacestode that develops from a procercoid larva; it usually shows little differentiation.

Plerocercus Tapeworm metacestode in the order Trypanorhyncha in which the posterior forms a bladder, the blastocyst, into which the rest of the body withdraws.

Plesiomorphic Ancestral characters; characters possessed by members of both ingroup and outgroup.

Pleura The membrane that lines each half of the thorax and covers the lungs.

Pleurite Lateral sclerite of a somite in an arthropod.

Plexus A network, especially of nerves or blood vessels.

Plica polonica Develops in untreated head louse (Pediculus humanus capitis) infection, hair matted together with exudate, fungal growth, fetid odor.

Plumule The shoot apex of a seedling, including embryonic leaves.

Pluteus Echinoid or ophiuroid larva with elongated processes like the supports of a desk; originally called "painter's easel larva."

Plywood A building material consisting of two or more thin sheets of wood bonded together.

P-N junction The boundary between P and N-type materials in a semiconductor crystal.

Pneumatic duct in fish, the connection between the esophagus and swim bladder

Pneumatic sacs Gas-filled sacs that arise from the esophagus, or another part of the digestive tract, of fishes. Pneumatic sacs are used in buoyancy regulation (swim bladders) or gas exchange (lungs).

Pneumatocyst A gas-filled bladder in seaweeds.

Pneumatophore A specialized root that grows upward into the air from roots growing in the mud; aids in gaseous exchange.

Pneumostome The opening of the mantle cavity (lung) of pulmonate gastropods to the outside.

Poachers Those who hunt wildlife illegally.

Pod 1. A dry dehiscent fruit that splits along two seams; a legume.

2. A school of cetaceans.

Podium A footlike structure, for example, the tube foot of echinoderms.

Podomere More or less cylindrical segment of a limb of an arthropod, generally articulated at both ends.

Podosoma Portion of the body of a tick or mite that bears the legs.

Poecilostome Describes mouthparts borne by members of the copepod order Poecilostomata (buccal cavity large, somewhat slitlike, with sickle-shaped mandibles); also, a member of the order Poecilostomata.

Pogonophora A phylum of protostomate, marine animals that are distributed throughout the world's oceans. Live in secreted, chitinous tubes in cold water at depths exceeding 100 m. Lack a mouth and digestive tract. Nutrients absorbed across the body wall and from endosymbiotic bacteria that they harbor. Beard worms.

Poikilotherm an organism whose body temperature varies with and is largely controlled by environmental temperatures

Poikilothermic Pertaining to animals whose body temperature is variable and fluctuates with that of the environment; cold blooded; compares with ectothermic.

Point mutations A change in the structure of a gene that usually arises from the addition, deletion, or substitution of one or more nitrogenous bases.

Point rule The sum of the currents toward a branch point is equal to the sum of the currents away from the same branch point.

Point sources Specific locations of highly concentrated pollution discharge, such as factories, power plants, sewage treatment plants, underground coal mines, and oil wells.

Poisson equation (s.d. poisson) The differential form of gauss' law, namely,

Div e = rho,

Poisson spot (s.d. poisson) Poisson originally predicted the existence of such a spot, and used the prediction to demonstrate how the wave theory of light must be in error to produce such a counterintuitive result. Subsequent observation of the arago spot provided a decisive confirmation of the wave nature of light.

Polar capsule Compartment bearing the polar filaments in myxozoans.

Polar covalent bond The type of bond that is formed by asymmetrical moving electrons.

Polar easterlies Variable winds that blow at high latitudes.

Polar effect For a reactant molecule RY, the polar effect of the group R comprises all the processes whereby a substituent may modify the electrostatic forces operating at the reaction centre Y, relative to the standard RoY. These forces may be governed by charge separations arising from differences in the electronegativity of atoms (leading to the presence of dipoles), the presence of unipoles, or electron delocalization. It is synonymous with electronic effect or "electrical effect" of a substituent as distinguished from other substituent effects, e.g. steric effects.

Sometimes, however, the term "polar effect" is taken to refer to the influence, other than steric, that non-conjugated substituents exert on reaction rates, i.e. effects connected with electron delocalization between a substituent and the molecular framework to which it is attached are excluded. Polar effect is then not synonymous with electronic effect.

Polar filament Threadlike organelles in Myxozoa and Microspora.

Polar granule Refractile granule within a coccidian oocyst.

Polar ice cap The portions of the globe close to the poles that are permanently covered with ice.

Polar nuclei Two of the eight haploid nuclei of a megagametophyte that migrate, one from each end, to the middle of the embryo sac. These two nuclei fuse with a sperm nucleus during double fertilization.

Polar ring Electron-dense organelles of unknown function, located under the cell membrane at the anterior tip of sporozoites and merozoites.

Polarity When applied to solvents, this rather ill-defined term covers their overall solvation capability (solvation power) for solutes (i.e. in chemical equilibria: reactants and products; in reaction rates: reactants and activated complex; in light absorptions: ions or molecules in the ground and excited state), which in turn depends on the action of all possible, nonspecific and specific, intermolecular interactions between solute ions or molecules and solvent molecules, excluding such interactions leading to definite chemical alterations of the ions or molecules of the solute. Occasionally, the term solvent polarity is restricted to nonspecific solute/solvent interactions only (i.e. to van der Waals forces).

Polarized A description for a membrane that has a potential difference due to an unequal distribution of ions across the membrane.

Polarized light Light radiations in which the vibrations of all light waves present are confined to planes parallel to each other.

Polaroplast Organelle, apparently a vacuole, near the polar filament of a microsporidean.

Polian vesicles Vesicles opening into ring canal in most asteroids and holothuroids.

Pollen The collective term for pollen grains, the male gametophytes.

Pollen grain The structure into which a haploid microspore develops; contains a halpoid tube nucleus and two haploid sperm nuclei at maturity.

Pollen tube A tube that develops from the pollen grain and carries the sperm to the ovule.

Pollex The thumb.

Pollination The transfer of pollen to a receptive surface; the stigma in angiosperms or the pollination droplet in most gymnosperms.

Pollinator An organism that effects pollination.

Pollution charges Fees assessed per unit of pollution based on the "polluter pays" principle.

Pollution The introduction of harmful substances or heat energy into

Polyandrous Having more than one male mate. Polyandry is advantageous when food is plentiful but, because of predation or other factors, the chances of successfully rearing young are low.

Polyandry Condition of having more than one male mate at one time.

Polycarpic Denoting a plant that flowers more than once during its lifetime.

Polycentric complex Cities with several urban cores surrounding a once dominant central core.

Polychaeta The class of annelids whose members are mostly marine and are characterized by a head with eyes and tentacles and a body with parapodia. Parapodia bear numerous setae. Examples: Nereis, Arenicola.

Polychlorinated biphenyls A group of nonbiodegradable pollutants.

Polychromatic light Light composed of several colors.

Polyembryony Development of a single zygote into more than one offspring.

Polygamy Condition of having more than one mate at a time.

Polygenes Genes at multiple loci that influence a trait in a quantitative fashion.

Polygenic Multiple genetic influences controlling a single trait, causes the trait to display continuous variation.

Polygenic trait Traits that are influenced by many genes and that are usually continuously distributed within a population.

Polygynous Having more than one female mate. Polygyny tends to occur in species whose young are relatively independent at birth or hatching.

Polygyny Condition of having more than one female mate at one time.

Polygyny threshold The point at which a female will benefit more by joining an already mated male possessing a good territory rather than an unmated male on a poor territory.

Polykinetid Rows or fields of kinetids in ciliates linked by fibrous networks.

Polymer A chemical compound composed of repeated structural units called monomers.

Polymerase chain reaction (PCR) A technique for preparing large quantities of DNA from tiny samples, making it easy to clone a specific gene as long as part of the sequence of the gene is known.

Polymerization The process of forming a polymer or polymeric compound.

Polymorphic A locus that contains two or more alleles within a population.

Polymorphism The presence in a species of more than one structural type of individual.

Polynomial A scientific name for an organism composed of more than two words.

Polynucleotide A nucleotide of many mononucleotides combined.

Polynucleotide chains Attachment of one nucleotide to another in a linear fashion.

Polyp Individual of the phylum Cnidaria, generally adapted for attachment to the substratum at the aboral end, often form colonies.

Polypeptide A molecule consisting of many joined amino acids, not as complex as a protein.

Polyphyletic group An assemblage of organisms that includes multiple evolutionary lineages. Polyphyletic assemblages usually reflect insufficient knowledge regarding the phylogeny of a group of organisms.

Polyphyly The condition that a taxon or other group of organisms does not contain the most recent common ancestor of all members of the group, implying that it has multiple evolutionary origins; such groups are not valid as formal taxa and are recognized as such only through error. Contrasts with monophyly and paraphyly.

Polyphyodont Having several sets of teeth in succession.

Polypide An individual or zooid in a colony, specifically in ectoprocts, which has a lophophore, digestive tract, muscles, and nerve centers.

Polyplacophora The class of molluscs whose members are elongate, dorsoventrally flattened, and have a shell consisting of eight dorsal plates.

Polyploid An organism possessing more than two full homologous sets of chromosomes.

Polyploidy Having more than two sets of chromosomes.

Polypod larva Caterpillar type of larva found in Lepidoptera and some Hymenoptera. It has thoracic appendages and abdominal locomotory processes (prolegs). Also called cruciform.

Polysaccharide A carbohydrate composed of many monosaccharide units, for example, glycogen, starch, and cellulose.

Polysome Two or more ribosomes connected by a molecule of messenger RNA.

Polytene chromosomes Chromosomes in the somatic cells of some insects in which the chromatin replicates repeatedly without undergoing mitosis.

Polyunsaturated fat A fat having several to many double bonds between carbon atoms.

Polyzoic A tapeworm forming a strobila of several to many proglottids; also, a colony of many zooids.

Pome A fleshy fruit derived from a compound inferior ovary; the fleshy edible part is the ripened tissue surrounding the ovary (derived from receptacle and perianth tissue); the ovary matures into the core and contains the seed.example: apple or pear.

Pongid Of or relating to the primate family Pongidae, comprising the anthropoid apes (gorillas, chimpanzees, gibbons, orangutans).

Pons A portion of the brain stem above the medulla oblongata and below the midbrain.

Population A group of organisms of the same species inhabiting a specific geographical locality.

Population crash A sudden population decline caused by predation, waste accumulation, or resource depletion; also called a dieback.

Population explosion Growth of a population at exponential rates to a size that exceeds environmental carrying capacity; usually followed by a population crash.

Population genetics The study of events occurring in gene pools.

Population hurdle The need for investment in infrastructure and social services in a rapidly growing population that prevents the capital investment necessary for real economic development.

Population momentum A potential for increased population growth as young members reach reproductive age.

Populational gradualism. The observation that new genetic variants become established in a population by increasing their frequencies across generations incrementally, initially from one or a few individuals and eventually characterizing a majority of the population.

Pore cell The tube-like cell of sponges that forms a pore.

Pore space Space between soil particles.

Porifera The animal phylum whose members are sessile and either asymmetrical or radially symmetrical. Body organized around a system of water canals and chambers. Cells are not organized into tissues or organs. Sponges.

Porocyte Type of cell found in asconoid sponges through which water enters the spongocoel.

Porose area Sunken areas on the basis capituli of certain mites and ticks.

Porosity The relative amount of pore space versus soil particulate space in soil.

Portal system System of large veins beginning and ending with a bed of capillaries; for example, hepatic portal and renal portal system in vertebrates.

Positive rays Rays coming through holes in a cathode on the side opposite the anode in a discharge tube Positively charged ions.

Postanal tail A tail that extends posterior to the anus; one of the four unique characteristics of chordates.

Posterior Situated at or toward the rear of the body; situated toward the back; in human anatomy the upright posture makes posterior and dorsal identical.

Posterior station Development of a protozoan in the hindgut or posterior midgut of its insect host, such as in the section Stercoraria of the Trypanosomatidae.

Postharvest physiology The study of the storage and perishability of fruits following harvest.

Post-kala-azar dermal leishmanoid Disfiguring dermal condition developing about one to two years after inadequate treatment of kala-azar.

Postmaterialist values A philosophy that emphasizes quality of life over acquisition of material goods.

Postmating isolation Isolation that occurs when fertilization is prevented even though mating has occurred.

Potential difference The work done per unit charge as a charge is moved between two points in an electric field

Potential energy Stored energy that is latent but available for use. A rock poised at the top of a hill or water stored behind a dam are examples of potential energy.

Potential-energy (reaction) surface A geometric hypersurface on which the potential energy of a set of reactants is plotted as a function of the coordinates representing the molecular geometries of the system.

For simple systems two such coordinates (characterizing two variables that change during the progress from reactants to products) can be selected, and the potential energy plotted as a contour map.

For simple elementary reactions, e.g. A-B + C$\longrightarrow$A + B-C, the surface can show the potential energy for all values of the A, B, C geometry, providing that the ABC angle is fixed.

For more complicated reactions a different choice of two coordinates is sometimes preferred, e.g. the bond orders of two different bonds. Such a diagram is often arranged so that reactants are located at the bottom left corner and products at the top right. If the trace of the representative point characterizing the route from reactants to products follows two adjacent edges of the diagram, the changes represented by the two coordinates take place in distinct succession; if the trace leaves the edges and crosses the interior of the diagram, the two changes are concerted. In many qualitative applications it is convenient (although not strictly equivalent) for the third coordinate to represent the standard Gibbs energy rather than potential energy.

Using bond orders is, however, an oversimplification, since these are not well-defined, even for the transition state.

The energetically easiest route from reactants to products on the potential-energy contour map defines the potential-energy profile.

Potential-energy profile A curve describing the variation of the potential energy of the system of atoms that make up the reactants and products of a reaction as a function of one geometric coordinate, and corresponding to the "energetically easiest passage" from reactants to products (i.e. along the line produced by joining the paths of steepest descent from the transition state to the reactants and to the products). For an elementary reaction the relevant geometric coordinate is the reaction coordinate; for a stepwise reaction it is the succession of reaction coordinates for the successive individual reaction steps. (The reaction coordinate is sometimes approximated by a quasi-chemical index of reaction progress, such as "degree of atom transfer" or bond order of some specified bond.)

Potocytosis Endocytosis of certain small molecules and ions bound to specific receptors limited to small areas on the cell surface. The areas of the receptors are invaginated and pinch off to form tiny vesicles.

Power The time rate of doing work Power factor The cosine of the phase angle between current and voltage in an a-c circuit.

Prairie An area of land dominated by grasses with occasional shrubby plants and small trees occurring where the grass cover is broken and with herbaceous perennials during certain seasons.

Praniza Parasitic larva of the isopod suborder Gnathiidea. It parasitizes fishes and feeds on blood.

Preadaptation Occurs when a structure or a process present in members of a species proves useful in promoting reproductive success when an individual encounters new environmental situations.

Pre-association A stcp on the reaction path of some stepwise reactions in which the molecular entity C is already present in an encounter pair or encounter complex with A during the formation of B from A. In this mechanism the chemical species C may but does not necessarily assist the formation of B from A, which may itself be a bimolecular reaction with some other reagent.

Pre-association is important when B is too short-lived to permit B and C to come together by diffusion.

Prebiotic synthesis. The chemical synthesis that occurred before the emergence of life.

Precambrian The geological era beginning with the formation of the earth, about 4700 million years ago and extending until about 590 million years ago. During this time the origin of life, the origin eukaryotes, and multicellular life-forms appeared.

Precession The motion that results from the application of a torque that tends to displace the axis of rotation of a rotating object.

Precious corals Gorgonians that secrete a red or pink skeleton consisting of fused calcareous spicules.

Precipitation Moisture falling from the atmosphere in the form of rain, sleet, snow, or hail.

Precise Sharply or clearly defined. Having small experimental uncertainty. A precise measurement may still be inaccurate, if there were an unrecognized determinate error in the measurement (for example, a miscalibrated instrument). Compare: accurate.

Precision The agreement between the numerical values of two or more measurements made in the same way and expressed in terms of deviation; the reproducibility of measured data

Precocial Having developed to a high degree of independence at the time of hatching or birth.

Precocial young Young that are born or hatched at a relatively advanced state of development and are capable of a more independent existence beginning at birth or hatching. Compare to altricial.

Precycling Making environmentally sound decisions at the store and reducing waste before we buy.

Predaceous, predacious Living by killing and consuming other animals; predatory.

Predation The derivation of an organism of elements essential for its existence from organisms of other species that it consumes and destroys. The ingestion of prey by a predator for energy and nutrients.

Predator An organism that preys on other organisms for its food.

Predictive model a somewhat simplified representation of a physical process that yields information with predictive value

Preformation The erroneous idea that gametes contain miniaturized versions of all of the elements present in an adult.

Prehensile Adapted for grasping.

Premating isolation When behaviors or other factors prevent animals from mating.

Premunition Resistance to reinfection or superinfection, conferred by a still existing infection, that does not destroy the organisms of the infection already present.

Prenymph Nonfeeding, quiescent stage in the life cycle of a chigger mite.

Preoptic nuclei Neurons in the hypothalamus that produce various neuropeptides.

Preparedness The genetically based predisposition to learn. Depending upon the learning situation, animals of a particular species may be prepared, unprepared, or contraprepared to learn.

Prepatent Developmental stage in an infection before agents produce evidence of their presence.

Presoma The proboscis, neck, and attached muscles and organs of an acanthocephalan.

Pressure The weight exerted over a unit area of surface. This is equal to 1 atmosphere (14.7 lbs per square inch) at the sea surface and 1 atmosphere plus the pressure exerted by the water column, which is 1 atmosphere per 10 m (33 ft) of depth.

Pressure Flow Hypothesis The theory that organic solutes move along a concentration gradient from source to sink through the phloem.

Pressure law The pressure of an ideal gas is directly proportional to the thermodynamic temperature at constant volume.

Pressure potential The water force created by a real pressure against a membrane.

Prevalance In epidemiology, the number of cases of a disease at a given time; that is, a static measurement. Contrast with incidence.

Prevention of significant deterioration A clause of the Clean Air Act that prevents degradation of existing clean air; opposed by industry as an unnecessary barrier to development.

Prey model A type of optimal foraging model that addresses the types of organisms an individual should try to catch and eat (i.e., its diet).

Priapulida A phylum of aschelminths commonly called priapulids.

Price elasticity A situation in which supply and demand of a commodity respond to price.

Primary A transformer winding that carries current and normally induces a current in one or more secondary windings.

Primary bilateral symmetry. Usually applied to a radially symmetrical organism descended from a bilateral ancestor and developing from a bilaterally symmetrical larva.

Primary cell An electrochemical cell in which the reacting materials must be replaced after a given amount of energy has been supplied to the external circuit.

Primary cell wall The cellulosic wall of all plant cells laid down at the time of mitosis and cytokinesis.

Primary colors Colors in terms of which all other colors may be described or from which all other colors may be evolved by mixtures.

Primary consumer An organism that consumes a producer organism as a food source; a herbivore.

Primary endosperm nucleus The product of the fusion of a sperm and two polar nuclei in the embryo sac of angiosperms; double fertilization.

Primary germ layers Blocks or layers of embryonic cells that give rise to tissues and organs of animals.

Primary growth Growth originating in the apical meristems of shoots and roots resulting in an increase in length of the axis.

Primary immune response The initial immune response following antigen exposure.

Primary pigments The complements of the primary colors.

Primary pit fields Regions within the primary cell wall in which plasmodesmata traverse the cell wall.

Primary pollutants Chemicals released directly into the air in a harmful form.

Primary producer An autotrophic organism; able to build its own complex organic molecules from simple inorganic substances in the environment.

Primary production The conversion of the inorganic carbon in carbon dioxide into organic carbon by autotrophs.

Primary productivity Synthesis of organic materials (biomass) by green plants using the energy captured in photosynthesis.

Primary radial symmetry.Usually applied to a radially symmetrical organism that did not have a bilateral ancestor or larva, in contrast to a secondarily radial organism.

Primary succession Plant successional events occurring in a pristine or newly forming habitat.

Primary tissue Any tissue derived from the apical meristem, either shoot or root.

Primary treatment A process that removes solids from sewage before it is discharged or treated further.

Primary wall The wall layer of a plant cell deposited during cell expansion, generally thin and elastic.

Primate Any mammal of the order Primates, which includes the tarsiers, lemurs, marmosets, monkeys, apes, and humans.

Priming pheromone A chemical signal that alters the physiology of another organism, eventually causing a change in its behavior. Contrast with releasing pheromone.

Primite Anterior member of a pair of gregarines in syzygy.

Primitive Primordial; ancient; little evolved; said of characteristics closely approximating those possessed by early ancestral types.

Primitive change One of the conceptually simpler molecular changes into which an elementary reaction can be notionally dissected. Such changes include bond rupture, bond formation, internal rotation, change of bond length or bond angle, bond migration, redistribution of charge, etc.

The concept of primitive changes is helpful in the detailed verbal description of elementary reactions, but a primitive change does not represent a process that is by itself necessarily observable as a component of an elementary reaction.

Primitive streak A medial thickening along the dorsal margin of an amniote embryo that forms during the migration of endodermal and mesodermal cells into the interior of the embryo.

Principal axis 1. A line drawn through the center of curvature and the vertex of a curved mirror

2. A line drawn through the center of curvature and the optical center of a lens.

Principal focus A point at which rays parallel to the principal axis converge or from which they diverge after reflection or refraction

Principle of competitive exclusion A result of natural selection whereby two similar species in a community occupy different ecological niches, thereby reducing competition for food.

Principle of independent assortment One of Mendel's observations on the behavior of hereditary units during gamete formation. A modern interpretation of this principle is that genes carried in one chromosome are distributed to gametes without regard to the distribution of genes in nonhomologous chromosomes.

Principle of parity For every process in nature there is a mirror-image process which is indistinguishable from the original process.

Principle of segregation One of Mendel's observations on the behavior of hereditary units during gamete formation. A modern interpretation of the principle of segregation is that genes exist in pairs, and during gamete formation, members of a pair of genes are distributed into separate gametes.

Pristine Denoting a natural and undisturbed state.

Proboscis A snout or trunk. Also, tubular sucking or feeding organ with the mouth at the end as in planarians, leeches, and insects. Also, the sensory and defensive organ at the anterior end of certain invertebrates.

Procaryotes Bacteria and cyanobacteria that lack the structural complexity and defined nucleus found in eucaryotes.

Procercoid Cestode metacestode developing from a coracidium in some orders. It usually has a posterior cercomer.

Procercoid larva Cestode developing from a coracidium in some orders; it usually has a posterior cercomer. Developmental stage between oncosphere and plerocercoid.300

Process To project from.

Procuticle Thicker layer beneath the epicuticle of arthropods that lends mass and strength to the cuticle. It contains chitin, sclerotin, and also inorganic salts in Crustacea. The layers within the procuticle vary in structure and composition.

Producer An organism that synthesizes food molecules from inorganic compounds by using an external energy source; most producers are photosynthetic.

Product development control The term is used for reactions under kinetic control where the selectivity parallels the relative (thermodynamic) stabilities of the products. Product development control is usually associated with a transition state occurring late on the reaction coordinate.

Product-determining step The step of a stepwise reaction, in which the product distribution is determined. The product-determining step may be identical to, or occur later than, the rate-controlling step on the reaction coordinate.

Production In ecology, the energy accumulated by an organism that becomes incorporated into new biomass.

Production frontier The maximum output of two competing commodities at different levels of production.

Productivity The amount of plant product resulting from a combination of water, nutrients and environmental factors.

Profile A graph that shows changes in temperature, salinity, or any other parameter with depth

Profundal zone The deepest portion of a lake.

Progametangia Swollen hyphae of fungi that fuse at the point of contact and eventually form gametangia.

Progesterone Hormone secreted by the corpus luteum and the placenta; prepares the uterus for the fertilized egg and maintains the capacity of the uterus to hold the embryo and fetus.

Proglottid One set of reproductive organs in a tapeworm strobila; usually corresponds to a segment. One of the linearly arranged segmentlike sections that make up the strobila of a tapeworm.

Progymnosperm An extinct group of plants, the Progymnospermophyta, which included woody plants bearing large plannated braching systems; the prototypes of the gymnospems (e.g., Archaeopteris).

Prohaptor Collective adhesive and feeding organs at the anterior end of a monogenetic trematode.

Prohormone A precursor of a hormone, especially a peptide hormone.

Prokaryote Organism in which the chromosomes are not contained within membrane-bound nuclei.

Prokaryotic Procaryotic Not having a membrane-bound nucleus or nuclei. Prokaryotic cells characterize the bacteria and cyanobacteria.

Prokaryotic cell A type of cell lacking a nucleus and membrane-bound organelles; found in the Kingdom Monera.

Prolegs Unjointed abdominal appendage in the larva of Lepidoptera and some other insects.

Promastigote Form of Trypanosomatidae with the free flagellum and the kinetoplast anterior to the nucleus, as in Leptomonas.

Promethean environmentalism A form of technological optimism that predicts that human ingenuity and enterprise will find cures for all our problems.

Promiscuity A mating system in which there is no prolonged association between the sexes and multiple mating by at least one sex.

Promoter. A region of DNA to which the RNA polymerase must have access for transcription of a structural gene to begin.

Pronatalist pressures Influences that encourage people to have children.

Pronephros Most anterior of three pairs of embryonic renal organs of vertebrates, functional only in adult hagfishes and larval fishes and amphibians, and vestigial in mammalian embryos.

Prop roots Adventitious roots arising on a stem above the ground and imparting some mechanical support to plants. The angled roots may provide for absorption of water and nutrients.

Propagate To travel through a material or space.

Propagation The act of propagating The action of traveling through a material or space.

Property A measurable aspect of matter, e.g., mass and inertia.

Prophase The stage of mitosis during which the chromosomes become visible under a light microscope. The first stage of mitosis during which the chromosomes are condensed but not yet attached to a mitotic spindle.

Proplastids Membrane-bound particles that develop some internal structure; may subsequently develop into chloroplasts, chromoplasts, or leucoplasts.

Propodeum First abdominal segment of hymenopterans, fused to the thorax.

Propodosoma Portion of the podosoma that bears the first and second pairs of legs of a tick or mite.

Propolar cells Anterior tier of cells in the calotte of a dicyemid mesozoan.

Proprioceptor Sensory receptor located deep within the tissues, especially muscles, tendons, and joints, that is responsive to changes in muscle stretch, body position, and movement.

Prosimian Any member of a group of arboreal primates including lemurs, tarsiers, and lorises, but excluding monkeys, apes, and humans.

Prosoma Anterior part of an invertebrate in which primitive segmentation is not visible; fused head and thorax of arthropod; cephalothorax.

Prosopyle Connections between the incurrent and radial canals in some sponges.

Prostaglandins A family of fatty-acid hormones, originally discovered in semen, known to have powerful effects on smooth muscle, nerves, circulation, and reproductive organs.

Prostate The prostate gland is "before" the bladder.

Prostate gland Gland located around the male urethra below the urinary bladder that adds its secretions to seminal fluid during ejaculation.

Prosthetic group Nonprotein groups that are attached to an enzyme or other protein and necessary for its function.

Prostomium Anterior closure of a metameric animal, anterior to the mouth.

Protandric hermaphroditism A pattern of sexuality in which a single individual functions as male and then female in sequence.

Protandrous Condition of hermaphroditic animals and plants in which male organs and their products appear before the corresponding female organs and products, thus preventing self-fertilization.

Protandry The condition in a monoecious organism in which male gonads mature before female gametes; prevents self-fertilization.

Protease An enzyme that digests proteins; includes proteinases and peptidases.

Protective coloration Coloration that benefits the individual by providing concealment from predators.

Protein A macromolecule of carbon, hydrogen, oxygen, and nitrogen and sometimes sulfur and phosphorus; composed of chains of amino acids joined by peptide bonds; present in all cells.

Protein synthesis The assemblage of protein as determined by the nucleotide sequence of a messenger RNA and the assistance of transfer RNA aligning amino acids in the proper arrangement.

Protelean parasite Organism parasitic during its larval or juvenile stages and free living as an adult, usually changing form with each stage.

Proterosoma Combination of the gnathosoma and propodosoma of the body of a tick or mite.

Prothallial cell Sterile cells present in the pollen grain of gymnosperms, believed to represent the last remnant of the vegetative male gametophyte thallus, which was, ancestrally,

Prothoracic gland An invertebrate endocrine gland located in the prothorax region directly behind the head, which secretes ecdysone, a steroid hormone that promotes molting.

Prothorax The first of the three thoracic segments of an insect; usually contains the first pair of walking appendages.

Prothrombin A constituent of blood plasma that is changed to thrombin by a catalytic sequence that includes thromboplastin, calcium, and plasma globulins; involved in blood clotting.

Protist A member of the kingdom Protista, generally considered to include the protozoa and eukaryotic algae.

Protista The kingdom whose members are characterized by being eukaryotic and unicellular or colonial.

Protochordates Chordates that lack a backbone.

Protocoel The anterior coelomic compartment in some deuterostomes, corresponds to the axocoel in echinoderms.

Protocooperation. A mutually beneficial interaction between organisms in which the interaction is not physiologically necessary to the survival of either.

Protoderm The dermal or outer tissue of an apical meristem that gives rise to the epidermis.

Protogenic (solvent) Capable of acting as a proton (hydron) donor strongly or weakly acidic (as a Brønsted acid). The term is preferred to the synonym "protic" or the more ambiguous expression "acidic" by itself. Also called HBD (hydrogen bond donor) solvent.

Protogynous Hermaphroditism in which the female gonads mature before the male gonads.

Protogyny A sequentially hermaphroditic species in which individuals change from females to males.

Protolysis This term has been used synonymously with proton (hydron)-transfer reaction. Because of its misleading similarity to hydrolysis, photolysis, etc., its use is discouraged.

Protomerite Anterior half of a cephaline gregarine protozoan.

Proton A positively charged subatomic particle having a mass of $1.6726485 \times 10^{-27}$ kg and a charge equal and opposite to that of the electron.

Proton affinity The negative of the enthalpy change in the gas phase reaction (real or hypothetical) between a proton (more appropriately hydron) and the chemical species concerned, usually an electrically neutral species to give the conjugate acid of that species.

Proton transfer reaction A chemical reaction, the main feature of which is the intermolecular or intramolecular transfer of a proton (hydron) from one binding site to another. For example,

$$CH_3CO_2H + (CH_3)_2C{=}O \longrightarrow CH_3CO_2^- + (CH_3)_2C{=}O^+H$$

In the detailed description of proton transfer reactions, especially of rapid proton transfers between electronegative atoms, it should always be specified whether the term is used to refer to the overall process (including the more-or-less encounter-controlled formation of a hydrogen bonded complex and the separation of the products.

Protonema The early, filamentous growth of the gametophyte in bryophytes and ferns.

Protonephridium Primitive osmoregulatory or excretory organ composed of a tubule terminating internally with a flame bulb or solenocyte; the unit of a flame bulb system. Protonephridia are specialized for ultrafiltration.

Protonymph Early, bloodsucking stage in the life cycle of some mesostigmatid mites.

Protophilic Capable of acting as proton acceptor, strongly or weakly basic (as a Brønsted base). Also called HBA (hydrogen bond acceptor) solvent.

Protoplasm Organized living substance; cytoplasm and nucleoplasm of the cell.

Protoplast All of a plant cell excluding the wall.

Protoplast fusion The technique of enzymatically digesting the cell wall of two distinctly different cells, then treating the plasma membrane so that protoplasts of the two cells fuse. The resulting hybrid may be difficult or impossible to achieve with traditional plant breeding.

Protopod, Basal portion of crustacean appendage, containing coxa and basis.

Protopod larva Larva found in some parasitic Hymenoptera and Diptera; limbs are rudimentary or absent; internal organs are incompletely differentiated; requires highly nutritive and sheltered environment for further development.

Protopodite The basal segment of a biramous appendage of a crustacean.

Protoscolex Juvenile scolex budded within a coenurus or a hydatid metacestode of a taeniid cestode.

Protostome Animal in which the embryonic blastopore becomes the mouth; often possesses a trochophore larva, schizocoelous coelom formation, and spiral embryonic cleavage.

Protostomia A group of phyla in which cleavage is determinate, the coelom (in coelomate forms) is formed by proliferation of mesodermal bands (schizocoelic formation), the mesoderm is formed from a particular blastomere (called 4d), and the mouth is derived from or near the blastopore. Includes the Annelida, Arthropoda, Mollusca, and a number of minor phyla. Compares with Deuterostomia.

Protostyle A rotating mucoid mass into which food is incorporated in the gut of a gastropod (phylum Mollusca).

Protoxylem The first xylem cells formed in the primary xylem.

Protozoa A subkingdom (formerly a phylum) comprising the simplest organisms called protista; divided into seven phyla.

Protozoans Animal-like protists; the various groups of unicellular and eukaryotic organisms that are mostly heterotrophic.

Protozoologist A person who studies protozoa.

Protract A muscle action that draws an element anteriorly.

Proventriculus In birds the glandular stomach between the crop and gizzard. In insects, a muscular dilation of foregut armed internally with chitinous teeth.

Proviral state The condition of a host cell after having been transformed.

Proximal Situated toward or near the point of attachment; opposite of distal, distant.

Proximate cause The factors that underlie the functioning of a biological system at a particular place and time, including those responsible for metabolic, physiological, and behavioral functions at the molecular, cellular, organs, and population levels.

Pruning The selective removal of parts of a plant, usually woody shrubs or trees.

Pseudo-catalysis If an acid or base is present in nearly constant concentration throughout a reaction in solution (owing to buffering or the use of a large excess), it may be found to increase the rate of that reaction and also to be consumed during the process. The acid or base is then not a catalyst and the phenomenon cannot be called catalysis according to the well-established meaning of these terms in chemical kinetics, although the mechanism of such a process is often intimately related to that of a catalysed reaction. It is recommended that the term pseudo-catalysis be used in these and analogous cases (not necessarily involving acids or bases). For example, if a Brønsted acid accelerates the hydrolysis of an ester to a carboxylic acid and an alcohol, this is properly called acid catalysis, whereas the acceleration, by the same acid, of hydrolysis of an amide should be described as pseudo-catalysis by the acid: the "acid pseudo-catalyst" is consumed during the reaction through formation of an ammonium ion. The terms "general acid pseudo-catalysis" and "general base pseudo-catalysis" may be used as the analogues of general acid catalysis and general base catalysis.

The term "base-promoted", "base-accelerated, or "base-induced" is sometimes used for reactions that are pseudo-catalysed by bases. However, the term "promotion" also has a different meaning in other chemical contexts.

Pseudocoel A body cavity not lined with peritoneum and not a part of the blood or digestive systems, embryonically derived from the blastocoel.

Pseudocoelom A body cavity between the mesoderm and endoderm; a persistent blastocoele that is not lined with peritoneum. Also pseudocoel.

Pseudoconditioning The strengthening of a response to a previously neutral stimulus by repeatedly eliciting the response with another stimulus without pairing the presentation of the two stimuli.

Pseudocyst Pocket of protozoa within a host cell but not surrounded by a cyst wall of parasite origin.

Pseudoforce A "force" which arises because an observer is naively treating an accelerating frame as an inertial one. Examples coriolis pseudoforce, centrifugal pseudoforce.

Pseudolabia Bilateral lips around the mouth of many nematodes of the order Spirurata; they are not homologous to the lips of most other nematodes but develop from the inner wall of the buccal cavity.

Pseudomolecular rearrangement The use of this awkwardly formed term is discouraged. It is synonymous with "intermolecular rearrangement".

Pseudomyiasis Presence within a host of a fly not normally parasitic.

Pseudopericyclic A concerted transformation is pseudopericyclic if the primary changes in bonding occur within a cyclic array of atoms at one (or more) of which nonbonding and bonding atomic orbitals interchange roles.

A formal example is the enol. Because the pi and sigma atomic orbitals that interchange roles are orthogonal, such a reaction does not proceed through a fully conjugated transition state and is thus not a pericyclic reaction and therefore not governed by the rules that express orbital symmetry restrictions applicable to pericyclic reactions.

Pseudopodia Temporary cytoplasmic extensions of amoebas that are used in feeding and locomotion.

Pseudopodium A temporary cytoplasmic protrusion extended out from a protozoan or ameboid cell, and serving for locomotion or for taking up food.

Pseudoreplication Results from either repeated testing of the same animal without proper statistical correction using a repeated measures design, or counting members of an interacting social group as separate data points. In both cases, problems with independence of data points arises with pseudoreplication.

Pseudo-unimolecular A term sometimes used as synonymous with pseudo-first order, but is inherently meaningless.

Psilophyta A division of leafless, rootless, vascular, homosporous plants, as exemplified by the genus Psilotum.

Psychoactive drug A drug that affects the central nervous system by influencing the release of neurotransmitters or mimicking their actions.

Psychobiology The study of the mechanism and function of the central nervous system from both psychological and biological perspectives.

Psychological dependence A condition marked by the strong desire to repeat the use of a drug to reexperience the feelings of well-being induced by the drug.

Psychopharmacology The study of the brain's behavior in terms of chemical, physiological, and psychological parameters.

Pterobranchia The class of hemichordates whose members lack gill slits and have two or more arms. Colonial, living in externally secreted encasements.

Pterophyta The division of vascular plants commonly referred to as the ferns.

Pteropods Pelagic gastropods in which the foot is modified for swimming and the shell is reduced or absent.

Ptilinum Balloonlike organ in the head of teneral dipterans that pushes off the operculum of the puparium.

P-type germanium "Hole-rich" germanium consisting of equal numbers of free positive holes and bound negative charges so that the net charge is zero

Puberty The age at which an organism can first reproduce.

Pubescence Having hairs or trichomes on the surface.

Public trust A doctrine obligating the government to maintain public lands in a natural state as guardians of the public interest.

Pudendo- pudenda = to be ashamed. External female genitalia.

Puff. Strands of DNA spread apart at certain locations on giant chromosomes of some flies where that DNA is being transcribed.

Pull factors (in urbanization) Conditions that draw people from the country into the city.

Pulmonary Relating to or associated with lungs.

Pulmonary circuit The system of blood vessels from the right ventricle of the heart to the lungs, transporting deoxygenated blood and returning oxygenated blood from the lungs to the left atrium of the heart.

Pulping The process of partially digesting and breaking up wood fibers to make paper.

Pulse A single nonrepeated disturbance.

Pumice A lightweight white, yellow, or gray stone formed from volcanic glass; used in polishing and cleaning.

Punctuated equilibrium model The idea that evolutionary change can occur rapidly over periods of thousands of years and that these periods of rapid change are interrupted by periods of constancy (stasis).

Punctuated equilibrium. A model of evolution in which morphological evolutionary change is discontinuous, being associated primarily with discrete, geologically instantaneous events of speciation leading to phylogenetic branching; morphological evolutionary stasis characterizes species between episodes of speciation; contrasts with phyletic gradualism.

Punnett square A tool used by geneticists to help predict the results of a genetic cross. Different kinds of gametes produced by each parent are placed on each axis of the square. Combining gametes in the interior of the square gives the results of random fertilization.

Pupa Inactive quiescent stage of the holometabolous insects. It follows the larval stages and precedes the adult stage.

Pupariation Formation of a puparium by the third-stage larvae of certain families of Diptera.

Puparium A pupal case formed from the last larval exoskeleton.

Purine A nitrogen-containing organic compound that contributes to the structure of a DNA or RNA nucleotide; uric acid is also derived from purines.

Push factors (in urbanization)Conditions that force people out of the country and into the city.

Pycnocline the ocean layer, usually near the bottom of the photic zone, marked by a sharp change in density.

Pycnogonida The class of chelicerate arthropods whose members have a reduced abdomen and four to six pairs of walking legs. Without special respiratory or excretory structures. Sea spiders.

Pygidium Posterior closure of a metameric animal, bearing the anus.

Pygostyle The fused posterior caudal vertebrae of a bird; helps support tail feathers that are important in steering.

Pyloric caecum (Each of the slender blind tubes found in the intestine of many bony fishes.

Pyogenic Pus producing.

Pyramid of biomass A representation showing the total mass of living matter in each trophic level of an ecosystem.

Pyramid of numbers The decrease in the number of individuals that is observed in each higher level of a food chain.

Pyrenoid Part of the chloroplast that synthesizes and stores polysaccharides.

Pyrimidine An organic base composed of a single ring of carbon and nitrogen atoms; parent substance of several bases found in nucleic acids.

Pyrogenic Substance that causes a rise in body temperature; causes fever.

Pyrolysis Thermolysis, usually associated with exposure to a high temperature.

Pyrrhophyta A division of protists commonly termed the dinoflagellates.

Pyruvic acid The end product 3-carbon compound resulting from glycolysis of glucose which enters the Krebs cycle in aerobic respiration.

Q

Quality The property of sound waves that depends on the number of harmonics and their prominence.

Quantitative structure-activity relationships (QSAR) The building of structure-biological activity models by using regression analysis with physicochemical constants, indicator variables or theoretical calculations. The term has been extended by some authors to include chemical reactivity, i.e. activity is regarded as synonymous with reactivity.

Quantitative trait A genetic trait that is determined by multiple interacting loci, and for which there is a range of phenotypes between phenotypic extremes.

Quantum An elemental unit of energy.

Quantum mechanics The branch of physics that heals with the behavior of particles whose specific properties are given by quantum numbers

Quantum number One of a set of notations used to characterize a discrete value that a quantized variable is allowed to assume.

Quantum theory A unifying theory based on the concept of the subdivision of radiant energy into discrete quanta (photons) and applied to the studies of structure at the atomic and molecular levels.

Quark Quarks are the particles which, when bound together form a hadron. The quarks are bound due to their reactions under the strong interaction much like the moon is bound to the Earth by gravity, like the Earth is bound to the sun by gravity, and like the electrons of an atom are bound to the nucleus of the atom by electromagnetism. There are currently six known quarks. They are distinguished by a property referred to as flavor. Each of these "flavors" of quark can have one of three "color charges".

Quartan malaria Malaria with fevers recurring every 72 hours. Caused by Plasmodium malariae.

Queen. In entomology, the single fully developed female in a colony of social insects such as bees, ants, and termites, distinguished from workers, nonreproductive females, and soldiers.

Quiescent A steady-state condition The operating condition of an electronic circuit when no input signal is applied.

Quotidian malaria Malaria with fevers recurring every 24 hours. Found in cases of overlapping infections.

R

R selection Selection favoring rapid rates of reproduction and growth, especially among species that specialize in colonizing short-lived habitats.

Raceme A vertical inflorescence with stalked flowers.

Rachis Central, longitudinal, supporting structure in the ovary of some nematodes.

Radial A spoke or ray.

Radial canals Canals along the ambulacra radiating from the ring canal of echinoderms; also choanocyte-lined canals in syconoid sponges.

Radial cleavage Embryonic development in which early cleavage planes are symmetrical to the polar axis, each blastomere of one tier lying directly above the corresponding blastomere of the next layer; indeterminate cleavage.

Radial cut A longitudinal section cut along a radius.

Radial micellation Reinforcement of the cell wall of guard cells in specific regions such that the cells curve outward when fully turgid.

Radial symmetry. A morphological condition in which the parts of an animal are arranged concentrically around an oral-aboral axis, and more than one imaginary plane through this axis yields halves that are mirror images of each other.

Radian A unit of angular measurement The angle that, when placed with its vertex at the center of a circle, subtends on the circumference an arc equal in length to the radius of the circle Approximately 57.3°.

Radian; rad The supplementary SI unit of angular measure, defined as the central angle of a circle whose subtended arc is equal to the radius of the circle.

Radiation A form of energy that includes visible light, ultraviolet light, and X rays; means by which body heat is lost in the form of infrared rays.

Radiation, alpha This radiation consists of two protons and two neutrons. (This is the same as the nucleus of a hydrogen atom.) Alpha particles travel about 7% of the speed of light (2.2 X 10^7 m/s) and require very mild shielding as protection against them — a few sheets of paper. Although they are easy to protect against externally, this radiation causes the most damage if it can reach beneath the skin. If one ingests or breathes alpha-radioactive dust (which isn't common) it can do some damage.

Radiation, beta This radiation consists of electrons or positrons. Beta particles travel anywhere from a quarter of to nearly the value of (99%) the speed of light and require mild shielding as protection against them — 15 meter of air or a millimeter or so of aluminum.

Radiation, gamma This radiation consists of photons (light). It is high-energy radiation in the gamma frequency. Since it is high energy, thick shielding (several feet of concrete or several centimeters of lead) is necessary as protection against it.

Radical (or free radical) A molecular entity such as.CH_3,.SnH_3, Cl. possessing an unpaired electron. (In these formulae the dot, symbolizing the unpaired electron, should be placed so as to indicate the atom of highest spin density, if this is possible.) Paramagnetic metal ions are not normally regarded as radicals. However, in the "isolobal analogy", the similarity between certain paramagnetic metal ions and radicals becomes apparent.

At least in the context of physical organic chemistry, it seems desirable to cease using the adjective "free" in the general name of this type of chemical species and molecular entity, so that the term "free radical" may in future be restricted to those radicals which do not form parts of radical pairs.

Depending upon the core atom that possesses the unpaired electron, the radicals can be described as carbon-, oxygen-, nitrogen-, metal-centred radicals. If the unpaired electron occupies an orbital having considerable s or more or less pure p character, the respective radicals are termed sigma or pi radicals.

In the past, the term "radical" was used to designate a substituent group bound to a molecular entity, as opposed to "free radical", which nowadays is simply calléd radical. The bound entities may be called groups or substituents, but should no longer be called radicals.

Radical centre(s) The atom (or group of atoms) in a polyatomic radical on which an unpaired electron is largely localized. Attachment of a monovalent atom to a radical centre gives a molecule for which it is possible to write a Lewis formula in which the normal stable valencies are assigned to all atoms.

Radical ion A radical that carries an electric charge. A positively charged radical is called a "radical cation" (e.g., the benzene radical cation $C_6H_6^{\cdot+}$); a negatively charged radical is called a "radical anion" (e.g., the benzene radical anion C6H6·⁻).

Commonly, but not necessarily, the odd electron and the charge are associated with the same atom.

Radical pair (or geminate pair) The term is used to identify two radicals in close proximity in solution, within a solvent cage. They may be formed simultaneously by some unimolecular process, e.g., peroxide decomposition, or they may have come together by diffusion. While the radicals are together, correlation of the unpaired electron spins of the two species cannot be ignored: this correlation is responsible for the CIDNP phenomenon.

Radicle The primary root of a germinating seed.

Radio waves Also called Hertzian waves Electromagnetic radiations produced by rapid reverses of current in a conductor.

Radioactive A word distinguishing radioactive materials from those which aren't. Usage: "U-235 is radioactive; He-4 is not." Radioactive is least misleading when used as an adjective, not as a noun. It is sometimes used in the noun form as an shortened stand-in for radioactive material, as in the example above.

Radioactive decay A change in the nuclei of radioactive isotopes that spontaneously emit high-energy electromagnetic radiation and/or subatomic particles while gradually changing into another isotope or different element.

Radioactive isotope An unstable form of an element that spontaneously emits either high-energy electromagnetic radiation or subatomic particles (or both). The decay rate describes the time necessary for a radioactive isotope to change to a stable element.

Radioactive material A material whose nuclei spontaneously give off nuclear radiation. Naturally radioactive materials (found in the earth's crust) give off alpha, beta, or gamma particles. Alpha particles are Helium nuclei, beta particles are electrons, and gamma particles are high energy photons.

Radioisotope An isotope that is radioactive, i.e., one having an unstable nucleus, which gives it the property of decay by one or more of several processes. Also called a radioactive isotope.

Radioisotope dating A technique for determining the age of fossil materials based on the rate of disintegration of the radioisotopes.

Radiolarian Members of the classes Acantharea, Phaeodarea, and Polycystinea (phylum Sarcomastigophora) with actinopodia and beautiful tests.

Radiolarian ooze A type of biogenous sediment that consists mostly of the silica shells of radiolarians.

Radioles Featherlike processes from the head of many tubicolous polychaete worms (phylum Annelida), used primarily for feeding.

Radiolysis The cleavage of one or several bonds resulting from exposure to high-energy radiation. The term is also often used loosely to specify the method of irradiation ("pulse radiolysis") used in any radiochemical reaction, not necessarily one involving bond cleavage.

Radionucleides Isotopes that exhibit radioactive decay.

Radula The rasping, tonguelike structure of most molluscs that is used for scraping food; composed of minute chitinous teeth that move over a cartilaginous odontophore.

Rain forest A forest with high humidity, constant temperature, and abundant rainfall (generally over 380 cm [150 in] per year); can be tropical or temperate.

Rain shadow Dry area on the downwind side of a mountain.

Ram ventilation The movement of water across gills as a fish swims through the water with its mouth open.

Random motion Molecular movement that is always present but nondirectional.

Range of optimum The range of values for a condition in the environment that is best able to support survival and reproduction of an organism.

Rangeland Grasslands and open woodlands suitable for livestock grazing.

Raphe a groove in the frustules of pennate diatoms through which cytoplasm extends for locomotion

Raprenox A process to reduce nitrogen oxide emissions by mixing combustion gases with isocyanic acid, which converts nitrogen oxides to dimolecular nitrogen gas.

Rarefaction The region of a longitudinal wave in which the vibrating particles are farther apart than their equilibrium distance.

Ras protein A protein that initiates a cascade of reactions leading to cell division when a growth factor is bound to the cell surface. The gene encoding Ras becomes an oncogene when a mutation produces a form of Ras protein that initiates the cascade even in the absence of the growth factor.

Rate A quantity of one thing compared to a quantity of another. In physics the comparison is generally made by taking a quotient. Thus speed is defined to be the dx/dt, the 'time rate of change of position'.

Rate law (or empirical differential rate equation) An expression for the rate of reaction of a particular reaction in terms of concentrations of chemical species and constant parameters (normally rate coefficients and partial orders of reaction) only. For examples of rate laws see equations (1.) to (3) under kinetic equivalence, and (1) under steady state.

Rate of reaction For the general chemical reaction

$$aA + bB \longrightarrow pP + qQ.$$

occurring under constant-volume conditions, without an appreciable build-up of reaction intermediates, the rate of reaction v is defined as

$$v = -\frac{1}{a}\frac{d[A]}{dt} = -\frac{1}{b}\frac{d[B]}{dt} = +\frac{1}{p}\frac{d[P]}{dt} = +\frac{1}{q}\frac{d[Q]}{dt}$$

where symbols placed inside square brackets denote amount (or amount of substance) concentrations (conventionally expressed in units of mol dm-3). The symbols R and r are also commonly used in place of v. It is recommended that the unit of time should always be the second.

In such a case the rate of reaction differs from the rate of increase of concentration of a product P by a constant factor (the reciprocal of its coefficient in the stoichiometric equation, p) and from the rate of decrease of concentration of the reactant A by 1/a.

For a stepwise reaction this definition of "rate of reaction" (and "extent of reaction",) will apply only if there is no accumulation of intermediate or formation of side products. It is therefore recommended that the term "rate of reaction" be used only in cases where it is experimentally established that these conditions apply. More generally, it is recommended that, instead, the terms "rate of disappearance" or "rate of consumption" of A (i.e. -d[A]/dt, the rate of decrease of concentration of A) or "rate of appearance" of P (i.e. d[P]/dt, the rate of increase of concentration of product P) be used, depending on the concentration change of the particular chemical species that is actually observed. In some cases reference to the chemical flux observed may be more appropriate.

The symbol v (without lettered subscript) should be used only for rate of reaction; v with a lettered subscript (e.g. vA) refers to a rate of appearance or rate of disappearance (e.g. of the chemical species A).

Rate-determining step, rate-limiting step These terms are best regarded as synonymous with rate-controlling step. However, other meanings that have been given to them should be mentioned, as it is necessary to be aware of them in order to avoid confusion:

Sometimes the term rate-determining is used as a special case of rate-controlling, being assigned only to an initial slow step which is followed by rapid steps. Such a step imposes an upper limit on the rate, and has also been called rate-limiting.

In view of the considerable danger of confusion when special meanings are applied to rate-determining and rate-limiting, it is recommended that they be regarded as synonymous, with the meaning explained under the entry rate-controlling step.

Ratio The quotient of two similar quantities. In physics, the two quantities must have the same units to be 'similar'. Therefore we may properly speak of the ratio of two lengths. But to say "the ratio of charge to mass of the electron" is improper. The latter is properly called "the specific charge of the electron."

Ratite Referring to birds having an unkeeled sternum; compares with carinate.

Rays Parenchymal cells found in secondary xylem and phloem in both woody angiosperms and conifers, which provide for lateral (radial) transport.

Rayleigh criterion; resolving power A criterion for determining how finely a set of optics may be able to distinguish. It begins with the assumption that central ring of one image should fall on the first dark ring of the other; for an objective lens with diameter d and employing light with a wavelength λ (usually taken to be 560 nm), the resolving power is approximately given by

$1.22\ \lambda/d$.

Rayleigh-jeans law For a blackbody at thermodynamic temperature t, the radiancy r over a range of frequencies between nu and nu + dnu is given by

$R = 2\ \text{pi}\ \text{nu}^2\ k\ t/c^2$.

Reactance The nonresistive opposition to current in an A-C circuit.

Reacting bond rules 1. For an internal motion of a molecular entity corresponding to progress over a transition state (energy maximum), any change that makes the motion more difficult will lead to a new molecular geometry at the energy maximum, in which the motion has proceeded further. Changes that make the motion less difficult will have the opposite effect.

2. For an internal motion of a molecular entity that corresponds to a vibration, any change that tends to modify the equilibrium point of the vibration in a particular direction will actually shift the equilibrium in that direction.

3. Effects on reacting bonds (bonds made or broken in the reaction) are the most significant. The bonds nearest the site of structural change are those most strongly affected.

Reaction Reaction forces are those equal and opposite forces of Newton's Third Law. Though they are sometimes called an action and reaction pair, one never sees a single force referred to as an action force.

Reaction coordinate A geometric parameter that changes during the conversion of one (or more) reactant molecular entities into one (or more) product molecular entities and whose value can be taken for a measure of the progress of an elementary reaction (for example, a bond length or bond angle or a combination of bond lengths and/or bond angles; it is sometimes approximated by a non-geometric parameter, such as the bond order of some specified bond).

In the formalism of "transition-state theory", the reaction coordinate is that coordinate in a set of curvilinear coordinates obtained from the conventional ones for the reactants which, for each reaction step, leads smoothly from the configuration of the reactants through that of the transition state to the configuration of the products. The reaction coordinate is typically chosen to follow the path along the gradient (path of shallowest ascent/deepest descent) of potential energy from reactants to products.

The term has also been used interchangeably with the term transition coordinate, applicable to the coordinate in the immediate vicinity of the potential energy maximum. Being more specific, the name transition coordinate is to be preferred in that context.

Reaction motor A heat engine whose acceleration is produced by the thrust of exhaust gases.

Reaction path 1. A synonym for mechanism.

2. A trajectory on the potential-energy surface.

3. A sequence of synthetic steps.

Reaction stage A set of one or more (possibly experimentally inseparable) reaction steps leading to and/or from a detectable or presumed reaction intermediate.

Reaction step An elementary reaction, constituting one of the stages of a stepwise reaction in which a reaction intermediate (or, for the first step, the reactants) is converted into the next reaction intermediate (or, for the last step, the products) in the sequence of intermediates between reactants and products.

Reaction wood Wood produced in response to a tree that has lost its vertical position, causing the tree to straighten.

Reactive, reactivity As applied to a chemical species, the term expresses a kinetic property. A species is said to be more reactive or to have a higher reactivity in some given context than some other (reference) species if it has a larger rate constant for a specified elementary reaction. The term has meaning only by reference to some explicitly stated or implicitly assumed set of conditions. It is not to be used for reactions or reaction patterns of compounds in general.

The term is also more loosely used as a phenomenological description not restricted to elementary reactions. When applied in this sense the property under consideration may reflect not only rate, but also equilibrium, constants.

Reactivity index Any numerical index derived from quantum mechanical model calculations that permits the prediction of relative reactivities of different molecular sites. Many indices are in use, based on a variety of theories and relating to various types of reaction. The more successful applications have been to the substitution reactions of conjugated systems where relative reactivities are determined largely by changes of pi-electron energy.

Reactivity-selectivity principle (RSP) This idea may be expressed loosely as: the more reactive a reagent is, the less selective it is.

Consider two substrates S_1 and S_2 undergoing the same type of reaction with two reagents R1 and R_2, S_2 being more reactive than S_1, and R_2 more reactive than R_1 in the given type of reaction. The relative reactivities for the four possible reactions may notionally be represented as follows:

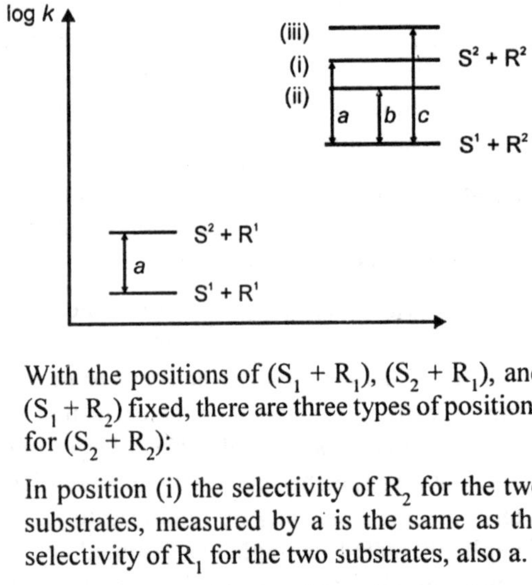

With the positions of $(S_1 + R_1)$, $(S_2 + R_1)$, and $(S_1 + R_2)$ fixed, there are three types of positions for $(S_2 + R_2)$:

In position (i) the selectivity of R_2 for the two substrates, measured by a is the same as the selectivity of R_1 for the two substrates, also a.

In position (ii) the selectivity of R_2 for the two substrates, measured by b, is less than the selectivity of R_1 for the two substrates, i.e. $b < a$. It is this situation which is in accord with the RSP.

In position (iii) the selectivity of R_2 for the two substrates, measured by c, is greater than the selectivity of R_1 for the two substrates, i.e. $c > a$. This situation may be described as anti-RSP.

There are many examples in which the RSP is followed, but there are also many examples corresponding to situations (i) and (iii). The RSP is in accord with intuitive feeling and certainly holds in the limiting case when reactivity is controlled by diffusion. However, the validity of the RSP is a matter of great controversy "..and diverse opinions have been expressed, from declaring the reactivity-selectivity principle as a universal law up to 'virtually useless in practice as a general rule'."

Real image The point(s) to which light rays converge as they emerge from a lens or mirror.

Real object The point(s) from which light rays diverge as they enter a lens or mirror.

Realized niche The multidimensional space that an animal actually occupies in the presence of competitors, predators, pathogens, and limited food. Contrast with fundamental niche.

Rearrangement stage The elementary reaction or reaction stage (of a molecular rearrangement) in which there is both making and breaking of bonds between atoms common to a reactant and a reaction product or a reaction intermediate. If the rearrangement stage consists of a single elementary reaction, this is a "rearrangement step".

Recapitulation. Summing up or repeating; hypothesis that an individual repeats its phylogenetic history in its development.

Receptacle The expanded tip of a pedicel or peduncle to which the floral organs are attached.

Receptive field The area of a group of sensory cells that, when stimulated, influences the activity of a given neuron.

Receptor-mediated endocytosis The type of endocytosis that involves a specific receptor on the plasma membrane that recognizes an extracellular molecule and binds with it.

Recessive A gene whose expression is masked when it is present in combination with a dominant allele.

Recessive gene Gene that is masked by its dominant allelic partner, having the recessive phenotype expressed only when both alleles for a given trait are recessive.

Recessive trait A trait caused by an allele, the phenotype of which is suppressed when it occurs with the dominant allele. Hence, two recessive alleles are necessary for the recessive phenotype to be displayed.

Recharge zone Area where water infiltrates into an aquifer.

Reciprocal altruism The trading of altruistic acts by individuals at different times, i.e., the payback to the altruist occurs some time after the receipt of the act.

Recognition surface The three-dimensional structure of a biological membrane surface that gives specificity due to macromolecules of various sizes and shapes that extend above the lipid bilayer.

Recolonization Revegetation; the reestablishment of a natural community following a natural or unnatural event that removed the existing community.

Recombination The formation of new combinations of alleles through exchange of sections of homologous chromosomes during the process of meiosis; sometimes refers to the combinations of new alleles that result from fertilization.

Rectal gland The excretory organ of elasmobranchs and the coelacanth located near the rectum; it excretes a hyperosmotic salt solution.

Rectifier A device for changing alternating current to direct current

Rectilinear propagation Traveling in a straight line.

Recycling Reprocessing of discarded materials into new, useful products; not the same as reuse of materials for their original purpose, but the terms are often used interchangeably.

Red algae Seaweeds that have a predominance of red pigments.

Red blood cell The type of blood cell that contains hemoglobin and no nucleus. During their formation in mammals, erythrocytes lose their nuclei; those of other vertebrates retain the nuclei.

Red clay A fine sediment that is the most common type of lithogenous sediment in the open ocean floor.

Red Queen hypothesis The hypothesis that sexual reproduction has evolved because the genetic variation that results is adaptive in the evolutionary "arms race" between hosts and their pathogens and parasites as well as between predators and prey.

Red tide A population explosion or bloom of minute, single-celled marine organisms called dinoflagellates. Billions of these cells can accumulate in protected bays where the toxins they contain can poison other marine life.

Redia A larval stage in the life cycle of flukes; it is produced by a sporocyst larva, and in turn gives rise to many cercariae.

Redox reactions Oxidation-reduction reaction; a chemical reaction involving the transfer of electrons from one molecule to another.

Reduced tillage systems Systems, such as minimum till, conserve-till, and no-till, that preserve soil, save energy and water, and increase crop yields.

Reduction The complete transfer of one or more electrons to a molecular entity (also called "electronation"), and, more generally, the reverse of the processes described under oxidation.

Reduction division The first nuclear division of meiosis during which the paired homologues migrate to opposite poles, resulting in cells with a reduction from a diploid to a haploid number of chromosomes.

Reductive elimination The reverse of oxidative addition.

Reef massive near-shore deposits of coral skeletal material

Reef crest The shallow outer edge of the reef slope of a coral reef.

Reef flat The wide and shallow upper surface of a coral reef.

Reef slope The outer, steep margin of a coral reef.

Reflectance The ratio of the light reflected from a surface to the light falling on it, expressed in percentage

Reflection The return of a wave from the boundary of a medium.

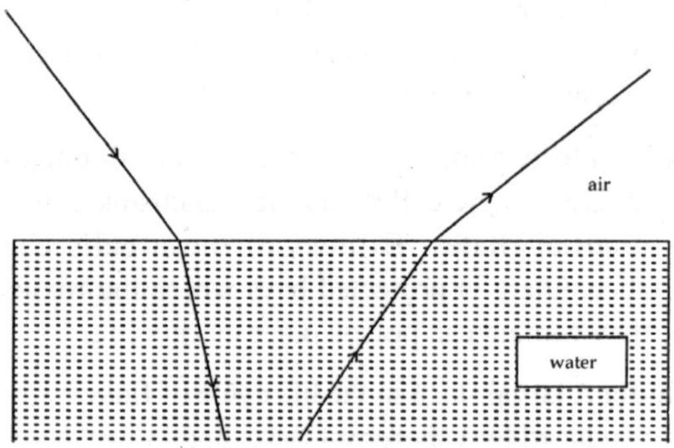

Fig. Refraction of light

Refraction law For a wavefront travelling through a boundary between two media, the first with a refractive index of N_1, and the other with one of N_2, the angle of incidence ø is related to the angle of refraction phi by

$N_1 \sin ø = N_2 \sin phi$.

Refraction The bending of a wave disturbance as it passes obliquely from one medium into another in which the disturbance has a different velocity

Refractory period In neurobiology, the period during which voltage-gated sodium channels in a neuron cannot be depolarized to threshold polarity by the influx of sodium in adjacent channels due to the hyperpolarizing effects of potassium exiting through potassium voltage-gated channels in that region.

Refuse-derived fuel Processing of solid waste to remove metal, glass, and other unburnable materials; organic residue is shredded, formed into pellets, and dried to make fuel for power plants.

Regelation The melting of a substance under pressure and the refreezing after the pressure is released.

Regeneration The ability of an organism to grow a body part that has been lost.

Regenerative farming Farming techniques and land stewardship that restore the health and productivity of the soil by rotating crops, planting ground cover, protecting the surface with crop residue, and reducing synthetic chemical inputs and mechanical compaction.

Regioselectivity A regioselective reaction is one in which one direction of bond making or breaking occurs preferentially over all other possible directions. Reactions are termed completely (100%) regioselective if the discrimination is complete, or partially (x%), if the product of reaction at one site predominates over the product of reaction at other sites. The discrimination may also semi-quantitatively be referred to as high or low regioselectivity.

In the past, the term "regiospecificity" was proposed for 100% regioselectivity. This terminology is not recommended owing to inconsistency with the terms stereoselectivity and stereospecificity.

Regular reflection Reflection from a polished surface in which scattering effects are negligible.

Regular symmetry Radial symmetry; having two or more planes through which the structure (flower) could be cut to result in mirror-image halves; actinomorphy.

Regulations Rules established by administrative agencies; regulations can be more important than statutory law in the day-to-day management of resources.

Regulative development Progressive determination and restriction of initially totipotent embryonic material.

Regulator genes Genes that control the activity of other genes that code for enzymes or structural proteins.

Rehabilitate land A utilitarian program to make an area useful to humans.

Reinforcement In learning, anything that alters the probability of behavior.

Relation A rule of correspondence between the set of values of one quantity to the values of another quantity, often (but not always) expressible as an equation.

Relative Colloquially "compared to". In the theory of relativity observations of moving observers are quantitatively compared. These observers obtain different values when measuring the same quantities, and these quantities are said to be relative. The theory, however, shows us how the differing measured values are precisely related to the relative velocity of the two observers. Some quantities are found to be the same for all observers, and are called invariant. One postulate of relativity theory is that the speed of light is an invariant quantity. When the theory is expressed in four dimensional form, with the appropriate choice of quantities, new invariant quantities emerge: the world-displacement $(x + y + z + ict)$, the energy-momentum four-vector, and the electric and magnetic potentials may be combined into an invariant four-vector. Thus relativity theory might properly be called invariance theory.

Relative deviation Percentage average deviation of a set of measurements

Relative humidity At any given temperature, a comparison of the actual water content of the air with the amount of water that could be held at saturation.

Relative uncertainty The uncertainty in a quantity compared to the quantity itself, expressed as a ratio of the absolute uncertainty to the size of the quantity. It may also be expressed as a percent uncertainty. The relative uncertainty is dimensionless and unitless.

Relativistic mass The mass of an object in motion with respect to the observer

Relativists Those who believe moral principles are always dependent on the particular situation.

Relativity principle The principle, employed by Einstein's relativity theories, that the laws of physics are the same, at least qualitatively, in all frames. That is, there is no frame that is better (or qualitatively any different) from any other. This principle, along with the constancy principle, constitute the founding principles of special relativity.

Relaxation Passage of an excited or otherwise perturbed system towards or into thermal equilibrium with its environment.

Releaser Simple stimulus that elicits an innate behavior pattern.

Releaser gland A gland in turbellarians that secretes a chemical that dissolves the attachment of the organism from a substrate.

Releasing factors Hormones or neurosecretions from the hypothalamus that travel either via the hypothalamic-pituitary portal system or along axons between the hypothalamus and the pituitary where they exert their effect in terms of production and release of hormones.

Remipedia A class of crustaceans whose members possess about 30 body segments and uniform, biramous appendages. This class contains a single species of cave-dwelling crustaceans from the Bahamas.

Renette An excretory structure found in some worms.

Renewable resources Resources normally replaced or replenished by natural processes; resources not depleted by moderate use; examples include solar energy, biological resources such as forests and fisheries, biological organisms, and some biogeochemical cycles.

Rennin A milk-clotting endopeptidase secreted by the stomach of some young mammals, including bovine calves and human infants.

Reorganization energy In a one-electron transfer reaction

$$A + D \rightleftharpoons A^{-} + D^{+}$$

the reorganization energy is the energy required for all structural adjustments (in the reactants and in the surrounding solvent molecules) which are needed in order that A and D assume the configuration required for the transfer of the electron.

Repolarized The reestablishment of polarity, especially the return of plasma membrane potential to resting potential after depolarization.

Repressible enzymes Enzymes whose production can be inhibited by a corepressor.

Repressor A compound that binds to and controls the regulator in gene regulation.

Reproductive barrier The factors that prevent one sexually propagating population from interbreeding and exchanging genes with another population.

Reproductive effort The energy expended and risk taken to reproduce, measured in terms of the decrease in ability of the organism to reproduce at a later time.

Reproductive isolation Occurs when individuals are prevented from mating, even though they may occupy overlapping ranges.

Reproductive strategy The reproductive patterns that are followed by a particular species.

Reproductive success The number of progeny born, or surviving progeny produced by an organism.

Reptiles Vertebrates with scales on their skin and leathery eggs that are laid on land. Marine reptiles include sea turtles, sea snakes, the marine iguana, and the saltwater crocodile.

Reptilia The class of vertebrates whose members have dry skin with epidermal scales and amniotic eggs that develop in terrestrial environments. Snakes, lizards, and alligators.

Repugnatorial glands Glands secreting a noxious substance for defense or offense, for example, as in the millipedes.

Reservoir Living or (rarely) nonliving means of maintaining an infectious agent in nature that can serve as source of infection for humans or domestic animals.

Residence time The length of time a component, such as an individual water molecule, spends in a particular compartment or location before it moves on through a particular process or cycle.

Residual magnetism Magnetism retained in a magnet after the magnetizing field has been removed.

Resilience The ability of a community or ecosystem to recover from disturbances.

Resilin The elastic protein found in some arthropod joints that stores energy and functions in jumping.

Resins Complex carbohydrates synthesized by certain plants in glands, canals, or ducts; insoluble in water. Used in various industrial products, including paints and varnishes. They may aid wood in resistance of decay.

Resin canal Long intercellular spaces present in the longitudinal system of vascular cells of pine, spruce, larch, and Douglas fir. Adjacent parenchymal cells secrete the resin into the canals.

Resistance The ratio of the potential difference across a conductor to the magnitude of current in it.

Resistivity A proportionality constant that relates the length and cross-sectional area of a given electric conductor to its resistance, at a given temperature.

Resolution of forces The resolving of a single force into component forces acting in given directions on the same point.

Resonance In the context of chemistry, the term refers to the representation of the electronic structure of a molecular entity in terms of contributing structures. Resonance among contributing structures means that the wavefunction is represented by "mixing" the wavefunctions of the contributing structures. The concept is the basis of the quantum mechanical valence bond methods. The resulting stabilization is linked to the quantum mechanical concept of "resonance energy". The term resonance is also used to refer to the delocalization phenomenon itself.

Resonance effect This is the term most commonly used to describe the influence (on reactivity, spectra, etc.) of a substituent through electron delocalization into or from the substituent. The use of the term obviates the need to attempt to distinguish between the operation of the mesomeric effect and the electromeric effect. (An alternative term with essentially the same meaning is "conjugative effect". At one time "tautomeric effect" was also used, but was abandoned because tautomerism implies reorganization of the atomic nuclei.) The effect is symbolized by R.

Resonance energy The difference in potential energy between the actual molecular entity and the contributing structure of lowest potential energy. The resonance energy cannot be measured, but only estimated, since contributing structures are not observable molecular entities.

Resource defense polygyny Males control access to females indirectly by monopolizing critical resources.

Resource partitioning In a biological community, various populations sharing environmental resources through specialization, thereby reducing direct competition.

Resource scarcity A shortage or deficit in some resource.

Respiration Gaseous interchange between an organism and its surrounding medium. In the cell, the release of energy by the oxidation of food molecules.

Respiratory fibrotic agents Special class of irritants, including chemical reagents and particulate materials, that damages the lungs, causing scar tissue formation that lowers respiratory capacity.

Respiratory pigment Organic compounds that have either metallic copper or iron with which oxygen combines.

Respiratory tree A pair of tubules attached to the rectum of a sea cucumber that branch through the body cavity and function in gas exchange.

Response to selection (R) The difference in the mean phenotypic value between a group of progeny and the mean value for the parents of those progeny.

Rest mass The mass of an object not in motion.

Resting membrane potential The potential difference that results from the separation of charges along the plasma membrane of a neuron.

Resting potential The unequal distribution of charges on either side of a neuron membrane that is not undergoing an action potential.

Restore ecosystems To reinstate an entire community of organisms to naturally occurring association.

Restriction endonuclease. An enzyme that cleaves a DNA molecule at a particular base sequence.

Resultant A vector representing the sum of several vector components Resultant force The single force that has the same effect as two or more forces applied simultaneously at the same point.

Rete mirabile A network of small blood vessels so arranged that the incoming blood runs countercurrent to the outgoing blood and thus makes possible efficient exchange between the two bloodstreams. Such a mechanism serves to maintain the high concentration of gases in the fish swim bladder.

Rete system Highly branched system of tubules in acanthocephalans, lying on longitudinal muscles or between longitudinal and circular muscles; thought to assist in contraction stimuli.

Reticular Resembling a net in appearance or structure.

Reticulate venation Netted venation; the arrangement of veins in a leaf that resembles a net; characteristic of dicot leaves.

Reticuloendothelial (RE) system Total complement of fixed macrophages in the body, especially reticular connective tissue and the lining epithelium of the blood vascular system. Some authorities also include the phagocytic white blood cells.

Reticulopodia Pseudopodia that branch and rejoin extensively.

Reticulopodium A pseudopodium that forms a threadlike branched mesh and contains axial microtubules.

Retina The posterior sensory membrane of the eye that receives images.

Retrofection Process of reinfection, whereby juvenile nematodes hatch on the skin and reenter the body before molting to third-stage larvae.

Retroperitoneal External to the peritoneum.

Retting The process that frees flax fibers by allowing microbial decomposition to break down the outer part of the stem.

Revegetate The natural or induced replacement of plants into a cleared area; the recurrence of the same plant community that existed prior to clearing.

Reverse bias Voltage applied to a semiconductor P-N junction that reduces the electron current across the junction.

Reversing thermometer A thermometer that permits the measurement of temperature at depth, since the mercury column breaks when the thermometer is turned upside down.

Rhabdite A rodlike structure in the cells of the epidermis or underlying parenchyma in certain tubellarians that are discharged in mucous secretions.

Rheoreceptor A sensory organ of aquatic animals that responds to water current.

Rheostat A variable resistance.

Rhinencephalon The part of the vertebrate brain that comprises the septum, cingulate gyrus, hypothalamus, amygdala, hippocampus, and the fornix; also called the limbic system.

Rhinophore Chemoreceptive tentacles in some molluscs (opisthobranch gastropods).

Rhizoid Rootlike absorptive structures on the underside of certain gametophytes.

Rhizome A fleshy, horizontal underground stem.

Rhodophyta The division of protists commonly called the red algae.

Rhodopsin Light-sensitive substance that occurs in the rods of the retina; visual purple.

Rhombogen Stage in the life cycle of a dicyemid mesozoan.

Rhopalium A sensory structure at the margin of the scyphozoan medusa. It consists of a statocyst and a photoreceptor (phylum Cnidaria).

Rhoptries Club-shaped bodies in Apicomplexa composing one of the structures of the apical complex; open at anterior and apparently functioning in penetration of host cell.

Rhynchocoel In nemerteans, the dorsal tubular cavity that contains the inverted proboscis. It has no opening to the outside.

Rhyniophyta An extinct division of early, vascular land plants characterized by their small stature, dichotomous branching, lack of leaves, roots, and possession of terminal sporangia (e.g., Rhynia, Cooksonia).

Ribbon (or nemertean) worms (phylum Nemertea)Invertebrates with a complete digestive tract, a true circulatory system, and a long proboscis to capture prey.

Ribonucleic acid (RNA) The nucleic acid formed from DNA and involved in protein synthesis; nucleotide of chain of phosphates, ribose sugars, and purine and pyrimidines.

Ribose A 5-carbon sugar important in RNA and many other compounds.

Ribosomal RNA A form of ribonucleic acid that makes up a portion of ribosomes.

Ribosome Subcellular structure composed of protein and ribonucleic acid. May be free in the cytoplasm or attached to the membranes of the endoplasmic reticulum; functions in protein synthesis.

Rickettsia A very small and specialized pathogenic bacterium transmitted by ticks, mites, and fleas.

Ridge and rise system the interconnecting chain of seafloor mountains that trace the edges of crustal plates and the sites of new oceanic crust production.

Rift A crack in the earth's crust formed as pieces of the crust separate.

Right-hand rule A trick for right-handed coordinate systems to determine which way the cross product of two 3-vectors will be directed. There are a few forms of this rule, and it can be applied in many ways. If u and v are two vectors which are not parallel, then u cross v is a vector which is directed in the following manner: orient your right hand so that your thumb is perpendicular to the plane defined by the vectors u and v. If you can curl your fingers in the direction from vector u to vector v, your thumb will point in the direction of u cross v. (if it doesn't, the vector is directed in the opposite direction.) This has immediate application for determining the orientation of the z-axis basis unit vector, k, in terms of the x- and y-axes' basis unit vectors; curl your right hand in the direction of i to j, and your thumb will point in the direction of i cross j = k.

The rule is also applicable in several practical applications, such as determining which way to turn a screw, etc. There is also a left-hand rule, which exhibits opposite chirality.

Rill erosion The removing of thin layers of soil as little rivulets of running water gather and cut small channels in the soil.

Ring-porous wood Secondary xylem characterized by larger vessels and tracheids being produced early in the season (or following favorable growing conditions). Thus each term of growth activity is seen as a ring.

Risk Probability that something undesirable will happen as a consequence of exposure to a hazard.

Risk assessment Evaluation of the short-term and long-term risks associated with a particular activity or hazard; usually compared to benefits in a cost-benefit analysis.

Risk-sensitive foraging A model that incorporates variation in prey encounter rates to predict where predators forage. Depending on their internal state and other factors, predators may be risk-prone, foraging where encounter rates are variable, or risk-averse, foraging where encounter rates are relatively constant.

Ritualization. In ethology, the evolutionary modification, usually intensification, of a behavior pattern to serve communication.

RNA Ribonucleic acid; nucleic acid used for transcription and translation of the genetic code found on DNA molecules.

RNA world. Hypothetical stage in the evolution of life on earth in which both catalysis and replication were performed by RNA, not protein enzymes and DNA.

Roche limit The position around a massive body where the tidal forces due to the gravity of the primary equal or exceed the surface gravity of a given satellite. Inside the roche limit, such a satellite will be disrupted by tides.

Rock cycle The process whereby rocks are broken down by chemical and physical forces; sediments are moved by wind, water, and gravity, sedimented and reformed into rock, and then crushed, folded, melted, and recrystallized into new forms.

Rockweeds Brown algae such as Fucus that are inhabitants of rocky shores in temperate zones.

Rod A class of vertebrate photoreceptors that, unlike cone cells, is usually insensitive to color.

Rod cell A type of light receptor that is responsible for color vision.

Romana's sign Symptoms of recent infection by Trypanosoma cruzi, consisting of edema of the orbit and swelling of the preauricular lymph node.

Romanovsky stain Complex stain, based on methylene blue and eosin, used to stain blood cells and hemoparasites. Wright's and Giemsa's stains are two common examples.

Root cap A thimble-shaped group of cells found at the tip of roots; it functions to protect the meristem.

Root hair A root epidermal cell that functions in water absorption.

Root nodule Gall-like structures on the roots of legumes that contain symbiotic nitrogen-fixing bacteria.

Root pressure A positive pressure in the xylem due to a negative solute potential and closed stomata.

Root system The portion of the plant formed from the growth of the radicle and its subsequent branching; generally the underground portion of the plant.

Rootbound The condition of a restricted root system caused by excess growth in too small a volume of soil; sometimes called potbound.

Root-mean-square (rms) current The effective value of an alternating current; the square root of the mean of the instantaneous values squared.

Rorquals Blue, fin, and the other baleen whales that display long grooves on the underside.

Rostellum Projecting structure on scolex of tapeworm, often with hooks.

Rostrum Dorsal part of capitulum projecting over chelicerae in acarines.

Rotary motion Motion of a body about an internal axis.

Rotate To rotate is to turn about a point which is (usually) held fixed. (It is possible to rotate about a translating point.) There are two contexts for this: First, an object can rotate either about another object (orbit) or about an axis through that object (spin). Second, if one rotates a coordinate system, then the origin of that coordinate system is held fixed and the axes are rotated around the origin.

Rotational equilibrium The state of a body in which the sum of all the clockwise torques in a given plane equals the sum of all the counterclockwise torques about a pivot point

Rotational inertia The property of a rotating object that resists changes in its angular velocity.

Rotifera The phylum of aschelminths that has members with a ciliated corona surrounding a mouth; muscular pharynx (mastax) present with jawlike features; nonchitinous cuticle; parthenogenesis common; both freshwater and marine species.

Rough ER A portion of the endoplasmic reticulum containing ribosomes.

R-selected Populations that are maintained below the carrying capacity of the environment.

R-strategist Species of organism that uses a survival and reproductive "strategy" characterized by high fecundity, high mortality, short longevity. Populations are controlled by density-independent factors.

RUBP carboxylase The enzyme that fixes CO_2 in the Calvin cycle of photosynthesis.

Rubpribulose 1, 5-bisphosphate, the 5-carbon compound that combines with CO_2 in the Calvin cycle.

Ruffles Slender projections of the exterior surface of a dicyemid mesozoan.

Rumen The large first compartment of the stomach of ruminant mammals.

Ruminant animals Cud-chewing animals, such as cattle, sheep, goats, and buffalo, with multichambered stomachs in which cellulose is digested with the aid of bacteria.

Runaway selection Selection for male ornaments that happens due to the genetic correlation, and resulting positive feedback relationship, between the trait and the preference for the trait.

Runner A horizontal stem that grows along the surface of the ground; also known as a stolon.

Runoff The excess of precipitation over evaporation; the main source of surface water and, in broad terms, the water available for human use.

Run-of-the-river flow Ordinary river flow not accelerated by dams, flumes, etc. Some small, modern, high-efficiency turbines can generate useful power with run-of-the-river flow or with a current of only a few kilometers per hour.

Rural area An area in which most residents depend on agriculture or the harvesting of natural resources for their livelihood.

Rydberg constant (rydberg) A constant which governs the relationship of the spectral line features of an atom through the rydberg formula. For hydrogen, it is approximately $1.097 \times 10^7\ m^{-1}$.

Rydberg formula (rydberg) A formula which describes all of the characteristics of hydrogen's spectrum, including the balmer, lyman, paschen, brackett, and pfund series.

For the transition between an electron in orbital m to one in orbital n (or the reverse), the wavelength λ involved is given by

$$1/\lambda = r\,(1/m_2 - {}^1/n_2).$$

S

S curve A curve that depicts logistic growth; called an S curve because of its shape.

Saccule Small chamber of the membranous labyrinth of the inner ear.

Sacrum Bone formed by fused vertebrae to which pelvic girdle is attached; pertaining to the sacrum.

Saefftigen's pouch Internal, muscular sac near the posterior end of a male acanthocephalan. It contains fluid that aids in manipulating the copulatory bursa.

Sagittal Pertaining to the median anteroposterior plane that divides a bilaterally symmetrical organism into right and left halves.

Salicylic acid A compound with pain-relieving characteristics, found in willow bark and other plants; the basic ingredient of aspirin.

Salience In the context of classical conditioning, the fact that animals can learn to make certain associations, depending upon which stimuli are involved, more rapidly than other associations.

Salinity Amount of dissolved salts (especially sodium chloride) in a given volume of water.

Salinization A process in which mineral salts accumulate in the soil, killing plants; occurs when soils in dry climates are irrigated profusely.

Saliva The enzyme-containing secretion of the salivary glands.

Salivarium Chamber in buccal cone of acarines into which salivary ducts open.

Salivary gland Any of the glands in molluscs, vertebrates, and other animals that release digestive enzymes into the mouth.

Salmon ranching Practice in which cultured juvenile salmon are released into fresh water and allowed to migrate to sea so they can be harvested later when they return as adults.

Salps Pelagic tunicates with a transparent, cylindrical body, sometimes forming long colonies.

Salt The reaction product of an acid and a base; dissociates in water solution to negative and positive ions, but not $H+$ or OH^-.

Salt gland An orbital gland of many reptiles and birds that secretes a hyperosmotic nacl or kcl solution an important osmoregulatory organ, especially for marine species.

Salt marsh A grassy area that extends along the shores of estuaries and sheltered coasts in temperate and subpolar regions.

Salt wedge A layer of denser, saltier seawater that flows along the bottom in estuaries.

Saltatory conduction type of nerve impulse conduction in which the impulse seems to jump from one neurofibril node to the next.

Saltwater intrusion Movement of saltwater into freshwater aquifers in coastal areas where groundwater is withdrawn faster than it is replenished.

Samara A simple, dry indehiscent fruit with the pericarp bearing winglike outgrowths; winged fruit of maple.

Sand Soil inorganic particles in the.02—2.0 mm size range.

Sand dollars Sea urchins with a flat, round test and short spines that live partly buried in soft sediments.

Sand fly Member of the dipteran subfamily Phlebotominae, family Psychodidae; sometimes also applied to Simuliidae (New Zealand) and Ceratopogonidae (Caribbean).

Sandstone A sedimentary rock composed of sand.

Saponin A glycoside with a steroid molecule as the active component, such as diosgenin from yams.

Saprobe An organism deriving its food from the dead body or nonliving products of another organism.

Saprophagous Feeding on decaying matter; saprobic; saprozoic.

Saprophyte A plant living on dead or decaying organic matter.

Saprophytic Plant living on dead organic matter.

Saprozoic nutrition Nutrition of an animal by absorption of dissolved salts and simple organic nutrients from surrounding medium. Also refers to feeding on decaying organic matter.

Sapwood The functional secondary xylem found between the vascular cambium and the nonfunctional heartwood in the center of the trunk or branch.

Sarcocystin Powerful toxin produced by zoitocysts of Sarcocystis.

Sarcodina The protozoan subphylum where members have pseudopodia for movement and food gathering; naked or with shell or test; mostly free living.

Sarcolemma The thin, noncellular sheath that encloses a striated muscle fiber.

Sarcomastigophora The protozoan phylum where members possess flagella, pseudopodia, or both for locomotion and feeding; single type of nucleus.

Sarcomere The contractile unit of a myofibril. The repeating units, delimited by the Z bands along the length of the myofibril.

Sarcoplasm The clear, semifluid cytoplasm between the fibrils of muscle fibers.

Sarcoptic mange Disease caused by mites of the genus Sarcoptes. Also called scabies.

Sargasso Sea The region of the Atlantic Ocean north of the West Indies that is characterized by floating masses of Sargasso weed, a brown alga.

Satellite Posterior member of a pair of gregarines in syzygy.

Saturated fat A fat in which all the carbons in the fatty acids are connected by single bonds, thereby having the maximum number of hygrogen atoms.

Saturation light intensity The light intensity that maximizes the photosynthetic rate.

Saturation point The maximum concentration of water vapor the air can hold at a given temperature.

Saturation transfer A term used in nuclear magnetic resonance. When a nucleus is strongly irradiated, its spin population may partly be transferred to another nucleus by an exchange process.

Sauropterygians Mesozoic marine reptiles.

Savanna Found between the tropics and deserts, this vegetative community is composed of extensive grasslands with scattered deciduous trees.

Saxitoxin a paralytic toxic produced by dinophytes that accumulates in the butter clam (Saxadoma)

Saytzeff rule Dehydrohalogenation of secondary- and tertiary-alkyl halides proceeds by the preferential removal of the -hydrogen from the carbon that has the smallest number of hydrogens. Originally formulated by A. Saytzeff (Zaitsev) to generalize the orientation in - elimination reactions of alkyl halides, this rule has been extended and modified, as follows: When two or more olefins can be produced in an elimination reaction, the thermodynamically most stable alkene will predominate. Exceptions to the Saytzeff rule are exemplified by the Hofmann rule.

Scabies Disease caused by mites of the genus Sarcoptes. Also called sarcoptic mange.

Scalar quantity A quantity that is completely specified by a magnitude

Scale The relative size of a quantity as compared to some other quantity in a given context.

Scale-limited A measuring instrument is said to be scale-limited if the experimental uncertainty in that instrument is smaller than the smallest division readable on its scale. Therefore the experimental uncertainty is taken to be half the smallest readable increment on the scale.

Scalid A set of complex spines found on the kinorhynchs, loriciferans, priapulans, and larval nematomorphs, with sensory, locomotor, food capture, or penetrant function.

Scan sampling A method for recording behavior observations that involves looking briefly at each animal at prescribed time intervals (e.g., every minute or every five minutes) to record their activity at the time of the sample.

Scanning electron microscope The type of microscope in which an electron beam, instead of light, forms a three-dimensional image for viewing, allowing much greater magnification and resolution.

Scaphopoda A class of molluscs whose members have a tubular shell that is open at both ends. Possess tentacles but no head. Dentalium.

Scarabaeiform Grublike larvae with lightly sclerotized cuticle; found in some coleopteran families.

Scarification A mechanical or chemical degradation of a hard surface, such as seed coats, so that oxygen and water can penetrate the hard layers.

Scavenger A substance that reacts with (or otherwise removes) a trace component (as in the scavenging of trace metal ions) or traps a reactive reaction intermediate.

Scent gland A gland located around the feet, face, or anus of many mammals; secretes pheromones, which may be involved with defense, species and sex recognition, and territorial behavior. Musk gland.

Schistosomule Juvenile stage of a blood fluke, between a cercaria and an adult; a migrating form taking the place of a metacercaria in the life cycle.

Schizeckenosy System of waste elimination found in some mites with a blindly ending midgut; the lobe breaks free from the ventriculus and is expelled through a split in the posterodorsal cuticle.

Schizocarp A dry indehiscent fruit that splits into two one-seeded halves at maturity.

Schizocoel A coelom formed by the splitting of embryonic mesoderm.

Schizocoelous mesoderm formation Embryonic formation of the mesoderm as cords of cells between ectoderm and endoderm; splitting of these cords results in the coelomic space.

Schizocoely Coelom formation accomplished by a split in the mesoderm during embryonic development of some animals (protostomes).

Schizogony Form of asexual reproduction in which multiple mitoses take place, followed by simultaneous cytokineses, resulting in many daughter cells at once.

Schizont Cell undergoing schizogony, in which nuclear divisions have occurred but cytokinesis is not completed; in its late phase sometimes called a segmenter.

School a well-defined social organization of marine animals consisting of a single species with all members of a similar size.

Schüffner's dots Small surface invaginations that appear as stippling on the membrane of an erythrocyte infected with Plasmodium vivax after Romanovsky staining.

Schwarzschild radius The radius R of the event horizon for a schwarzschild black hole of mass m is given by (in geometrized units) R= 2 m. In conventional units,

$R = 2\ g\ m/c^2$.

Science A methodical, precise, objective way to study the natural world.

Scientific method The objective process of approaching a problem. Involves hypothesis establishment, testing and observing the results, reevaluation of the hypothesis in light of new knowledge, and retesting to seek repeatability and thus validity of the hypothesis.

Scientific name A Latinized binomial (genus and species) unique to each identified organism.

Scientific notation A style of expressing numbers in a convenient manner. The number 524 is written as 5.24e2 or 5.24×10^2 which is 5.24 multiplied by 10 squared: 5.24 x 10 x 10 = 52.4 x 10 = 524. While this may not seem convenient for numbers near one or ten, it is more obviously convenient for numbers like 0.000000524, which is 5.24e-7, or 5.24×10^{-7}, or 5.24 divided by 10, seven times. (The negative in the exponent means that you should divide instead of multiply.)

Scintillation counter A device that counts the impacts of charged subatomic particles on a fluorescent screen by means of a photomultiplier tube.

Scion A small twig or bud that is grafted to a stock.

Sclereid A sclerenchyma cell with a thick, lignified secondary wall having many pits; variable in form but not usually elongated.

Sclerenchyma The tissue type characterized by thick, sclerified cell walls; includes both fibers and sclereids (stone cells).

Sclerite A hard chitinous or calcareous plate or spicule; one of the plates making up the exoskeleton of arthropods, especially insects.

Scleroblast An amebocyte specialized to secrete a spicule, found in sponges.

Sclerocyte An amebocyte in sponges that secretes spicules.

Sclerophyllous Vegetation characterized by thick leathery leaves with abundant sclerenchyma cells; vegetation of a chaparral.

Sclerosponges Sponges with a massive calcareous skeleton.

Sclerotic Pertaining to the tough outer coat of the eyeball.

Sclerotin Highly resistant and insoluble protein occurring in the cuticle of arthropods; also thought to occur in structures secreted by various other animals, such as in the eggshells of some trematodes, in which stabilization of the protein is achieved by orthoquinone crosslinks between free imino or amino groups of the protein molecules.

Sclerotium A fungal resting body resistant to unfavorable conditions; a firm, hardened mass of hyphae (or a hardened plasmodium of a slime mold) that will germinate on the return of favorable conditions.

Sclerotization Process of hardening of the cuticle of arthropods by the formation of stabilizing cross linkages between peptide chains of adjacent protein molecules.

Scolex The attachment or holdfast organ of a tapeworm, generally considered the anterior end; it is used to adhere to the host.

Scoliosis Lateral curvature of the spine.

Scramble polygyny A mating system in which males actively search for mates without overt competition.

Scrotum The pouch that contains the testes in most mammals.

Scrub typhus Rickettsial disease transmitted by certain chigger mites.

Scuba (self-contained underwater breathing apparatus) The use of tanks of compressed air for breathing underwater.

Scutellum The single cotyledon in grass seeds.

Scutum Large, anteriodorsal sclerite on a tick or mite.

Scyphistoma The polyp stage of a scyphozoan (phylum Cnidaria); develops from a planula and produces ephyrae by budding.

Scyphozoa A class of Cnidarians whose members have prominent medusae. Gametes are gastrodermal in origin and are released to the gastrovascular cavity. Nematocysts are present in the gastrodermis. Polyps are small. Aurelia.

Sea A wave that has a sharp peak and a relatively flat trough. Seas are found in areas where waves are generated by the wind.

Sea anemones Anthozoans that consist of one large polyp.

Sea cucumbers Echinoderms with a soft, elongate body that lacks spines.

Sea spiders Arthropods that have a reduced body and four pair of legs.

Sea squirts Tunicates with a sac-like, attached body as adults.

Sea stars Echinoderms with five or more radiating arms and tube feet that are used in locomotion.

Sea urchins Echinoderms with a round or flattened test and movable spines.

Sea-floor spreading The process by which new sea floor is formed as it moves away from spreading centers in mid-ocean ridges.

Seagrasses Grass-like flowering plants such as eelgrass that are adapted to live at sea.

Seamount A submarine volcano in the abyssal plain.

Search image The hypothetical mental image of a prey species that a predator forms as it improves in its ability to capture that particular species.

Seasonal Fluctuations in environmental factors such as temperature over an annual cycle.

Seaweed A common name for various macroalgae.

Sebaceous A type of mammalian epidermal gland that produces a fatty substance.

Sebum Oily secretion from the sebaceous gland.

Secernentea The class of nematodes formerly called Phasmidea. Examples: Ascaris, Enterobius, Necator, Wuchereria.

Second; s The fundamental SI unit of time, defined as the period of time equal to the duration of 9192631770 periods of the radiation corresponding to the transition between two hyperfine levels of the ground state of the Cesium-133 atom.

Second law of photoelectric emission The kinetic energy of photoelectrons is independent of the intensity of the incident light.

Second law of thermodynamics It is not possible for an engine to transfer heat from one body to another at a higher temperature unless work is done on the engine.

Second World The industrialized, socialist, centrally planned economy nations of Eastern Europe and the former Soviet Union and its allies.

Secondary A transformer output winding in which the current is due to inductive coupling with another winding called the primary.

Secondary axis Any line other than the principal axis drawn through the center of curvature of a mirror or the optical center of a lens.

Secondary cell wall A cellulosic wall, often impregnated.

Secondary compounds Organic molecules synthesized by certain species of plants and not thought to be directly involved in essential metabolism.

Secondary consumer An animal that feeds on other consumers.

Secondary emission Emission of electrons as a result of the bombardment of an electrode by high-velocity electrons.

Secondary growth Growth derived from secondary or lateral meristems, the vascular and cork cambiums; secondary growth results in an increase in diameter or circumference; contrasted with primary growth, which results in an increase in length.

Secondary immune response The immune response that follows a second exposure to a specific antigen.

Secondary lamella a small extension of a gill filament containing blood capillaries for gas exchange

Secondary palate A plate of bone that separates the nasal and oral cavities of mammals and some reptiles.

Secondary phloem All phloem tissue formed by the vascular cambium in woody tissues.

Secondary pollutants Chemicals modified to a hazardous form after entering the air or that are formed by chemical reactions as components of the air mix and interact.

Secondary product Chemical compounds synthesized by plants or fungi but not critical for the basic metabolic functions of that organism; often functioning to deter predators or attract pollinators; a secondary metabolite.

Secondary recovery technique Pumping pressurized gas, steam, or chemical- containing water into a well to squeeze more oil from a reservoir.

Secondary succession Revegetation of cleared land; return to previous community structure.

Secondary treatment Bacterial decomposition of suspended particulates and dissolved organic compounds that remain after primary sewage treatment.

Secondary wall The innermost layer of a cell wall formed after cell elongation has ceased; often characterized by the deposition of lignin.

Secondary xylem All xylem tissue formed by the vascular cambium in woody plants.

Secure landfill A solid waste disposal site lined and capped with an impermeable barrier to prevent leakage or leaching. Drain tiles, sampling wells, and vent systems provide monitoring and pollution control.

Sedatives Compounds that produce a relaxed, euphoric state—opium and its derivatives, morphine and heroin, are sedatives.

Sedentary Stationary, sitting, inactive; staying in one place.

Sediment Loose material such as sand and mud that settles on the bottom.

Sedimentary rock Deposited material that remains in place long enough or is covered with enough material to compact into stone; examples include shale, sandstone, breccia, and conglomerates.

Sedimentation The deposition of organic materials or minerals by chemical, physical, or biological processes. Sediments can be transported from their source to their place of deposition by gravity, wind, water, or ice. If subjected to sufficient heat, pressure, or chemical reactions, sediments can solidify into sedimentary rock.

Seed A matured ovule containing an embryo and food supply and covered by a seed coat.

Seed bank Storage facilities for seeds of domesticated plants and wild relatives; facilities for preserving genetic diversity.

Seed coat The outer layer of a seed that is developed from the integuments of the ovule; the testa.

Seed ferns Fossil plants with fernlike foliage but producing seeds belonging to the division Pteridospermophyta.

Seed plant Common term for gymnosperms and angiosperms.

Seedling The embryonic product of the germination of a seed; the young shoot and root axis.

Seed-scale complex The spirally arranged scales on a female strobilus in gymnosperms.

Segmentation 1. In many animal species, a series of body units that may be externally similar to, or quite different from, one another.

2. The oscillating back-and-forth movement in the small intestine that mixes food with digestive secretions and increases the efficiency of absorption.

Segmented worms Invertebrates that display an elongate body with distinct segmentation and a digestive tract that lies in a coelom.

Seisonidea A class of rotifers containing members that are commensals of crustaceans; large and elongate body with rounded corona. Example: Seison.

Seive cell The organic solute-conducting cell of the phloem in gymnosperms.

Selection differential (S) The difference in mean phenotypic value for a selected group of breeding animals and the mean value in the population from which they are drawn.

Selection pressure The tendency for natural selection to occur; natural selection occurs whenever some genotypes are more fit than other genotypes.

Selective cutting Harvesting only mature trees of certain species and size; usually more expensive than clear-cutting, but it is less disruptive for wildlife and often better for forest regeneration.

Selective permeability The ability of the plasma membrane to let some substances in and keep others out.

Selectively permeable. Permeable to small particles, such as water and certain inorganic ions, but not to larger molecules.

Selectively permeable membrane a membrane that is permeable to small molecules, usually H_2O, O_2, and CO_2, but not permeable to larger molecules or ions.

Selectivity factor A quantitative representation of selectivity in aromatic substitution reactions (usually electrophilic, for monosubstituted benzene derivatives). If the partial rate factor, f, expresses the reactivity of a specified position in the aromatic compound PhX relative to that of a single position in benzene, then the selectivity factor Sf (expressing discrimination between p- and m-positions in PhX) is defined as

$$S_f = \lg (f_p^X / f_m^X).$$

Selectivity The discrimination shown by a reagent in competitive attack on two or more substrates or on two or more positions in the same substrate. It is quantitatively expressed by ratios of rate constants of the competing reactions, or by the decadic logarithms of such ratios.

Self-incompatibility Condition of a flower that cannot successfully complete the reproductive process with pollen produced by its own stamens.

Self-inductance The ratio of the induced emf across a coil to the rate of change of current in the coil.

Self-pollinating plant A plant that has its own pollen fall on its own stigma.

Self-pollination Transfer of pollen from stamen to stigma within the same plant.

Self-propagating A description of the events occurring during an action potential, with each regional depolarization by sodium voltage-gated channels causing a similar event at an adjacent area downstream.

Self-regulating population A population with a growth rate that is dependent on its own numbers.

Self-shading Reduction in the amount of light available to phytoplankton that live below other phytoplankton.

Self-stimulation A situation in which a subject animal is able to introduce stimuli to its nervous system.

Semantic communication Of or relating to meaning of signals; specifically used to denote the use of different alarm signals to warn about different predators.

Semelparity The production of offspring by an organism once in its life.

Semen The thick, whitish secretion of the reproductive organs in the male; composed of sperm and secretions from the prostate, seminal vesicles, and various other glands and ducts.

Semicircular canals Bony canals in the vertebrate inner ear; the hair cells within the canals detect changes in the direction of movement of the head.

Semi-conductor A material which has a medium-sized energy gap is a semi-conducting material. A semi-conductor may conduct electricity; while it conducts better than an insulator, it does not conduct as well as a conductor.

Semidiurnal tide A tidal pattern with two high and two low tides each day.

Seminal receptacle A structure in the female reproductive system that stores sperm received during copulation (e.g., many insects and annelids).

Seminal vesicle 1. One of the paired accessory glands of the reproductive tract of male mammals. It secretes the fluid medium for sperm ejaculation (phylum Chordata).

2. A structure associated with the male reproductive tract that stores sperm prior to its release.

Seminiferous Pertains to the tubules that produce or carry semen in the testes.

Seminiferous tubule The male duct that conveys semen.

Semipermeable Permeable to small particles, such as water and certain inorganic ions, but not to larger molecules.

Senescence The aging process, usually characterized by the loss of some functional capacity, including reproduction.

Sensation The process of transducing environmental stimuli or energy into action potentials.

Sensilla Modifications of the exoskeleton of an arthropod that, along with nerve cells, form sensory receptors.

Sensillum, A small sense organ, especially in the arthropods.

Sensitive period The time interval when an animal can develop an imprinting attachment. More broadly, in behavior development, this refers to time intervals when particular events must occur for proper ontogenetic sequencing.

Sensitization 1. Enhanced responsiveness to a repeated stimulus.

2. Strengthening of a response that was initially produced via a CS resulting from a pairing with a US and UR.

3. A stimulus priming the animal to pay particular attention to what follows.

Sensory (afferent) neuron or nerve A neuron or nerve that conducts an impulse from a receptor organ to the central nervous system.

Sensory adaptation A process that occurs at the level of the sensory receptors and that consists of a slowing down or cessation of nerve impulses transmitted to the central nervous system.

Sensory deprivation Withholding all or a specified portion of the sensory input that an animal would normally be receiving.

Sensory filter Neural circuits that selectively transmit some features of a sensory input and ignore other features.

Sensory neuron A neuron that is modified to respond to a particular set of stimuli.

Sepal The flower part attached outside the others, enclosing the flower when in bud.

Septa Peritoneal (mesodermal) sheets separating adjacent segments, as in annelids, or body divisions, as in chaetognaths.

Septate Divided by crosswalls into cells.

Septicemia Systemic infection where a pathogen is present in the circulating blood.

Septum, A wall between two cavities.

Sequential hermaphroditism The type of hermaphroditism that occurs when an animal is one sex during one phase of its life cycle and an opposite sex during another phase.

Seral stage One stage of the communities in an ecological succession.

Sere An entire successional sequence in an ecosystem (e.g., the sequence of stages in the succession of a lake to a climax forest).

Serially homologous Metameric structures that have evolved from a common form; the biramous appendages of crustaceans are serially homologous.

Series circuit An electric circuit in which the components are arranged to provide a single conducting path for current.

Series resonance A condition in which the impedance of a series circuit containing resistance, inductance, and capacitance is equal to the resistance of the circuit and the voltage across the circuit is in phase with the current

Seriously undernourished Those who receive less than 80 percent of their minimum daily caloric requirements; are likely to suffer permanently stunted growth, mental retardation, and other social and developmental disorders.

Serosa The outer embryonic membrane of birds and reptiles; chorion. Also, the peritoneal lining of the body cavity.

Serotonin A phenolic amine, found in the serum of clotted blood and in many other tissues, that possesses several poorly understood metabolic, vascular, and neural functions; 5-hydroxytryptamine.

Serous Watery, resembling serum; applied to glands, tissue, cells, fluid.

Serum The liquid that separates from the blood after coagulation; blood plasma from which fibrinogen has been removed. Also, the clear portion of a biological fluid separated from its particulate elements.

Sesamoid Resembling a sesame grain. A small bone formed in tendons.

Sessile Attached at the base; fixed to one spot, not able to move about.

Seta Hairlike modifications of an arthropod's exoskeleton that may be set into a membranous socket. Displacement of a seta initiates a nerve impulse in an associated cell.

Sex chromosome one of a pair of chromosomes whose composition determines gender.

Sex hormone A hormone that controls the timing of reproduction and sexual characteristics in vertebrates.

Sex-influenced traits Traits that behave as if they are dominant in one sex but recessive in the other sex.

Sexual Pertaining to the fusion of gametes; sexual reproduction.

Sexual imprinting The process by which young of many species establish an attachment to opposite sex conspecifics; the effect manifests itself later in life during the process of finding mates.

Sexual reproduction Reproduction involving fusion of gametes.

Sexual selection Selection in relation to mating. It is composed of intrasexual competition among members of one sex (usually males) for access to the other sex and intersexual choice of members of one sex by members of the other sex (usually females).

Shale A sedimentary rock composed of mud.

Shallow ecology A critical term applied to superficial environmentalists who claim to be green but are quick to compromise and who do little to bring about fundamental change. Many of those accused of being shallow ecologists counter that they are merely pragmatic, progressive reformers who prefer to work within the established social contract rather than try to overthrow it.

Shantytowns Settlements created when people move onto undeveloped lands and build their own shelter with cheap or discarded materials; some are simply illegal subdivisions where a landowner rents land without city approval; others are land invasions.

Shaping The procedure used in operant conditioning wherein the experimenter can reinforce particular responses or behavioral actions by the subject and not others, thus enhancing the frequency of some actions and extinguishing other actions.

Shard A piece of broken clay or ceramic pot placed over the drainage hole of a pot to prevent the loss of soil during watering.

Shattering A trait found in wild plants in which the fruiting head breaks apart to scatter the seeds over a wide area.

Shear boundary The boundary between two plates that move past each other on the earth's surface.

Shear strain The ratio of the amount of deformation of the side of a body to the length of the side.

Sheath The base of a leaf in monocots, usually wrapping around the stem.

Sheet erosion Peeling off thin layers of soil from the land surface; accomplished primarily by wind and water.

Shelf break The section of the continental shelf where the slope abruptly becomes steeper, usually at a depth of 120 to 200 m (400 to 600 ft).

Shell The calcium carbonate outer layer of Cnidarians, molluscs, and other animals. Produced by mucous glands.

Shielding In the context of NMR spectroscopy shielding is the effect of the electron shells of the observed and the neighbouring nuclei on the external magnetic field. The external field induces circulations in the electron cloud. The resulting magnetic moment is oriented in the opposite direction to the external field, so that the local field at the central nucleus is weakened, although it may be strengthened at other nuclei (deshielding).

Shoot system Collective term for the stem and its leaves; generally the aboveground portion of the plant.

Short circuit An electric circuit through a negligible resistance that usually shunts a normal load and overloads the circuit.

Short grass prairie A grassland biome characterized by short grass and low rainfall; also known as the plains.

Short-day plant A plant that flowers when the length of day is shorter than some critical value.

Sickle cell anemia A condition that causes the red blood cells to collapse (sickle) under oxygen stress. The condition becomes manifest when an individual is homozygous for the gene for hemoglobin-S (hbs).

Siemens The derived SI unit of electrical conductance equal to the conductance of an element that has a resistance of 1 ohm.

Sieve plate The part of the wall of sieve-tube members bearing one or more highly differentiated sieve areas or modified primary pit fields.

Sieve pores Holes or modified plasmodesmata found within a sieve area either on a sieve plate or sieve cell.

Sieve tube A long tube specialized for the conduction of food materials (products of photosynthesis) and consisting of several-

Sievert; sv The derived SI unit of dose equivalent, defined as the absorbed dose of ionizing radiation multiplied by internationally-agreed-upon dimensionless weights, since different types of ionizing radiation cause different types of damage in living tissue. The sv, like the gy, has units of j/kg.

Sieve-tube member One of the cells of a sieve tube; found primarily in flowering plants and typically associated with a companion cell; also called sieve-tube element.

Sigmatropic rearrangement A molecular rearrangement that involves both the creation of a new sigma bond between atoms previously not directly linked and the breaking of an existing sigma bond. There is normally a concurrent relocation of pi bonds in the molecule concerned, but the total number of pi and sigma bonds does not change. The term was originally restricted to intramolecular pericyclic reactions, and many authors use it with this connotation. It is, however, also applied in a more general, purely structural, sense.

Sign stimuli Specific external stimuli that trigger stereotyped responses from conspecifics, usually called fixed action patterns.

Signal Any behavior pattern that conveys information. Patterns modified through evolution to convey information are called displays.

Signaling pheromone A chemical signal that is quickly perceived and causes an immediate response. Also called releasing pheromone. Contrast with priming pheromone.

Significant figures Those digits in an observed quantity (measurement) that are known with certainty plus the first digit that is uncertain.

Silent mutation A permanent genetic change, but one that is never expressed by the phenotype.

Silica (SiO_3) A mineral similar to glass that is the major component of the cell wall, shell, or skeleton of many marine organisms.

Siliceous Containing silica.

Siliceous ooze A type of biogenous sediment that consists mostly of the silica shell and skeletons of marine organisms.

Silicoflagellates Unicellular and eukaryotic members of the phytoplankton that have a star-shaped silica skeleton.

Silique A simple fruit that develops from a two-carpeled ovary; at maturity the two halves fall away, leaving the seeds attached to the persistent central wall.

Silt Soil inorganic particles in the.002-.02 size range.

Silurian The geological period beginning about 435 million years ago during which time the first land plants occurred.

Silylene 1. Generic name for H_2Si: and substitution derivatives thereof, containing an electrically neutral bivalent silicon atom with two non-bonding electrons. (The definition is analogous to that given for carbene.)

(2) The silanediyl group (H_2Si), analogous to the methylene group (H_2C).

Simian Pertaining to monkeys or apes.

Simphyla A class of arthropods whose members are characterized by having long antennae, 10 to 12 pairs of legs, and centipedelike bodies. Occupy soil and leaf mold.

Simple Without complexity. In the case of a tissue, without layers.

Simple diffusion The process of molecules spreading out randomly from where they are more concentrated to where they are less concentrated until they are evenly distributed.

Simple fruit A fruit that develops from a single ovary.

Simple harmonic motion Motion in which the acceleration is proportional to the displacement from an equilibrium position and is directed toward that position.

Simple leaf A leaf having a single blade portion; may be highly lobed or dissected.

Simple pistil A pistil that contains a single carpel.

Simple sugar A sugar, such as glucose, that cannot be broken down into simpler sugar molecules.

Simultaneity principle The principle that all frames of reference will have invariant simultaneity; that is, two events perceived as simultaneous (i.e., having the same time coordinate) in one frame will be perceived as simultaneous in all other frames. According to special relativity, however, this is not the case; simultaneity is frame-dependent.

Since blackbody power emission is proportional to the area of the hole and the fourth power of its thermodynamic temperature, the emitted power scales as m^{-2} — that is, as the inverse square of the mass.

Single-electron transfer mechanism (SET) A reaction mechanism characterized by the transfer of a single electron between the species occurring on the reaction coordinate of one of the elementary steps.

Single-step reaction A reaction that proceeds through a single transition state.

Singularity The center of a black hole, where the curvature of spacetime is maximal. At the singularity, the gravitational tides diverge; no solid object can even theoretically survive hitting the singularity. Although singularities generally predict inconsistencies in theory, singularities within black holes do not necessarily imply that general relativity is incomplete so long as singularities are always surrounded by event horizons.

Sinistral Pertaining to the left; in gastropods, shell is sinistral if opening is to left of columella when held with spire up and facing observer.

Sink A site of collection of metabolites, such as sugar; metabolic sinks may exist anywhere in the plant where organic solutes are being transported by the phloem and stored.

Sinkholes A large surface crater caused by the collapse of an underground channel or cavern; often triggered by groundwater withdrawal.

Sinus A cavity or space in tissues or in bone.

Siphon A tubular structure through which fluid flows; siphons of some molluscs allow water to enter and leave the mantle cavity.

Siphonogamy A reproductive process in seed plants in which a pollen tube carries the sperm cells to the egg located within the integumented megasporangium.

Siphonoglyph Ciliated furrow in the gullet of sea anemones.

Siphonophores Hydrozoans that exist as drifting colonies.

Siphuncle Cord of tissue running through the shell of a nautiloid, connecting all chambers with body of animal.

Sipuncula A phylum of protostomate worms whose members burrow in soft marine substrates throughout the world's oceans. Range in length from 2 mm to 75 cm.

Sister chromatid One of the two identical parts of a duplicated chromosome in a eukaryotic cell. Sister chromatids consist of exact copies of a long coiled DNA molecule with associated proteins. Sister chromatids are joined at the centromere of a duplicated chromosome.

Sister group The relationship between a pair of species or higher taxa that are each other's closest phylogenetic relatives.

Skeletal muscle Type of muscle tissue found in muscles attached to skeletal parts.

Skin The outer integument or covering of an animal body, consisting of the dermis and the epidermis and resting on the subcutaneous tissues.

Slater-type orbital An approximate atomic orbital that attempts to allow for electron-electron repulsion by scaling the nuclear charge for each orbital.

Sleeping sickness African trypanosomiasis and mosquito-borne, virus-induced encephalitis.

Slime ball Mass of mucus-covered cercariae of dicrocoeliid flukes, released from land snails. Also a term of derogation applied to really disgusting persons.

Slime mold Members of the Myxomycetes, characterized by a creeping, plasmodial stage.

Sludge Semisolid mixture of organic and inorganic materials that settles out of wastewater at a sewage treatment plant.

Slums Legal but inadequate multifamily tenements or rooming houses; some are custom built for rent to poor people, others are converted from some other use.

Small intestine The part of the digestive system consisting of the duodenum, jejunum, and ileum.

Smog The term used to describe the combination of smoke and fog in the stagnant air of London; now often applied to photochemical pollution products or urban air pollution of any kind.

Smolt Ayoung salmon just before it migrates downstream and out to sea.

Smooth ER The portion of endoplasmic reticulum that lacks ribosomes.

Smooth muscle Type of muscle tissue found in the walls of the hollow organs; visceral muscle.

Social deprivation Withholding or removal of contact with any form of stimulation from conspecifics.

Social ecology A socialist/humanist philosophy based on the communitarian anarchism of the

Russian geographer Peter Kropotkin. It shares much with deep ecology except that it is more humanist in its outlook.

Social justice Equitable access to resources and the benefits derived from them; a system that recognizes inalienable rights and adheres to what is fair, honest, and moral.

Social organization The species-typical pattern of relationships among all members of a group. This would include spatial distribution patterns, interindividual relationships involving dominance hierarchies or territoriality, mating systems, parenting, and dispersal.

Society A group of individuals belonging to the same species and organized in a cooperative manner. Usually assumed to extend beyond sexual behavior and parental care of offspring.

Sociobiology A study that involves the application of the principles of evolution to the study of the social behavior and social systems of animals.

Sociopolitical The awareness by society of how the political process can be influenced by well-informed individuals.

Sodium-potassium ATPase pump The active transport mechanism that functions to concentrate sodium ions on the outside of a plasma membrane and potassium ions on the inside of the membrane.

Soft corals Colonial anthozoans with no hard skeleton.

Softwood General term for the wood (secondary xylem) of conifers.

Soil A complex mixture of weathered mineral materials from rocks, partially decomposed organic molecules, and a host of living organisms.

Soil horizons Horizontal layers that reveal a soil's history, characteristics, and usefulness.

Soil profile A vertical section of soil showing the zones of particle sizes from surface down to bedrock.

Solar spectrum The band of colors produced when sunlight is dispersed by a prism.

Solenia Channels through the coenenchyme connecting the polyps in an alcyonarian colony (phylum Cnidaria).

Solenocyte Special type of flame bulb in which the bulb bears a flagellum instead of a tuft of flagella.

Solenoid A long helical wound coil of insulated wire.

Solenophage Blood-feeding arthropod that introduces its mouthparts directly into a blood vessel to feed.

Soleus The sole of the foot.

Solid state detector A device used to detect the passage of charged subatomic particles by their crystal-distorting or ionizing effects on a nonconducting or nonconducting solid.

Solidification The change of phase from a liquid to a solid.

Solstice As the orientation of the Earth (relative to the sun) varies through the year, the sun appears to move across the sky a little higher in the sky or a little lower. During the winter, the sun is lower; in the summer, higher. During the northern hemisphere summer, the southern hemisphere sees the sun lower in the sky and experiences winter. During the northern hemisphere winter, the southern hemisphere sees the sun higher in the sky and experiences summer. The word solstice came to Middle English from the Latin solstitium (via Old French) combining sol- (sun) and -stitium (a stoppage), meaning "the sun stands still." This is the time of the year when the Sun stops drifting north or south and begins moving in the opposite direction. The solstice, thus marks the half-way point for the long days of summer and the short days for winter. The northern hemisphere's summer solstice (when the Earth's north pole is tilted its maximum amount towards the Sun) is the southern hemisphere's winter solstice. And vica versa (when the south pole is tilted its maximum amount towards the Sun).

Solubility The relative ability of a solute to be dissolved.

Solute Any material dissolved in a solution.

Solute potential The water potential component caused by the presence of solutes in water.

Solvation Any stabilizing interaction of a solute (or solute moiety) and the solvent or a similar interaction of solvent with groups of an insoluble material (i.e., the ionic groups of an ion-exchange resin). Such interactions generally involve electrostatic forces and van der Waals forces, as well as chemically more specific effects such as hydrogen bond formation.

Solvatochromism The change in position and sometimes intensity of an electronic absorption or emission band, accompanying a change in the polarity of the medium. Negative (positive) solvatochromism corresponds to a hypsochromic (bathochromic) shift with increasing solvent polarity.

Solvent The liquid matrix in which a solute is dissolved.

Solvent parameter Quantitative measures of the capability of solvents for interaction with solutes. Such parameters have been based on numerous different physico-chemical quantities, e.g. rate constants, solvatochromic shifts in ultraviolet/visible spectra, solvent-induced shifts in infrared frequencies, etc. Some solvent parameters are purely empirical in nature, i.e. they are based directly on some experimental measurement. It may be possible to interpret such a parameter as measuring some particular aspect of solvent-solute interaction or it may be regarded simply as a measure of solvent polarity. Other solvent parameters are based on analysing experimental results. Such a parameter is considered to quantify some particular aspect of solvent capability for interaction with solutes.

Solvolysis Generally, reaction with a solvent, or with a lyonium ion or lyate ion, involving the rupture of one or more bonds in the reacting solute. More specifically the term is used for substitution, elimination and fragmentation reactions in which a solvent species is the nucleophile ("alcoholysis" if the solvent is an alcohol, etc.).

Solvophobicity parameter A solvent parameter defined by

$$Sp = 1 - M/M(\text{hexadecane})$$

derived from the Gibbs energy of transfer (ΔGo) of a series of solutes from water to numerous aqueous-organic mixtures and to pure solvents:

$$\Delta G^o(\text{to solvent}) = MR_T + D$$

where R_T is a solute parameter, and M and D characterize the solvent. The M values are used to define a solvent solvophobic effect so that S_p values are scaled from unity (water) to zero (hexadecane).

Soma The whole of an organism except the germ cells (germ plasm).

Somaclonal variant A plant showing a mutation that developed asexually during the tissue culture of a single callus.

Somatic Either pertaining to body cells (as opposed to gametes) or to the body wall (as opposed to the viscera).

Somatic cell Ordinary body cell; pertaining to or characteristic of a body cell. Any cell other than a germ cell or germ-cell precursor.

Somatic mutation A mutation that occurs in cells of leaves, stems, or roots; a mutation occurring in any cells that are not involved in gamete formation.

Somatocoel Posterior coelomic compartment of echinoderms; left somatocoel gives rise to oral coelom, and right somatocoel becomes aboral coelom.

Somatoplasm The living matter that makes up the mass of the body as distinguished from germ plasm, which makes up the reproductive cells. The protoplasm of body cells.

Somite One of the blocklike masses of mesoderm arranged segmentally (metamerically) in a longitudinal series beside the neural tube of the embryo; metamere.

SOMO A singly occupied molecular orbital (such as the half-filled HOMO of a radical).

Sonar A system that uses sound at sonic or ultrasonic frequencies to detect and locate objects.

Sonication Irradiation with (often ultra)sound waves, e.g. to increase the rate of a reaction or to prepare vesicles in mixtures of surfactants and water.

Sonometer A device, consisting of two or more wires or strings stretched over a sounding board, used for testing the frequency of strings and for showing how they vibrate.

Sorting. Differential survival and reproduction among varying individuals; often confused with natural selection which is one possible cause of sorting.

Sorus A cluster of sporangia found on a fern leaf.

Sound The series of disturbances in matter to which the human ear is sensitive Also similar disturbances in matter above and below the normal range of human hearing.

Sound intensity The rate at which sound energy flows through a unit area

Sound window The use of frequencies for communication that are transmitted through the environment with little loss of strength (attenuation).

Space charge The negative charge in the space between the cathode and plate of a vacuum tube.

Spacetime, Euclidean This description of spacetime treats only the geometry of flat surfaces such as a tabletop. The surface of a globe, for example, is not a two-dimensional Euclidean surface; it is a three-dimensional Euclidean surface.

Spacetime, non-Euclidean This description of spacetime treats a more general case than that of Euclidean spacetime. In non-Euclidean geometry, one can treat the surface of a sphere, for example, as a two-dimensional surface.

Sparganum Cestode plerocercoid of unknown identity.

Spark chamber A device used to detect the passage of charged subatomic particles by the light flashes they trigger

Spawning The release of gametes or eggs into the water.

Special salt effect The initial steep rate increase observed in the kinetic electrolyte effect on certain solvolysis reactions, upon addition of some non-common ion salts, especially $LiClO_4$.

Specialists Species that occupy a narrow range of habitats and eat a narrow range of foods. Contrast with generalists.

Specialized niche A species occupying a niche with a narrow range of tolerance.

Speciation The process by which two or more species are formed from a single ancestral stock.

Species A group of interbreeding individuals of common ancestry that are reproductively isolated from all other such groups; a taxonomic unit ranking below a genus and designated by a binomen consisting of its genus and the species name.

Species diversity The number and relative abundance of species present in a community.

Species recovery plan A plan for restoration of an endangered species through protection, habitat management, captive breeding, disease control, or other techniques that increase populations and encourage survival.

Species-typical behavior Actions and displays that are broadly characteristic of a species.

Specific In physics and chemistry the word specific in the name of a quantity usually means divided by an extensive measure that is, divided by a quantity representing an amount of material. Specific volume means volume divided by mass, which is the reciprocal of the density. Specific heat capacity is the heat capacity divided by the mass.

Specific catalysis The acceleration of a reaction by a unique catalyst, rather than by a family of related substances. The term is most commonly used in connection with specific hydrogen-ion or hydroxide-ion (lyonium ion or lyate ion) catalysis.

Specific gravity The ratio of the mass density of a substance to that of water

Specific heat The amount of heat required to raise the temperature of 1 gm of any substance 1 degree Celsius.

Specific-Mate Recognition System (SMRS) Signals that increase the likelihood that members of the same population will mate with one another. Contrast with species isolating mechanisms.

Spectator mechanism A pre-association mechanism in which one of the molecular entities, C, is already present in an encounter pair with A during formation of B from A, but does not assist the formation of B, e.g.

$$A + C \xrightleftharpoons{\text{Pre-association}} \underset{\text{encounter complex}}{(A\cdots C)} \longrightarrow \underset{\text{encounter complex}}{(B\cdots C)} \xrightarrow{\text{rapid}} \text{Product}$$

The formation of B from A may itself be a bimolecular reaction with some other reagent. Since C does not assist the formation of A, it is described as being present as a spectator, and hence such a mechanism is sometimes referred to as a spectator mechanism.

Spectroscope Optical instrument used for the study of spectra

Speed of light The speed at which electromagnetic radiation propagates in a vacuum; it is defined as 299792458 m/s.

Sperm A male gamete.

Sperm competition A situation in which one male's sperm fertilize a disproportionate number of eggs when a female copulates with more than one male.

Sperm nuclei Each pollen grain produces two sperm nuclei, which effect double fertilization in angiosperms.

Spermaceti organ a large organ in the forehead of sperm whales that is filled with a fine-quality liquid or waxy spermaceti oil.

Spermagonium A structure that produces spermatia in the rust fungi.

Spermalege Organ that receives the sperm in the female cimicid bug during copulation.

Spermatheca A sac in the female reproductive organs for the reception and storage of sperm.

Spermatid A growth stage of a male reproductive cell arising by division of a secondary spermatocyte; gives rise to a spermatozoon.

Spermatium Minute, nonmotile male gametes that occur in the rust fungi.

Spermatocyte A growth stage of a male reproductive cell; gives rise to a spermatid.

Spermatodactyl Modification in some Acari of chelicera, which functions in transfer of sperm from male's gonopore to copulatory receptacles between third and fourth coxae of female.

Spermatogenesis Formation and maturation of spermatozoa.

Spermatogonium Precursor of mature male reproductive cell; gives rise directly to a spermatocyte.

Spermatophore Formed "container" or packet of sperm that is placed in or on the body of a female, in contrast to the sperm in copulation which are conducted directly from male reproductive structures into the female's body.

Sphenoid The sphenoid bone.

Sphenophyta The vascular plant division commonly termed the horsetails to which the genus Equisetum belongs.

Spherical aberration The failure of parallel rays to meet at a single point on a spherical surface after reflection or refraction.

Sphincter A ringlike band of muscle fibers that constricts a passage or closes a natural orifice.

Spice A pungent, aromatic plant product derived from plants native to tropical regions and used to flavor foods.

Spicule One of the minute calcareous or siliceous skeletal bodies found in sponges, radiolarians, soft corals, and sea cucumbers.

Spike An inflorescence in which the main axis is elongated and the flowers are sessile.

Spikelet A small group of grass flowers; a unit of the inflorescence in grasses.

Spin density The unpaired electron density at a position of interest, usually at carbon, in a radical. It is often measured experimentally by electron paramagnetic resonance (EPR, ESR (electron spin resonance)) spectroscopy through hyperfine coupling constants of the atom or an attached hydrogen.

Spin label A stable paramagnetic group (typically a nitryl radical) that is attached to a part of a molecular entity whose microscopic environment is of interest and may be revealed by the electron spin resonance (ESR) spectrum of the spin label.

When a simple paramagnetic molecular entity is used in this way without covalent attachment to the molecular entity of interest it is frequently referred to as a "spin probe".

Spin trapping In certain reactions in solution a transient radical will interact with a diamagnetic reagent to form a more persistent radical. The product radical accumulates to a concentration where detection and, frequently, identification are possible by EPR/ESR spectroscopy. The key reaction is usually one of attachment; the diamagnetic reagent is said to be a "spin trap", and the persistent product radical is then the "spin adduct". The procedure is referred to as spin trapping, and is used for monitoring reactions involving the intermediacy of reactive radicals at concentrations too low for direct observation. Typical spin traps are C-nitroso compounds and nitrones, to which reactive radicals will rapidly add to form nitryl radicals. A quantitative development, in which essentially all reactive radicals generated in a particular system are intercepted, has been referred to as "spin counting". Spin trapping has also been adapted to the interception of radicals generated in both gaseous and solid phases. In these cases the spin adduct is in practice transferred to a liquid solution for observation in order to facilitate interpretation of the EPR/ESR spectra of the radicals obtained.

Spindle The aggregation of microtubules that is involved in the movement and separation of chromosomes during mitosis and meiosis.

Spindle fibers The protein fibers formed during prophase of nuclear division; chromosomes attach to these fibers at the centromere.

Spine A sharp projection from the body or main part of a bone.

Spin-orbit effect An effect that causes atomic energy levels to be split because electrons have intrinsic angular momentum (spin) in addition to their extrinsic orbital angular momentum.

Spinthariscope A device used to detect subatomic particles by the light flashes they produce on a zinc sulfide screen.

Spiracle An opening for ventilation. The opening(s) of the tracheal system of an arthropod or an opening posterior to the eye of a shark, skate, or ray.

Spiral cleavage A type of embryonic cleavage in which cleavage planes are diagonal to the polar axis and unequal cells are produced by the alternate clockwise and counterclockwise cleavage around the axis of polarity; determinate cleavage.

Spiral valve A spiral portion in the intestine of cartilaginous fishes.

Spirillum A spirally coiled bacterium.

Spondylosis Degeneration of a vertebra.

Sponges (phylum Porifera) Invertebrates that consist of a complex aggregation of cells, including collar cells, and have a skeleton of fibers and/or spicules.

Spongin Fibrous, collagenous material making up the skeletal network of horny sponges.

Spongioblast Cell in a sponge that secretes spongin, a protein.

Spongocoel Central cavity in sponges.

Spongocyte A cell in sponges that secretes spongin.

Spongy parenchyma A tissue composed of loosely packed, irregular parenchymatous cells containing chloroplasts; commonly found in leaves.

Spontaneous ovulators Species in which females release eggs whether they have copulated or not.

Spontaneous recovery A process in which a conditioned response has been extinguished and then is followed by some time interval (generally one minute up to one day or more depending upon the species and experimental conditions), after which the animal may immediately exhibit a nearly normal correct response rate upon reintroduction to the test situation.

Sporadin Mature trophozoite of a gregarine protozoan.

Sporangiospore A spore that develops within a sporangium.

Sporangium A structure in which spores are produced.

Spore A reproductive unit (often unicellular) that is capable of developing into a new organism without fusion with another cell.

Sporoblast Cell mass that will differentiate into a sporocyst within an oocyst.

Sporocyst 1. Stage of development of a sporozoan protozoan, usually with an enclosing membrane, the oocyst.

2. An asexual stage of development in some digenean trematodes that arises from a miracidium and gives rise to rediae.

Sporocyst residuum Cytoplasmic material "left over" within a sporocyst after sporozoite formation; seen as an amorphous mass.

Sporogony Multiple fission that produces sporozoites after zygote formation. Occurs in the class Sporozoea.

Sporont Undifferentiated cell mass within an unsporulated oocyst.

Sporophyll A leaf that bears sporangia.

Sporophyte The diploid, spore-producing plant in the alternation of generations; undergoes meiosis to produce the haploid spores.

Sporoplasm Amebalike portion of a microsporan or myxosporan cyst that is infective to the next host.

Sporozoite A stage in the life history of many sporozoan protozoa; released from oocysts.

Spout (blow)The water vapor and seawater that is observed when whales surface and exhale.

Spring overturn Springtime lake phenomenon that occurs when the surface ice melts and the surface water temperature warms to its greatest density at 4C and then sinks, creating a convection current that displaces nutrient-rich bottom waters.

Spring tides The tides with a large tidal range; they occur around the times of full or new moon. Compare neap tides.

Springwood The cells in the secondary xylem that are formed early in the season, usually with wide vessels (angiosperms) or wide tracheids (gymnosperms); also called

Squalene A liquid acyclic triterpene hydrocarbon found especially in the liver oil of sharks.

Squama Prominent lobe in the anal angle of a dipteran wing.

Squamous epithelium Simple epithelium of flat, nucleated cells.

Squamous Flat or scalelike.

Squatter towns Shantytowns that occupy land without owner's permission; some are highly organized movements in defiance of authorities; others grow gradually.

Stability In ecological terms, a dynamic equilibrium among the physical and biological factors in an ecosystem or a community; relative homeostasis.

Stabilizing selection Natural selection that results in the decline of both extremes in a phenotypic range; results in a narrowing of the phenotypic range.

Stable As applied to chemical species, the term expresses a thermodynamic property, which is quantitatively measured by relative molar standard Gibbs energies. A chemical species A is more stable than its isomer B if $\Delta G^o > 0$ for the (real or hypothetical) reaction A $\longrightarrow$B, under standard conditions. If for the two reactions

$P \longrightarrow X + Y\ (\Delta rG_1{}^o)$

$Q \longrightarrow X + Z\ (\Delta rG_2{}^o)$

$\Delta_r G_1{}^o > \Delta_r G_2{}^o$, P is more stable relative to the product Y than is Q relative to Z. Both in qualitative and quantitative usage the term stable is therefore always used in reference to some explicitly stated or implicitly assumed standard.

The term should not be used as a synonym for unreactive or "less reactive" since this confuses thermodynamics and kinetics. A relatively more stable chemical species may be more reactive than some reference species towards a given reaction partner.

Stable runoff The fraction of water available year-round; usually more important than total runoff when determining human uses.

Stalk-and-rush Predators that approach prey as closely as possible, then close with a sudden burst of speed.

Stamen The floral organ that produces pollen; consisting of an anther and filament.

Staminate A unisexual flower having stamens but no pistil.

Staminate flower A flower having stamens but no carpels.

Staminodium A sterile stamen; nonfunctional anthers and often with petaloid filaments.

Standard deviation A statistical measure of the degree of variation from the mean value among the individual measurements in a series of values.

Standard error of the mean (SEM) A statistical measure of variation most properly restricted to use with a group of means, though often reported as a measure of variation around the mean value in a series of individual measurements.

Standard quantum limit The limit imposed on standard methods of measurement by the uncertainty principle within quantum mechanics.

Standing crop total amount of plant or animal material in an area at any one time

Standing stock The total amount, or biomass, of an organism at a given time.

Standing wave The resultant of two wave trains of the same wavelength, frequency, and

Starch A polysaccharide composed of a thousand or more glucose molecules; the chief food storage material of most plants.

Static electricity Electricity at rest

Static limit The distance from a rotating black hole where no observer can possibly remain at rest (with respect to the distant stars) because of inertial frame dragging; this region is outside of the event horizon, except at the poles where it meets the horizon at a point. The region between the event horizon and the static limit is called the ergosphere.

Stationary state In quantum mechanics: A state that does not evolve with time.

Statistics, Bose-Einstein Statistics, which describe the motion of particles with integral spin, ie bosons.

Statistics, Fermi-Dirac Statistics which describe the motion of particles with half-integral spin, ie fermions.

Statistics, Maxwell-Boltzmann statistics which describe the motion of particles without accounting for their spin.

Statoblast Biconvex capsule containing germinative cells and produced by most freshwater ectoprocts by asexual budding. Under favorable conditions it germinates to give rise to new zooid.

Statocyst Sense organ of equilibrium; a fluid-filled cellular cyst containing one or more granules (statoliths) used to sense direction of gravity.

Statolith Small calcareous body resting on tufts of cilia in the statocyst.

Statutory law Rules passed by a state or national legislature.

Steam point The boiling point of water at standard atmospheric pressure

Stefan-Boltzmann law The radiated power p (rate of emission of electromagnetic energy) of a hot body is proportional to the radiating surface area, a, and the fourth power of the thermodynamic temperature, t. The constant of proportionality is the stefan-boltzmann constant.

Stele The central vascular cylinder of roots and stems of vascular plants.

Stenohaline Pertaining to aquatic organisms that have restricted tolerance to changes in environmental saltwater concentration.

Stenophagous Eating few kinds of foods.

Stenotopic Refers to an organism with

Stepwise reaction A chemical reaction with at least one reaction intermediate and involving at least two consecutive elementary reactions.

Steradian The supplementary SI unit of solid angle defined as the solid central angle of a sphere that encloses a surface on the sphere equal to the square of the sphere's radius.

Stereocilia The short, modified cilia at the apex of a hair cell.

Stereoelectronic Pertaining to the dependence of the properties (especially the energy) of a molecular entity in a particular electronic state (or of a transition state) on relative nuclear geometry. The electronic ground state is usually considered, but the term can apply to excited states as well. Stereoelectronic effects are ascribed to the different alignment of electronic orbitals in different arrangements of nuclear geometry.

Stereoelectronic control Control of the nature of the products of a chemical reaction (or of its rate) by stereoelectronic factors. The term is usually applied in the framework of an orbital approximation. The variations of molecular orbital energies with relative nuclear geometry (along a reaction coordinate) are then seen as consequences of variations in basis-orbital overlaps.

Stereogastrula A solid type of gastrula, such as the planula of Cnidarians.

Stereom Meshwork structure of endoskeletal ossicles of echinoderms.

Stereospecificity, stereospecific 1. A reaction is termed stereospecific if starting materials differing only in their configuration are converted into stereoisomeric products. According to this definition, a stereospecific process is necessarily stereoselective but not all stereoselective processes are stereospecific. Stereospecificity may be total (100%) or partial. The term is also applied to situations where reaction can be performed with only one stereoisomer. For example the exclusive formation of trans-1,2-dibromocyclohexane upon bromination of cyclohexene is a stereospecific process, although the analogous reaction with (E)-cyclohexene has not been performed.

2. The term has also been applied to describe a reaction of very high stereoselectivity, but this usage is unnecessary and is discouraged.

Stereoselectivity, stereoselective Stereoselectivity is the preferential formation in a chemical reaction of one stereoisomer over another. When the stereoisomers are enantiomers, the phenomenon is called enantioselectivity and is quantitatively expressed by the enantiomer excess; when they are diastereoisomers, it is called diastereoselectivity and is quantitatively expressed by the diastereomer excess. Reactions are termed (100%) stereoselective if the discrimination is complete or partially (x%) stereoselective if one product predominates. The discrimination may also be referred to semiquantitatively as high or low stereoselectivity.

Stereotyped behavior A pattern of behavior repeated with little variation in performance.

Steric effect The effect on a chemical or physical property (structure, rate or equilibrium constant) upon introduction of substituents having different steric requirements. The steric effect in a reaction is ascribed to the difference in steric energy between, on the one hand, reactants and, on the other hand, a transition state, (or products). A steric effect on a rate process may result in a rate increase ("steric acceleration") or a decrease ("steric retardation"). (The adjective "steric" is not to be confused with stereochemical.)

Steric effects arise from contributions ascribed to strain as the sum of 1. non-bonded repulsions,

2. bond angle strain, and

3. bond stretches or compressions.

For the purpose of correlation analysis or linear free-energy relations various scales of steric parameters have been proposed, notably A values, Taft's Es and Charton's scales.

In a reactant molecule RY and an appropriate reference molecule R0Y, the "primary steric effect" of R is the direct result of differences in compressions which occur because R differs from R0 in the vicinity of the reaction centre Y. A "secondary steric effect" involves the differential moderation of electron delocalization by non-bonded compressions.

Some authors make a distinction between "steric" effects attributed to van der Waals repulsions alone, and "strain" effects, attributed to deviations of bond angles from "ideal" values.

Steric hindrance The original term for a steric effect arising from crowding of substituents.

Steric-approach control Control of stereoselectivity of a reaction by steric hindrance towards attack of the reagent, which is directed to the less hindered face of the molecule. Partial bond making is strong enough at the transition state for steric control to take place. This suggests that the transition state should not be close to products.

Stern-Gerlach experiment An experiment that demonstrates the features of spin (intrinsic angular momentum) as a distinct entity apart from orbital angular momentum.

Sternite Main ventral sclerite of a somite of an arthropod.

Sternum Ventral plate of an arthropod body segment; breastbone of vertebrates.

Steroid An organic substance whose molecules include four complex rings of carbon and hydrogen atoms. Examples are estrogen, cholesterol, and testosterone.

Sterol One of a class of organic compounds containing a molecular skeleton of four fused carbon rings; it includes cholesterol, sex hormones, adrenocortical hormones, and vitamin D.

Stewards Protectors, careful managers of land and nature's resources.

Stewardship A philosophy that holds that humans have a unique responsibility to manage, care for, and improve nature.

Stichosome Column of large, rectangular cells called stichocytes, supporting and secreting into the esophagus of most nematodes of the family Trichuridae.

Stieda body Plug in the inner wall of one end of a coccidian oocyst.

Stigma The mass of bright red photoreceptor granules found in certain flagellated protozoa (Euglena) that serves as a shield for the photoreceptor. Also the spiracle of certain terrestrial arthropods.

Stimulant A psychoactive compound that excites and enhances mental alertness and physical activity; often reduces fatigue and suppresses hunger.

Stimulus Any form of energy an animal is able to detect with its receptors.

Stimulus filter The ability of the nervous system to block incoming stimuli that are unimportant for the animal.

Stipe A supporting stalk; such as those in mushrooms and brown algae.

Stipule A small appendage found in pairs at the base of leaves.

Stock The size of a population.

Stolon A rootlike extension of the body wall giving rise to buds that may develop into new zooids, thus forming a compound animal in which the zooids remain united by the stolon. Found in some colonial anthozoans, hydrozoans, ectoprocts, and ascidians.

Stoma The epidermal complex consisting of two guard cells and the pore between them.

Stomach The expansion of the alimentary canal between the esophagus and duodenum.

Stomates The small openings in leaves, herbaceous stems, and fruits through which gases and water vapor pass.

Stomochord Anterior evagination of the dorsal wall of the buccal cavity into the proboscis of hemichordates; the buccal diverticulum.

Stony corals Anthozoans, often colonial, that secrete a calcareous skeleton.

Stopped flow A technique for following the kinetics of reactions in solution (usually in the millisecond time range) in which two reactant solutions are rapidly mixed by being forced through a mixing chamber. The flow of the mixed solution along a uniform tube is then suddenly arrested. At a fixed position along the tube the solution is monitored (as a function of time following the stoppage of the flow) by some method with a rapid response (e.g. photometry).

Storage cell An electrochemical cell in which the reacting materials are regenerated by the use of a reverse current from an external source.

Strain Strain is present in a molecular entity or transition structure if the energy is enhanced because of unfavourable bond lengths, bond angles, or dihedral angles ("torsional strain") relative to a standard.

It is quantitatively defined as the standard enthalpy of a structure relative to a strainless structure (real or hypothetical) made up from the same atoms with the same types of bonding. (The enthalpy of formation of cyclopropane is 53.6 kJ mol-1, whereas the enthalpy of formation based on three "normal" methylene groups, from acyclic models, is -62 kJ mol-1. On this basis cyclopropane is destabilized by ca. 115 kJ mol-1 of strain energy.)

Strata Layers of sedimentary rock, the oldest rocks occurring at the bottom.

Strategic minerals Materials a country cannot produce itself but that it uses for essential materials or processes.

Stratification The separation of the water column into layers, with the densest at the bottom and the least dense at the surface. A stratified water column is said to be stable. An unstable column results when the surface water becomes more dense than the water below.

Stratosphere The zone in the atmosphere extending from the tropopause to about 50 km (30 mi) above the earth's surface; temperatures are stable or rise slightly with altitude; has very little water vapor but is rich in ozone.

Streambank erosion Washing away of soil from banks of established streams, creeks, or rivers, often as a result of the removal of trees and brush along streambanks or cattle damage to the banks.

Stress Physical, chemical, or emotional factors that place a strain on an animal. Plants also experience physiological stress under adverse environmental conditions.

Stress shock A loose set of physical, psychological, and/or behavioral changes thought to result from the stress of excess competition and extreme closeness to other members of the same species.

Stretch receptor Sensory receptor that responds to stretch; found in muscle tissue, lungs, and other organs that undergo changes in position or size.

Stridulation The production of sound by an insect rubbing one body part against another (e.g., in male crickets).

Strike Deposition of fly eggs or larvae on a living host.

Strip-farming Planting different kinds of crops in alternating strips along land contours; when one crop is harvested, the other crop remains to protect the soil and prevent water from running straight down a hill.

Strip-mining Removing surface layers over coal seams using giant, earth-moving equipment; creates a huge open-pit from which coal is scooped by enormous surface-operated machines and transported by trucks; an alternative to deep mines.

Strobila A stage in the development of the scyphozoan jellyfish. Also, the chain of proglottids of a tapeworm.

Strobilation Repeated, linear budding of individuals, as in scyphozoans (phylum Cnidaria), or sets of reproductive organs, as in tapeworms (phylum Platyhelminthes).

Strobilocercoid Cysticercoid that undergoes some strobilation; found only in Schistotaenia.

Strobilus A number of sporangia- or ovule-bearing stuctures (sporophylls, sporangiophores, scales, etc.) Grouped together on an axis.

Stroma The matrix between the grana in chloroplasts and site of the dark reactions of photosynthesis.

Stroma thylakoid A thylakoid that does not occur in a granum; connects separate grana.

Stromatolites Massive calcareous skeletons formed by cyanobacteria (blue-green algae).

Strong nuclear interaction The interaction that holds the particles of the nucleus together and is independent of charge.

Structural color A color that results when light is reflected by a particular surface.

Structural genes Genes that code for enzymes or structural proteins rather than proteins that control the function of other genes.

Structural molecule A complex molecule, such as cellulose, that provides support and protection.

Structure Patterns of organization, both spatial and functional, in a community.

Style Terminal segment of the antenna of a brachyceran dipteran. It is drawn into a sharp point.

Styloid stylos = pillar or stalk.

Stylops Member of the insect order Strepsiptera.

Stylostome Hardened, tubelike structure secreted by a feeding chigger mite.

Subchelate Condition of an arthropod appendage in which the terminal podomere can fold back like a pincer against the subterminal podomere.

Subduction The downward movement of a plate into the mantle that occurs in trenches, which are also known as subduction zones.

Suberin A fatty material found in the cell walls of cork cells and the Casparian strip of the endodermis.

Subjacent orbital The next-to-highest occupied molecular orbital ("NHOMO", also called "HOMO-") and the second lowest unoccupied molecular orbital (SLUMO). Subjacent orbitals are sometimes found to play an important role in the interpretation of molecular interactions according to the frontier orbital approach.

Sublimation The process by which water can move between solid and gaseous states without ever becoming liquid.

Submarine canyon A narrow, deep depression in the continental shelf formed by the erosion of rivers or glaciers before the shelf was submerged.

Submergent plant community a marine plant community restricted to subtidal environments

Submetacentric Pertaining to a chromosome with a centromere located between the center and one end of the chromosome.

Subnivean Applied to environments beneath snow, in which snow insulates against a colder atmospheric temperature.

Subsidence A settling of the ground surface caused by the collapse of porous formations that result from withdrawal of large amounts of groundwater, oil, or other underground materials.

Subsoil A layer of soil beneath the topsoil that has lower organic content and higher concentrations of fine mineral particles; often contains soluble compounds and clay particles carried down by percolating water.

Substiedal body Additional plug material underlying a Stieda body.

Substituent An atom or group of bonded atoms that can be considered to have replaced a hydrogen atom (or two hydrogen atoms in the special case of bivalent groups) in a parent molecular entity (real or hypothetical).

Substitution reaction A reaction, elementary or stepwise, in which one atom or group in a molecular entity is replaced by another atom or group. For example.

$$CH_3Cl + OH \longrightarrow CH_3OH + Cl^-$$

Substrate A chemical species, the reaction of which with some other chemical reagent is under observation (e.g., a compound that is transformed under the influence of a catalyst). The term should be used with care. Either the context or a specific statement should always make it clear which chemical species in a reaction is regarded as the substrate.

Substrate-level phosphorylation The generation of ATP by coupling strongly exergonic reactions with the synthesis of ATP from ADP and phosphate.

Subtidal zone The bottom above the continental shelf.

Succession the gradual replacement, through time, of one group of species in a community by other groups.

Successor complex The radical ion pair which forms by the transfer of an electron from the donor D to the acceptor A after these species have diffused together to form the precursor or encounter complex:

$$A + D \longrightarrow (A\,D) \longrightarrow (A^- D^+)$$

Succulent A fleshy plant that accumulates water.

Sucrose A disaccharide (glucose + fructose) found in many plants; the primary form in which sugar produced by photosynthesis is translocated.

Sudoriferous gland A sweat gland.

Sugar A monosaccharide; a carbohydrate with the general formula $C_nH_{2n}O_n$.

Sulcus A furrow or groove.

Sulfide One of the minerals that is abundant in the hot water that seeps through hydrothermal vents.

Sulfur dioxide A colorless, corrosive gas directly damaging to both plants and animals.

Sulfur oxide Produced from burning coal, this gas is a pollutant that can produce sulfuric acid in rain droplets.

Summerwood The cells in the secondary xylem that are formed late in the season, usually with few vessels (angiosperms) or narrow tracheids (gymnosperms); also called late wood.

Sun compass The mechanism in which animals use the sun and an internal clock that allows them to adjust for movement in the sun as they navigate.

Superacid A medium having a high acidity, generally greater than that of 100 wt.-% sulfuric acid. The common superacids are made by dissolving a powerful Lewis acid (e.g. SbF_5) in a suitable Brønsted acid, such as HF or HSO_3F. (An equimolar mixture of HSO_3F and SbF_5 is known by the trade name "magic acid".).

Superconductivity The phenomena by which, at sufficiently low temperatures, a conductor can conduct charge with zero resistance.

Supercooling The process of cooling a substance below its normal phase change point without a change of phase.

Superfluidity The phenomena by which, at sufficiently low temperatures, a fluid can flow with zero viscosity. Its causes are associated with superconductivity.

Superior Above a point of reference (e.g., the neck is superior to the chest of humans).

Superior ovary Attachment of the ovary to the receptacle above and free from the attachment of the other floral parts.

Superposition Combining the displacements of two or more waves vectorially to produce a resultant displacement

Superposition principle The general idea that, when a number of influences are acting on a system, the total influence on that system is merely the sum of the individual influences; that is, influences governed by the superposition principle add linearly.

Superposition principle of forces The net force on a body is equal to the sum of the forces impressed upon it.

Superposition principle of states The resultant quantum mechnical wavefunction due to two or more individual wavefunctions is the sum of the individual wavefunctions.

Supply The quantity of that product being offered for sale at various prices, other things being equal.

Suprachiasmtic nucleus (SCN) A brain nucleus involved in the visual pathway that has been clearly identified for its involvement in biological rhythms mediated by photoperiod.

Supramolecule A system of two or more molecular entities held together and organized by means of intermolecular (noncovalent) binding interactions.

Suprapopulation of parasites All individuals of a single parasite species at all stages in the life cycle in all hosts in an ecosystem.

Surf A wave that becomes so high and steep as it approaches the shoreline that it breaks.

Surface (mixed) layer The upper layer of water that is mixed by wind, waves, and currents.

Surface current long-term directional flow of water at the sea surface

Surface mining Some minerals are also mined from surface pits.

Surface of a cell. Eukaryotic flagella are longer versions of cilia. Flagellar undulations drive a cell through a fluid medium.

Surface tension the mutual attraction of water molecules at the surface of a water mass that creates a flexible molecular "skin" over the water surface

Surface-to-volume ratio (S/V ratio) The amount of surface area relative to the total volume of an organism.

Surra Disease of large mammals caused by Trypanosoma evansi.

Survivorship The percentage of a population reaching a given age or the proportion of the maximum life span of the species reached by any individual.

Suspension feeder The type of feeding whereby an animal obtains its nutrients by the removal of suspended food particles from the surrounding water by some sort of capturing, trapping, or filtering structure.

Sustainable agriculture An ecologically sound, economically viable, socially just, and humane agricultural system. Stewardship, soil conservation, and integrated pest management are essential for sustainability.

Sustainable development A real increase in well-being and standard of life for the average person that can be maintained over the long-term without degrading the environment or compromising the ability of future generations to meet their own needs.

Sustained yield Utilization of a renewable resource at a rate that does not impair or damage its ability to be fully renewed on a long-term basis.

Swain-Lupton equation A dual parameter approach to the correlation analysis of substituent effects, which involves a field constant (F) and a resonance constant (R). The original treatment was modified later.

Swarmer Daughter trophozoites resulting from multiple fissions of Ichythophthirius multifiliis and a few other protozoa.

Sweeper tentacle A type of tentacle in corals that is used to sting neighboring colonies.

Swell A wave with a flatter, rounded crest and trough. Swells are found away from the area where waves are generated by the wind.

Swim bladder A gas-filled sac, which is usually located along the dorsal body wall of bony fishes; it is an outgrowth of the digestive tract and regulates buoyancy of a fish.

Sycon A type of canal system in certain sponges. Sometimes called syconoid.

Sylvatic Existing normally in the wild, not in the human environment.

Symbiology Study of symbioses.

Symbiont Any organism involved in a symbiotic relationship with another organism, the host.

Symbiosis The term was originally applied to describe the maximum flocking of either hard or soft ligands in the same complexes. For hydrocarbon molecules, symbiosis implies that those containing a maximum number of C-H bonds (e.g. CH_4) or C-C bonds (e.g. Me_4C) are the most stable.

Symbiotic An association of two or more different organisms in which both benefit.

Symmetrogenic fission Mitotic fission between the rows of flagella of protozoa.

Symmetry A balanced arrangement of similar parts on either side of a common point or axis.

Sympathetic nervous system Portion of the autonomic nervous system that arises from the thoracic and lumbar regions of the spinal cord; also called thoracolumbar division.

Sympatric Having the same or overlapping regions of geographical distribution.

Sympatric speciation Speciation that occurs in populations that have overlapping ranges.

Sympatry Populations or species with overlapping geographic distributions.

Symplast The interconnected protoplasm of all cells in a plant.

Symplastic Pertaining to the movement of water and solutes through tissues by passing through biological membranes.

Symplesiomorphies Taxonomic characters that are common to all members of a group of organisms. These characters indicate common ancestry but cannot be used to describe relationships within the group.

Symproportionation Synonymous with comproportionation.

Synangium Fused sporangia.

Synanthropism Habit of an organism of living in or around human dwellings.

Synapomorphies Characters that have arisen within a group since it diverged from a common ancestor. Synapomorphies are used to indicate degrees of relatedness within a group. Also called shared, derived characters.

Synapomorphy Shared, evolutionarily derived character states that are used to recover patterns of common descent among two or more species.

Synapse The place at which a nerve impulse passes between neuron processes, typically from an axon of one nerve cell to a dendrite of another nerve cell.

Synapsids An amniote lineage comprising the mammals and the ancestral mammal-like reptiles, having a skull with a single pair of temporal openings.

Synapsis The pairing of homologous chromosomes that occurs in Prophase I of meiosis.

Synaptic cleft The narrow space between the terminal ending of a neuron and the receptor site of the postsynaptic cell.

Synaptic processes The threadlike extensions at the end of an axon.

Synaptonemal complex The structure that holds homologous chromosomes together during synapsis in prophase of meiosis I.

Synchronization (principle of nonperfect synchronization) This principle applies to reactions in which there is a lack of synchronization between bond formation or bond rupture and other primitive changes that affect the stability of products and reactants, such as resonance, solvation, electrostatic, hydrogen bonding and polarizability effects. The principle states that a product-stabilizing factor whose development lags behind bond changes at the transition state, or a reactant-stabilizing factor whose loss is ahead of bond changes at the transition state, increases the intrinsic barrier and decreases the "intrinsic rate constant" of a reaction. For a product-stabilizing factor whose development is ahead of bond changes, or reactant factors whose loss lags behind bond changes, the opposite relations hold. The reverse effects are observable for factors that destabilize a reactant or product.

Synchronous A concerted process in which the primitive changes concerned (generally bond rupture and bond formation) have progressed to the same extent at the transition state is said to be synchronous. The term figuratively implies a more or less synchronized progress of the changes. However, the progress of the bonding change (or other primitive change) has not been defined quantitatively in terms of a single parameter applicable to different bonds or different bonding changes. The concept is therefore in general only qualitatively descriptive and does not admit an exact definition except in the case of concerted processes involving changes in two identical bonds.

Synchrotron A particle accelerator in which the oscillating frequency varies.

Syncytial hypothesis The idea that multicellular organisms could have arisen by the formation of cell boundaries within a large multinucleate protist.

Syncytium A multinucleated cell.

Syndrome A group of symptoms characteristic of a particular disease or abnormality.

Synergid Two haploid cells on either side of the egg cell at the micropylar end of the embryo sac.

Synergism The interaction of two or more agents or forces so that their combined effect is greater than the sum of their individual effects, as when two hormones combine to affect target tissues.

Synergistic effects When an injury caused by exposure to two environmental factors together is greater than the sum of exposure to each factor individually.

Syngamy Fertilization of one gamete with another individual gamete to form a zygote, found in most animals with sexual reproduction.

Synkaryon Zygote nucleus resulting from fusion of pronuclei.

Synsacrum The fused posterior thoracic vertebrae, all lumbar and sacral vertebrae, and anterior caudal vertebrae of a bird; helps maintain proper flight posture.

Syntax Information provided by the sequence in which signals are transmitted.

Syrinx The vocal organ of birds located at the base of the trachea.

System A group of organs specialized for one function.

System, electrical A collection of electrical devices connected with electrical circuits to achieve some end. A communication system can contain a microphone (transducer), an oscillator, a modulator, a transmitting antenna, a receiving antenna, a detector, amplifiers, and a speaker (transducer).

Systematic variation In learning, a procedure that involves examining qualitative differences in learning by investigating a wide variety of species from diverse taxonomic groups.

Systematics The study of the classification and phylogeny of organisms.

Système Internationale d'unités (si) The coherent and rationalized system of units, derived from the M.K.S. System (which itself is derived from the metric system) in common use in physics today.

Systemic A condition or process that affects the whole body; many metabolic poisons are systemic.

Systemic circuit The portion of the circulatory system concerned with blood flow from the left ventricle of the heart to the entire body and back to the heart via the right atrium.

Systole Phase of the cardiac cycle during which a heart chamber wall is contracted; also systolic pressure.

Systolic pressure The portion of blood pressure measurement that represents the highest pressure reached during ventricular ejection; it is the first number shown in a blood pressure reading.

Syzygy Stage during sexual reproduction of some gregarines in which two or more sporadins connect end to end.

T

T cell A type of lymphocyte derived from bone marrow stem cells that matures into an immunologically competent cell under the influence of the thymus. T cells are involved in a variety of cell-mediated immune reactions; also known as a T lymphocyte.

Tachyon A purely speculative particle, which is presumed to travel faster than light. According to Einstein's equations of special relativity, a particle with an imaginary rest mass and a velocity greater than c would have a real momentum and energy. Ironically, the greater the kinetic energy of a tachyon, the slower it travels, approaching c asymptotically (from above) as its energy approaches infinity. Alternatively, a tachyon losing kinetic energy travels faster and faster, until as the kinetic energy approaches zero, the speed of the tachyon approaches infinity; such a tachyon with zero energy and infinite speed is called transcendent.

Special relativity does not seem to specifically exclude tachyons, so long as they do not cross the lightspeed barrier and do not interact with other particles to cause causality violations. Quantum mechanical analyses of tachyons indicate that even though they travel faster than light they would not be able to carry information faster than light, thus failing to violate causality. But in this case, if tachyons are by their very nature indetectable, it brings into question how real they might be.

Tachyon paradox The argument demonstrating that tachyons (should they exist, of course) cannot carry an electric charge. For a (imaginary-massed) particle travelling faster than c, the less energy the tachyon has, the faster it travels, until at zero energy the tachyon is travelling with infinite velocity, or is transcendent. Now a charged tachyon at a given (non-infinite) speed will be travelling faster than light in its own medium, and should emit cherenkov radiation. The loss of this energy will naturally reduce the energy of the tachyon, which will make it go faster, resulting in a runaway reaction where any charged tachyon will promptly race off to transcendence.

Although the above argument results in a curious conclusion, the meat of the tachyon paradox is this: in relativity, the transcendence of a tachyon is frame-dependent. That is, while a tachyon might appear to be transcendent in one frame, it would appear to others to still have a nonzero energy. But in this case we have a situation where in one frame it would have come to zero energy and would stop emitting cherenov radiation, but in another frame it would still have energy left and should be emitting cherenkov radiation on its way to transcendence. Since they cannot both be true, by relativistic arguments, tachyons cannot be charged.

This argument naturally does not make any account of quantum mechanical treatments of tachyons, which complicate the situation a great deal.

Tachyzoite Small, merozoitelike stages of Toxoplasma. They develop in the host cells' parasitophorous vacuole by endodyogeny.

Tactile (touch) receptor A sensory receptor in the skin that detects light pressure; formerly called Meissner's corpuscle.

Tadpole larva The larva of tunicates.

Tagma, A compound body section of an arthropod resulting from embryonic fusion of two or more segments; for example, head, thorax, abdomen.

Tagmatization The specialization of body regions of a metameric animal for specific functions. The head of an arthropod is specialized for feeding and sensory functions, the thorax is specialized for locomotion, and the abdomen is specialized for visceral functions.

Taiga Areas of coniferous forests north of 50degree latitude having extended cold and snow cover winter conditions.

Tailings Mining waste left after mechanical or chemical separation of minerals from crushed ore.

Taking Unconstitutional confiscation of private property.

Tall grass prairie A grassland biome characterized by many tall grasses up to 5 meters tall.

Tannin A secondary product found in many plants that have been widely utilized as stains, dyes, inks, or tanning agents for leather; believed to function in plants by discouraging herbivores.

Tantulus Larva of a tantulocaridan (subphylum Crustacea).

Tapetum The nutritive somatic tissue surrounding the microsporocyte.

Tapeworms Parasitic flatworms, typically consisting of a chain of repeated units.

Taproot system A system of roots characterized by the presence of one dominant axis (the primary root) and several lateral, subordinate roots.

Tardigrada A phylum of animals whose members live in marine and freshwater sediments and in water films on terrestrial lichens and mosses. Possess four pairs of unsegmented legs and a proteinaceous cuticle. Water bears.

Tardon A particle which has a positive real mass and travels at a speed less than c in all inertial frames.

Target cell The cell influenced by a specific hormone.

Tarsus Most distal podomere of the insect or acarine limb; articulates proximally with the tibia and usually is subdivided into two to five subsegments in insects.

Taste buds Structures in the mouth and other locations of fishes that are sensitive to chemical stimuli.

Tau-theta paradox (1950s) When two different types of kaons, tau and theta (today tau refers to a completely different particle) decay, tau decays into three particles, while the theta decays into two. The tau and theta differ only in parity; and at the time, it was thought that parity was strictly conserved, and that particles differing only in parity should behave exactly the same. Since the two decay differently, a paradox ensued. The paradox was resolved when experiments carried out according to f. Yang and T.D. Lee's theoretical calculations indeed indicate that parity is not conserved in weak interactions.

Tautomerism Isomerism of the general form

$$G\text{-}X\text{-}Y{=}Z \rightleftharpoons X{=}Y\text{-}Z\text{-}G$$

where the isomers (called tautomers) are readily interconvertible; the atoms connecting the groups X,Y,Z are typically any of C, H, O, or S, and G is a group which becomes an electrofuge or nucleofuge during isomerization. The commonest case, when the electrofuge is H^+, is also known as "prototropy".

Examples:

Keto-enol tautomerism, such as

$H\text{-}O\text{-}C(CH_3){=}CH\text{-}CO_2Et$ (enol) $\rightleftharpoons (CH_3)C({=}O)\text{-}CH_2\text{-}CO_2Et$ (keto)

(G = H, X = O, Y = CCH_3, Z = $CHCO_2Et$)

$ArCH_2$-N=CHAr' ⇌ ArCH=N-CH_2Ar'

(G = H, X = CHAr, Y = N, Z = CHAr')

The grouping Y may itself be a three-atom (or five-atom) chain extending the conjugation, as in

(G = H, X = O, Y = C–CH=CH, Z = N)

The double bond between Y and Z may be replaced by a ring, when the phenomenon is called ring-chain tautomerism.

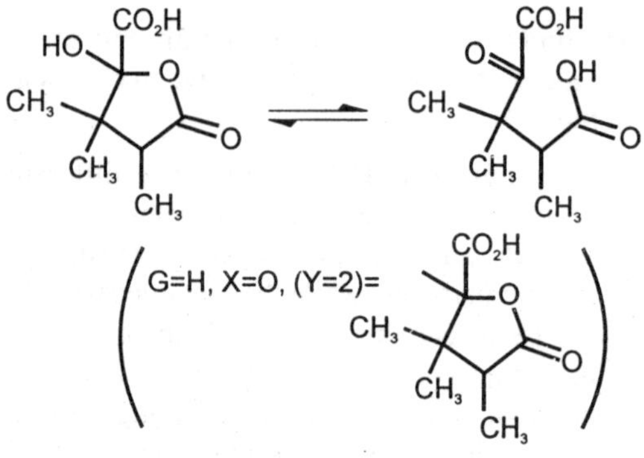

Taxis Directed reactions to a stimulus involving an orientation of the long axis of the body in line with the stimulus source.

Taxon A group of organisms that are genetically (evolutionarily) related.

Taxonomy The description of species and the classification of organisms into groups that reflect evolutionary relationships.

Technological optimists Those who believe that technology and human enterprise will find cures for all our problems. Also called Promethean environmentalism.

Technology The application of science to human needs and goals.

Technopolis Also called a vertical city; this model of city development proposes that cities grow vertically instead of horizontally.

Tectonic estuary An estuary that results from the sinking of land due to movements of the crust.

Tectonic plates Huge blocks of the earth's crust that slide around slowly, pulling apart to open new ocean basins or crashing ponderously into each other to create new, larger landmasses.

Tectum A rooflike structure, for example, dorsal part of capitulum in ticks and mites.

Tegument Surficial covering of a multicellular organism, an integument.

Tele A prefix meaning at a distance, as in telescope, telemetry, television.

Telencephalon The most anterior vesicle of the brain; the anterior-most subdivision of the prosencephalon that becomes the cerebrum and associated structures.

Teleomorphic The sexual phase, or perfect stage, in a fungal life cycle.

Tele-substitution A substitution reaction in which the entering group takes up a position more than one atom away from the atom to which the leaving group was attached.

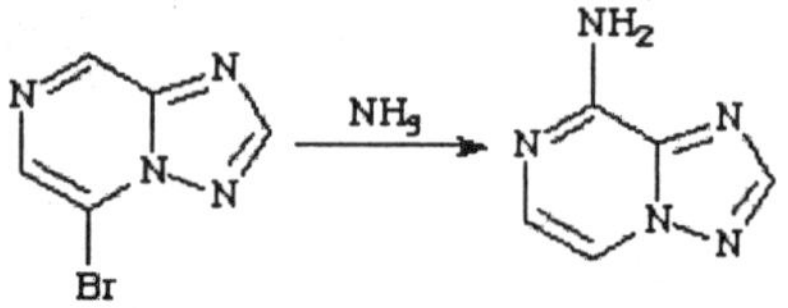

Teliospore A thick-walled spore found in the rust and smut fungi; karyogamy occurs within the teliospore and it gives rise to the basidium.

Telmophage Blood-feeding arthropod that cuts through skin and blood vessels to cause a small hemorrhage of blood from which it feeds.

Telocentric Chromosome with centromere at the end.

Telolecithal Having the yolk concentrated at one end of an egg.

Telome theory A theory that regards the primitive vascular plant as consisiting of upright, dichotomously branched axes, some of which bear terminal sporangia and from which, through a series of geometric, architectural modifications, the reproductive and vegetative organs of other vascular land plants evolved.

Telomerization The formation of an addition oligomer, having uniform end groups X'..X", by a chain reaction in which a chain transfer limits the length of the polymer ("telomer") produced. An example is the polymerization of styrene in bromotrichloromethane solution (X' = CCl_3, X" = Br), where Cl_3C. radicals are formed in the initiation step to produce $Cl_3C[CH_2CHPh]nBr$, with n greater than 1 and often less than 10.

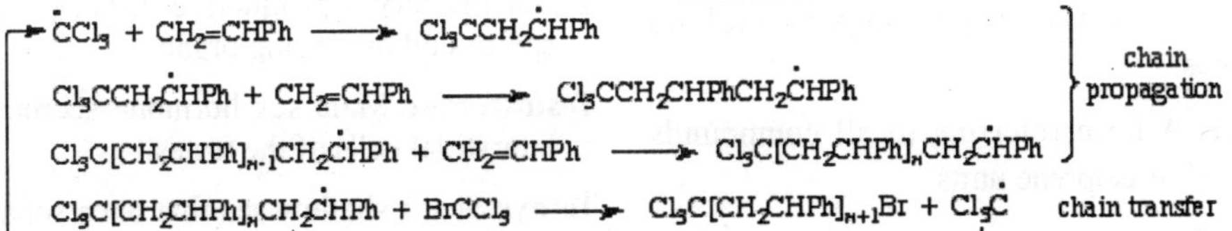

Telophase Stage in mitosis during which daughter cells become separate structures; the two sets of separated chromosomes decondense, and become enclosed by nuclear envelopes.

Telson Posterior projection of the last body segment in many crustaceans.

Temnospondyls A large lineage of amphibians that extended from the Carboniferous to the Triassic.

Temperate Characterized by a mild or moderate temperature.

Temperate deciduous forest An ecosystem that occurs in regions with moderate climate and well-defined seasons; deciduous trees predominate; animals include insects, white-tailed deer, and wolves.

Temperate rain forest A biome dominated by coniferous trees, high rainfall, and high humidity; moist coniferous forest.

Temperature-compensated rhythm The relative insensitivity of biological rhythms to the effects of temperature; this contrasts with the fact that many chemical reactions double in rate for every 1°C increase in temperature.

Template A pattern or mold guiding the formation of a duplicate; often used with reference to gene duplication.

Temporary magnet A magnet produced by induction.

Temporary parasite Parasite that contacts its host only to feed and then leaves. Also called an intermittent parasite or micropredator.

Tendon Fibrous band connecting muscle to bone or other movable structure.

Tendril A modified leaf or stem in which only a slender strand of tissue constitutes the entire structure.

Teneral Newly emerged adult arthropod that is soft and weak.

Tensile strength The force required to break a rod or wire of unit cross-sectional area.

Tension wood The reaction wood produced along the upper side of leaning woody trees, straightening the trunk by contracting and "pulling" the tree upright.

Tentacle A flexible, elongate appendage.

Tentaculata The class of ctenophorans with tentacles that may or may not be associated with sheaths into which tentacles can be retracted. Pleurobranchia.

Tentaculocyst One of the sense organs along the margin of medusae; a rhopalium.

Tentorium An extension of the parietal bone between the cerebral and cerebellar hemispheres; actually an ossification of the meninges.

Tepal Members of the perianth that are not differentiated into sepals and petals.

Teratogens Chemicals or other factors that specifically cause abnormalities during embryonic growth and development.

Terebra Functional unit of a hymenopteran ovipositor, formed from first and second valvulae.

Tergite Main dorsal sclerite of a somite of an arthropod.

Tergum Dorsal part of an arthropod body segment.

Termination The steps in a chain reaction in which reactive intermediates are destroyed or rendered inactive, thus ending the chain.

Terpenes A group of secondary compounds composed of two to many isoprene units in a chain or ring; sometimes categorized as hydrocarbons only, sometimes to include terpenoids.

Terpenoids A term referring to all compounds composed of isoprene units.

Terracing Shaping the land to create level shelves of earth to hold water and soil; requires extensive hand labor or expensive machinery, but it enables farmers to farm very steep hillsides.

Terrarium A closed biological system in which plants and animals coexist without external inputs or discharges; H_2O, CO_2, O_2, and nutrients cycle in the closed system.

Terrestrial Pertaining to the land.

Territoriality An intense form of intraspecific competition in which organisms define an area surrounding their home site or nesting site and defend it, primarily against other members of their own species.

Territory A restricted area preempted by an animal or pair of animals, usually for breeding purposes, and guarded from other individuals of the same species.

Tertian malaria Malaria in which fevers recur every 48 hours. Caused by Plasmodium vivax, P. Ovale, and P. Falciparum.

Tertiary treatment The removal of inorganic minerals and plant nutrients after primary and secondary treatment of sewage.

Tesla The derived SI unit of magnetic flux density, defined the magnetic flux density of a magnetic flux of 1 Wb through an area of 1 m^2; it thus has units of Wb/m^2.

Test A shell or hardened outer covering.

Testcross Crossing an organism having a dominant expression for a trait with an organism having a homozygous recessive genotype for that trait to determine the genotype of the organism expressing the dominant phenotype.

Testing Application of theories to examine repeatability and accuracy, should have comparisons to known information.

Testis Primary reproductive organ of a male; a sperm-cell producing organ.

Testosterone Male sex hormone secreted by the interstitial cells of the testes.

Tethys Sea A shallow sea that once separated the Eurasian and African sections of the supercontinent Pangaea. It eventually gave rise to the modern Mediterranean Sea.

Tetracotyle Strigeoid metacercaria in the family Strigeidae.

Tetrad A pair of homologous chromosomes during synapsis (prophase I of meiosis). A tetrad consists of four chromatids.

Tetrahedral intermediate A reaction intermediate in which the bond arrangement around an initially double-bonded carbon atom (typically a carbonyl carbon atom) has been transformed from trigonal to tetrahedral. For example, aldol in the condensation reaction of acetaldehyde (but most tetrahedral intermediates have a more fleeting existence).

Tetrapods A nontaxonomic designation used to refer to amphibians, reptiles, birds, and mammals.

Tetrathyridium Only metacestode form known in the tapeworm cyclophyllidean genus Mesocestoides. A large, solid-bodied cysticercoid.

Thalamus An oval mass of gray matter within the diencephalon that serves as a sensory relay area.

Thaliacea A class of urochordates whose members are planktonic. Adults are tailless and barrel shaped. Oral and atrial openings are at opposite ends of the tunicate. Water currents are produced by muscular contractions of the body wall and result in a weak form of jet propulsion.

Thallus The complete body of a seaweed.

Thecodonts A large assemblage of Triassic archosaurian diapsids of the order Thecodontia and characterized by having teeth set in sockets.

Theileriosis Disease of cattle and other ruminants, caused by Theileria parva. Also called East Coast fever.

Thelyotoky, thelytoky Type of parthenogenesis in which all individuals are uniparental and essentially no males are produced.

Theoretical Describing an idea which is part of a theory, or a consequence derived from theory.

Theory A well-tested mathematical model of some part of science. In physics a theory usually takes the form of an equation or a group of equations, along with explanatory rules for their application. Theories are said to be successful if 1. they synthesize and unify a significant range of phenomena;

2. they have predictive power, either predicting new phenomena, or suggesting a direction for further research and testing. Compare: hypothesis, and law.

Theory of evolution by natural selection A theory conceived by Charles Darwin and Alfred Russell Wallace on how some evolutionary changes occur.

Theory of inheritance of acquired characteristics The mistaken idea that organisms develop new organs, or modify existing organs as environmental problems present themselves, and that these traits are passed on to offspring.

Therapsids Extinct Mesozoic mammal-like reptiles from which true mammals evolved.

Thermal energy The total potential and kinetic energy associated with the random motions of the particles of a material.

Thermal plume A plume of hot water discharged into a stream or lake by a heat source, such as a power plant.

Thermal pollution Pollution by heated water.

Thermal stratification The layering of different temperatures of water or air caused by different densities, less dense floating on more dense layers.

Thermionic emission The liberation of electrons from the surface of a hot body.

Thermocline Layer of water separating upper warmer and lighter water from lower colder and heavier water in a lake or sea; a stratum of abrupt change in water temperature.

Thermoconformer To conform to the temperature of one's external environment.

Thermocouple An electric circuit composed of two dissimilar metals whose junctions are maintained at different temperatures.

Thermodynamic control (of product composition) The term characterizes conditions that lead to reaction products in a proportion governed by the equilibrium constant for their interconversion and/or for the interconversion of reaction intermediates formed in or after the rate-limiting step. (Some workers prefer to describe this phenomenon as "equilibrium control".).

Thermodynamics The branch of science that deals with heat, energy, and the interconversion of these; the study of energy transformations.

Thermodynamic laws The change in internal energy of a system is the sum of the heat transferred to or from the system and the work done on or by the system.

Second law of thermodynamics The entropy — a measure of the unavailability of a system's energy to do useful work — of a closed system tends to increase with time.

Third law of thermodynamics For changes involving only perfect crystalline solids at absolute zero, the change of the total entropy is zero.

Zeroth law of thermodynamics If two bodies are each in thermal equilibrium with a third body, then all three bodies are in thermal equilibrium with each other.

Thermoelectric effect The production of an electron current in a closed circuit consisting of two dissimilar metals as a result of the emf developed when the two junctions are maintained at different temperatures.

Thermogenesis The generation of heat by muscle contraction.

Thermolysis The uncatalysed cleavage of one or more covalent bonds resulting from exposure of a compound to a raised temperature, or a process in which such cleavage is an essential part.

Thermonuclear reaction Nuclear fusion

Thermoplastics Soft plastics composed of single-chain, unlinked polymers, such as polyethylene, polypropylene, polyvinylchloride, polystyrene, and polyester, that can be remelted and reformed to make useful products.

Thermoreceptor A sensory receptor that is sensitive to changes in temperature; a heat receptor.

Thermoregulation Heat regulation.

Thermoset polymers Hard plastics composed of cross-linked molecular networks, such as acrylic, phenolic, or epoxy resins, that cannot be remelted or recycled.

Thermosphere The highest atmospheric zone; a region of hot, dilute gases above the mesosphere extending out to about 1600 km from the earth's surface.

Third law of photoelectric emission Within the region of effective frequencies, the maximum kinetic energy of photoelectrons varies directly with the difference between the frequency of the incident light and the cut-off frequency.

Third World Less-developed countries that are not capitalistic and industrialized (First World) or centrally-planned socialist economies (Second World); not intended to be derogatory.

Thomson experiment When an electric current flows through a conductor whose ends are maintained at different temperatures, heat is released at a rate approximately proportional to the product of the current and the temperature gradient.

Thoracic Pertaining to the thorax or chest.

Thorn A modified stem terminating in a sharp point.

Thorn forest The dry end of a savanna community; hotter and drier than a typical savanna, these areas are characterized by dense, thorny small trees.

Thought experiment An idealized experiment that cannot be performed under actual conditions.

Threatened species While still abundant in parts of its territorial range, this species has declined significantly in total numbers and may be on the verge of extinction in certain regions or localities.

Threshold of hearing The intensity of the faintest sound audible to the average human ear, 10-16 w/cm^2 at 10^3 hz

Threshold of pain For audible frequencies of sound, an intensity level above which pain results in the average human ear Rock concert.

Threshold The minimum stimulus necessary to initiate an all-or-none response.

Thrombin. Enzyme catalyzing fibrinogen transformation into fibrin. Percursor is prothrombin.

Thrombus Blood clot in a blood vessel or in one of the cavities of the heart.

Thylakoid The lamellar structure of the grana of chloroplasts.

Thylakoid membrane A saclike photosynthetic membrane in chloroplasts; stacks of thylakoids form the grana.

Thymine A pyrimidine base occurring in DNA but not in RNA.

Thymus gland A ductless mass of flattened lymphoid tissue situated behind the top of the sternum; it forms antibodies in the newborn and is involved in the development of the immune system.

Thyroid An endocrine gland near the shield-shaped cartilage of the larynx.

Ti plasmid The tumor-inducing plasmid from the bacterium, Agrobacterium tumefaciens;

commonly used as a vector for recombinant DNA studies in plants.

Tibia Podomere of an insect or acarine leg that articulates proximally with the femur in insects and patella in acarines and distally with the tarsus in insects or with the metatarsus or tarsus in acarines.

Tidal bore A steep wave generated as high tides move up some estuaries and rivers.

Tidal current A current that is generated by tides.

Tidal energy Energy that can be harnessed as a result of the movement of tides.

Tidal range The difference in water level between successive high and low tides.

Tidal station A dam built across a narrow bay or estuary traps tide water flowing both in and out of the bay. Water flowing through the dam spins turbines attached to electric generators.

Tide The periodic, rhythmic rise and fall of the sea surface.

Tide pool A depression that holds seawater at low tide.

Tide table A table that gives the predicted time and height of tides for particular points along a coast.

Tied ridges Series of ridges running at right angles to each other so that runoff is blocked in all directions and water is allowed to soak into the soil.

Tiedemann's bodies Four or five pairs of pouchlike bodies attached to the ring canal of sea stars, apparently functioning in production of coelomocytes.

Tight junction Region of actual fusion of cell membranes between two adjacent cells.

Tillering The production of lateral buds and shoots near the ground to result in a plant with several shoots instead of one; particularly important in grasses and grain crops.

Timberline In mountains, the highest-altitude edge of forest that marks the beginning of the treeless alpine tundra.

Tintinnids Ciliates that secrete vase-like cases, or loricas.

Tipler machine A solution to Einstein's equations of general relativity that allows time travel. An extremely dense (on the order of the density of neutron star matter), infinitely-long cylinder which rotates very rapidly can form closed timelike curves in its vicinity, which will allow time travel and possible subsequent violations of causality.

Tissue A group of similar cells that performs a specialized function.

Titer Concentration of a substance in a solution as determined by titration.

Tolerance Degree of precision obtainable with a measuring instrument

Tolerance range The range of variation in an environmental parameter that is compatible with the life of an organism.

Tonicity The state of tissue tone or tension; in body fluid physiology, the effective osmotic pressure equivalent.

Tonoplast The membrane surrounding the vacuole.

Tool-making revolution The invention of tools—probably first of stone, wood, or bone_ sometime in the early Paleolithic period. Knowledge of tool-making and tool-using technologies may have predated modern humans, as evidenced by tool utilization by other primates.

Top carnivore The consumer at the end of a food chain or web; a carnivore that ordinarily has no predator under those ecological conditions.

Topography The surface condition of an area of land; relief features.

Topsoil The first true layer of soil; layer in which organic material is mixed with mineral particles; thickness ranges from a meter or more under virgin prairie to zero in some deserts.

Tornado A violent storm characterized by strong swirling winds and updrafts; tornadoes form when a strong cold front pushes under a warm, moist air mass over the land.

Tornaria A free-swimming larva of enteropneusts that rotates as it swims; resembles somewhat the bipinnaria larva of echinoderms.

Torpor A time of decreased metabolism and lowered body temperature that occurs in daily activity cycles.

Torque Product of a force and the effective length of its torque arm

Torque arm The perpendicular distance between the line of action of the torque producing force and the axis of rotation.

Torquoselectivity The term refers to the preference for "inward" or "outward" rotation of substituents in conrotatory or disrotatory electrocyclic ring opening reactions.

Torsion A developmental twisting of the visceral mass of a gastropod mollusc that results in an anterior opening of the mantle cavity and a twisting of nerve cords and the digestive tract.

Tortoiseshell The polished shell of hawksbill turtles.

Total fertility rate The number of children born to an average woman in a population during her entire reproductive life.

Total growth rate The net rate of population growth resulting from births, deaths, immigration, and emigration.

Total reflection The reflection of light at the boundary of two transparent

Toxic colonialism Shipping toxic wastes to a weaker or poorer nation.

Toxicyst Structures possessed by predatory ciliate protozoa, which on stimulation expel a poison to subdue the prey.

Toxins Poisonous chemicals that react with specific cellular components to kill cells or to alter growth or development in undesirable ways; often harmful, even in dilute concentrations.

Trabecula In general anatomical usage, a septum extending from an envelope through an enclosed substance, which, together with other trabeculae, forms part of the framework of various organs; here referring specifically to the cell processes connecting the perikarya of cestode and trematode tegumental cells with the distal cytoplasm. Also called internuncial process.

Trabecular net Network of living tissue formed by pseudopodia of amebocytes in Hexactinellida (phylum Porifera).

Trace element an element needed for normal metabolism but available only in minute amounts from the environment

Trace mineral Dietary minerals that are required in minute quantities.

Tracheae The small tubes that carry air from spiracles through the body cavity of an arthropod; arthropod tracheae are modifications of the exoskeleton.

Tracheal system System of cuticle-lined tubes in many insects and acarines that functions in respiration; opens to outside through spiracles.

Tracheid An elongated, tapering xylem cell that is specialized for conducting water and support with lignified pitted walls.

Tracheophytes Plants with a well-defined vascular system.

Tract A bundle of nerve fibers within the central nervous system.

Transcription The formation of a messenger RNA molecule that carries the genetic code from the nucleus to the cytoplasm of a cell.

Transducer A receptor that converts one form of energy into another.

Transduction. Condition in which bacterial DNA (and the genetic characteristics it bears) is transferred from one bacterium to another by the agent of viral infection.

Transfer RNA (tRNA). A form of RNA of about 70 or 80 nucleotides, which are adapter molecules in the synthesis of proteins. A specific amino acid molecule is carried by transfer RNA to a ribosome-messenger RNA complex for incorporation into a polypeptide.

Transferability Transferability assumes invariance of properties, associated conceptually with an atom or a fragment present in a variety of molecules. The property, such as electronegativity, nucleophilicity, NMR chemical shift, etc. is held as retaining a similar value in all these occurrences.

Transform fault A large horizontal displacement in the mid-ocean ridge.

Transformation A transformation is a change, generally in the context of a coordinate system. In this context, the transformation will change the basis set which defines the dimensions of the coordinate system. The two basic types of transformations are translations and rotations. An example of a rotation transformation is to switch from considering North-South versus East-West to considering Northeast-Southwest versus Northwest-southeast. (The rotation is usually about the origin, but it can be about any point.) An example of a translation transformation is to switch from considering Chicago as the origin to considering New York as the origin.

Transformer A device for changing an alternating voltage from one potential to another.

Transforming factor A substance that can be passed from one cell to another and cause a permanent change in heredity.

Transgenic Cells or organisms that contain genes that were inserted into them from other organisms using the techniques of genetic engineering.

Transient (chemical) species Relating to a short-lived reaction intermediate. It can be defined only in relation to a time scale fixed by the experimental conditions and the limitations of the technique employed in the detection of the intermediate. The term is a relative one.

Transient species are sometimes also said to be "metastable". However, this latter term should be avoided, because it relates a thermodynamic term to a kinetic property, although most transients are also thermodynamically unstable with respect to reactants and products.

Transistor A semiconductor device used as a substitute for vacuum tubes in electronic applications.

Transition coordinate The reaction coordinate at the transition state corresponding to a vibration with an imaginary frequency. Motion along it in the two opposite senses leads towards the reactants or towards the products.

Transition state In theories describing elementary reactions it is usually assumed that there is a transition state of more positive molar Gibbs energy between the reactants and the products through which an assembly of atoms (initially composing the molecular entities of the reactants) must pass on going from reactants to products in either direction. In the formalism of "transition state theory" the transition state of an elementary reaction is that set of states (each characterized by its own geometry and energy) which an assembly of atoms, when randomly placed there, would have an equal probability of forming the reactants or of forming the products of that elementary reaction. The transition state is characterized by one and only one imaginary frequency. The assembly of atoms at the transition state has been called an activated complex.

It may be noted that the calculations of reaction rates by the transition state method and based on calculated potential-energy surfaces refer to the potential energy maximum at the saddle point, as this is the only point for which the requisite separability of transition state coordinates may be assumed. The ratio of the number of assemblies of atoms that pass through to the products to the number of those that reach the saddle point from the reactants can be less than unity, and this fraction is the "transmission coefficient".

Transition structure A saddle point on a potential-energy surface. It has one negative force constant in the harmonic force constant matrix.

Transition temperature The temperature (dependant on the substance involved) below which a superconducting substance conducts electricity with zero resistance; consequently, the temperature above which a superconductor loses its superconductive properties.

Transitional mutants A type of mutation in which a single purine-pyrimidine base pair is replaced by another.

Transitional zone A zone in which populations from two or more adjacent communities meet and overlap.

Translation The second stage of protein synthesis in which the codon of mRNA pairs with the anticodon of tRNA at the surface of the ribosome.

Translational equilibrium The state of a body in which there are no unbalanced forces acting on it.

Translocation Moving animals from one location to another, for instance to determine whether and how soon they shift their activity cycle to match the photoperiod and/or other features in their new location.

Transmission electron microscope The type of microscope that produces highly magnified images of ultrathin tissue sections or other specimens.

Transpiration The evaporation of water from plant surfaces, especially through stomates.

Transpirational pull Water molecules being "pulled" up the xylem are triggered by transpirational water loss at the leaf surface.

Transpiration-Cohesion Theory The theory that explains water movement in the xylem; the driving force is the pull of transpiration and the cohesion of water molecules.

Transplantation The movement of neural or hormonal tissue from one area of an animal to another or from one animal to another. Also, the movement of individuals or colonies from one location to another.

Transport host Paratenic host.

Transuranium elements Elements with atomic number greater than 92.

Transverse cut A section cut at right angles to the long axis. Also termed a cross section.

Transverse plane A plane or section that lies or passes across a body or structure.

Transverse wave A wave in which the vibrations are at right angles to the direction of propagation of the wave

Transversional mutant A type of mutation in which a purine-pyrimidine base pair is replaced by a pyrimidine-purine base pair.

Trapping The interception of a reactive molecule or reaction intermediate so that it is removed from the system or converted into a more stable form for study or identification.

Trauma Injury caused by accident or violence.

Trematoda The class of platyhelminthes that has members that are all parasitic; several holdfast devices present; have complicated life cycles involving both sexual and asexual reproduction.

Trench deep area in the ocean floor, generally deeper than 6,000 m

Triactinomyxon Stage in the life cycle of a myxozoan, formerly assigned to a separate class.

Tribocytic organ Glandular, padlike organ behind the acetabulum of a strigeoid trematode.

Trichinosis A disease resulting from infection by Trichinella spiralis (Nematoda) larvae by eating undercooked meat; characterized by muscular pain, fever, edema, and other symptoms.

Trichocysts An anchoring structure present in the ectoplasm of some ciliates. A bottle-shaped extrusible organelle of the ciliate pellicle.

Trichogon Spiny male larva of a rhizocephalan cirripede that comes to lie within a special receptacle in the female.

Trichome An epidermal appendage, such as a hair or scale.

Triglyceride A type of lipid formed from three fatty acids bonded to a molecule of glycerol; a fat or oil.

Trilobitamorpha The subphylum of arthropods whose members had bodies divided into three longitudinal lobes. Head, thorax, and abdomen present. One pair of antennae and biramous appendages. Entirely extinct.

Trimerophytophyta An extinct division of early vascular plants characterized by a monopodial axis, dichotomously branched lateral axes bearing terminal sporangia. Believed to have been the ancestor of the ferns and progymnosperms.

Trimerous Body in three main divisions, as in lophophorates and some deuterostomes.

Triode Vacuum tube consisting of a plate, grid, and cathode.

Triple point The single condition of temperature and pressure at which the solid, liquid, and vapor phases of a substance can coexist in stable equilibrium.

Triplet A group of three nucleotides on a nucleic acid that codes for a particular amino acid.

Triploblastic Animals whose body parts are organized into layers that are derived embryologically from three tissue layers: ectoderm, mesoderm, and endoderm. Platyhelminthes and all coelomate animals are triploblastic.

Triploid A cell or nucleus that contains three sets of chromosomes; common in endosperm.

Triquetrum A carpal bone.

Tritonymph Third nymphal stage in most acarines.

Tritosternum Ventral, bristlelike sensory organ just behind the gnathosoma of a mesostigmatid mite.

Triungulin First instar larva of some parasitic, hypermetamorphic Neuroptera and Coleoptera and of the Strepsiptera, which is an active, campodeiform oligopod.

Trochanter Podomere of an insect or acarine leg that articulates basally with the coxa and distally with the femur; usually fixed to the femur in insects.

Trochophore A free-swimming ciliated marine larva characteristic of most molluscs and certain ectoprocts, brachiopods, and marine worms; an ovoid or pyriform body with preoral circlet of cilia and sometimes a secondary circlet behind the mouth.

Trochophore larva A larval stage characteristic of many molluscs, annelids, and some other protostomate animals.

Trojan satellites Satellites which orbit a body at one or the other trojan points relative to a secondary body. There are several examples of this in our own solar system: a group of asteroids which orbit in the the trojan points of jupiter; daughter satellites which orbit in the trojan points of the saturn-tethys system, and an additional satellite (helene) which orbits in the forward trojan point of saturn and dione.

Trombe wall An interior, heat-absorbing wall; may be water-filled glass tubes that absorb heat rays and let light into interior rooms.

Trophallaxis Exchange of food between young and adults, especially certain social insects.

Trophi Jaw-like structures in the mastax of rotifers.

Trophic Pertaining to feeding and nutrition.

Trophic hormones Hormonal or neurosecretory products from endocrine glands or neurosecretory cells that influence the production and release of other hormone products from endocrine glands.

Trophic levels The feeding level of an organism in an ecosystem; green plants and other autotrophs function at producer trophic levels; animals function at the consumer trophic levels.

Trophoblast Outer ectodermal nutritive layer of blastodermic vesicle; in mammals it is part of the chorion and attaches to the uterine wall.

Trophont Stage in the life cycle of gregarines.

Trophosome Organ in poganophorans bearing mutualistic bacteria, derived from midgut.

Trophozoite Active, feeding stage of a protozoan, in contrast to a cyst. Also called the vegetative stage.

Tropic Related to the tropics (tropical); in endocrinology, a hormone that influences the action of another hormone or endocrine gland.

Tropical rain forest An ecosystem characterized by very high rainfall and temperatures between 20 and 25C; very diverse life-forms; broad-leaved, nondeciduous trees; highly stratified forest.

Tropical seasonal forest Semievergreen or partly deciduous forests tending toward open woodlands and grassy savannas dotted with scattered, drought-resistant tree species; distinct wet and dry seasons, hot year-round.

Tropomyosin Low-molecular weight protein surrounding the actin filaments of striated muscle.

Troponin Complex of globular proteins positioned at intervals along the actin filament of skeletal muscle; thought to serve as a calcium-dependent switch in muscle contraction.

Tropopause The boundary between the troposphere and the stratosphere.

Troposphere The layer of air nearest to the earth's surface; both temperature and pressure usually decrease with increasing altitude.

Trough A region of downward displacement in a transverse wave.

True language Communication that includes the use of symbols to represent abstract objects or ideas, and syntax, where those symbols convey different messages depending on their sequence.

True navigation The ability to maintain or establish reference to a goal without the use of landmarks.

Truth This is a word best avoided entirely in physics except when placed in quotes, or with careful qualification. Its colloquial use has so many shades of meaning from 'it seems to be correct' to the absolute truths claimed by religion, that it's use causes nothing but misunderstanding. Someone once said "Science seeks proximate (approximate) truths." Others speak of provisional or tentative truths. Certainly science claims no final or absolute truths.

Trypomastigote Form of Trypanosomatidae with an undulating membrane and the kinetoplast located posterior to the nucleus. An example is Trypanosoma.

Tsetse fly Bloodsucking fly of the genus Glossina.

Tsunami Giant seismic sea swells that move rapidly from the center of an earthquake; they can be 10 to 20 meters high when they reach shorelines hundreds or even thousands of kilometers from their source.

Tube feet (podia). Numerous small, muscular, fluid-filled tubes projecting from body of echinoderms; part of water-vascular system; used in locomotion, clinging, food handling, and respiration.

Tube nucleus One division of the microscope nucleus in a pollen grain that is responsible for the formation of a pollen tube from the stigma through the style to the ovule.

Tuber An enlarged, fleshy, underground stem tip, such as the potato.

Tubercle Small protuberance, knob, or swelling.

Tuberculum Process on some bones.

Tuberosity A normal raised area on a bone, such as the deltoid tuberosity, which is the site of insertion of the deltoid muscle.

Tuberous roots Modified fibrous roots that have become fleshy and enlarged with food reserves.

Tubular eyes Specialized eyes of many midwater animals that allow upward or downward vision.

Tubular nerve cord A hollow nerve cord that runs middorsally along the back of chordates; one of four unique chordate characteristics; also called the neural tube and, in vertebrates, the spinal cord.

Tubulin Globular protein forming the hollow cylinder of microtubules.

Tumor A spherical mass of cells in which cell divisions occur at random and often in an uncontrolled fashion.

Tumor necrosis factor A cytokine, the most important source of which is macrophages, that is a major mediator of inflammation.

Tundra Terrestrial habitat zone, located between taiga and polar regions; characterized by absence of trees, short growing season, and mostly frozen soil during much of the year.

Tunic In tunicates, a cuticular, cellulose-containing covering of the body secreted by the underlying body wall.

Tunicates Chordates that show the three basic chordate characteristics only in the larva.

Tuning fork A metal two-prong fork that produces a sound of a definite pitch.

Tunnelling The process by which a particle or a set of particles crosses a barrier on its potential energy surface without having the energy required to surmount this barrier. Since the rate of tunnelling decreases with increasing reduced mass, it is significant in the context of isotope effects of hydrogen isotopes.

Turbellaria The class of Platyhelminthes that has members that are mostly free living and aquatic; external surface usually ciliated; predaceous; possess rhabdites; protrusable proboscis; mostly hermaphroditic. Examples: Convoluta, Notoplana, Dugesia.

Turbellarians Mostly free-living flatworms.

Turbinate Folded parts of the ethmoid bone.

Turbulence random, nonlaminar flow of a fluid

Turf The intertwined fibrous roots of grasses forming a mass with the soil just below ground level.

Turgid A swollen, distended cell that is firm due to water uptake.

Turgor pressure The real pressure developed in living cells by pressing against a membrane.

Turning over The mixing of thermal zones in a standing body of freshwater due to cold, more dense water "sinking" from the top through the other layers.

Turnover rate The rate at which members of a population or community replace themselves

Turpentine A solvent that includes two terpenes—camphor and pinene.

Tusk shells Molluscs that have an elongate, tapered shell that is open at both ends.

Twin paradox One of the most famous "paradoxes" in history, predicted by a. Einstein's special theory of relativity. Take two twins, born on the same date on earth. One, leaves home for a trip around the universe at very high speeds (very close to that of light), while the other, stays at home at rests. Special relativity predicts that when first one returns, he will find himself much younger than second one.

That is actually not the paradox. The paradox stems from attempting to naively analyze the situation to figure out why. From second one's point of view (and from everyone else on earth), first one seems to speed off for a long time, linger around, and then return. Thus he should be the younger one, which is what we see. But from first one's point of view, it's second one (and the whole of the earth) that are travelling, not he. According to special relativity, if second one is moving relative to first one, then first one should measure his clock as ticking slower — and thus second one is the one who should be younger. But this is not what happens.

So what's wrong with our analysis? The key point here is that the symmetry was broken. First one did something that second one did not — first one accelerated in turning around. second one did no accelerating, as he and all the other people on the earth can attest to (neglecting gravity). So first one broke the symmetry, and when he returns, he is the younger one.

Tympanal (tympanic) organs Auditory receptors present on the abdomen or legs of some insects.

Type specimen. A specimen deposited in a museum that formally defines the name of the species that it represents.

Typhlosole A longitudinal fold projecting into the intestine in certain invertebrates such as the earthworm.

Typology A classification of organisms in which members of a taxon are perceived to share intrinsic, essential properties, and variation among organisms is regarded as uninteresting and unimportant.

U

Ulcer Area of inflammation that opens out to the skin or a mucous surface.

Ultimate causation (factors) Those aspects of behavior that are concerned with why the behavior evolved and its functional significance in an ecological context.

Ultimobranchial The last embryonic gill slit.

Ultimobranchial gland In jawed fishes, primitive tetrapods, and birds, the small gland(s) that forms ventral to the esophagus; produces the hormone calcitonin that helps regulate calcium concentrations.

Ultradian rhythm A cyclical rhythm of less than 24 hours.

Ultraplankton phytoplankton with cells between 2 and 5 micrometers in width.

Ultrasonic range Vibrations in matter above 20,000 vibrations/second

Ultrastructure Structure of an organism or cell at the electron microscopic level.

Ultraviolet catastrophe A shortcoming of the Rayleigh-Jeans formula, which attempted to describe the radiancy of a blackbody at various frequencies of the electromagnetic spectrum. It was clearly wrong because as the frequency increased, the radiancy increased without bound; something quite not observed; this was dubbed the "ultraviolet catastrophe." It was later reconciled and explained by the introduction of the Planck radiation law.

Ultraviolet light Electromagnetic radiations of shorter wavelength than visible light but longer than X rays.

Ultraviolet That portion of the sun's total range of radiation having wavelengths immediately shorter than the shortest of the visible spectrum (purple); between approximately 380 and 100 nm.

Umbel An inflorescence in which the individual pedicels all arise from the apex of the peduncle.

Umbilical cord Cordlike structure that connects the fetus to the placenta.

Umbilical Refers to the navel, or umbilical cord.

Umbo One of the prominences on either side of the hinge region in a bivalve mollusc shell. Also, the "beak" of a brachiopod shell.

Umbra The part of a shadow from which all light rays are excluded.

Umpolung Any process by which the normal alternating donor and acceptor reactivity pattern of a chain, which is due to the presence of O or N heteroatoms, is interchanged. Reactivity umpolung is most often achieved by temporary exchange of heteroatoms (N, O) by others, such as P, S, and Se.

The original meaning of the term has been extended to the reversal of any commonly accepted reactivity pattern. For example, reaction of R-C≡CX (X = halide) as a synthon for "R-C≡C^+ (i.e. electrophilic acetylene) is an umpolung of the normal more common acetylide, R-C≡C^- (i.e. nucleophilic) reactivity.

Umwelt The sensory and perceptual world of the animal, dependent upon the types of sensory receptors that it possesses and the internal nervous system processes for receiving and interpreting those stimuli.

Uncertainty A measure of the inherent variability of repeated measurements of a quantity. A prediction of the probable variability of a result, based on the inherent uncertainties in the data, found from a mathematical calculation of how the data uncertainties would, in combination, lead to uncertainty in the result. This calculation or process by which one predicts the size of the uncertainty in results from the uncertainties in data and procedure is called error analysis. Uncertainties are always present; the experimenter's job is to keep them as small as required for a useful result. We recognize two kinds of uncertainties: indeterminate and determinate. Indeterminate uncertainties are those whose size and sign are unknown, and are sometimes (misleadingly) called random. Determinate uncertainties are those of definite sign, often referring to uncertainties due to instrument miscalibration, bias in reading scales, or some unknown influence on the measurement.

Unconditioned response (UCR) In classical conditioning, the animal's response that is initially given to the unconditioned stimulus.

Unconditioned stimulus (US) In classical conditioning, the stimulus that produces what is initially the unconditioned response.

Unconventional air pollutants Toxic or hazardous substances, such as asbestos, benzene, beryllium, mercury, polychlorinated biphenyls, and vinyl chloride, also called noncriteria pollutants.

Unconventional oil Resources such as shale oil and tar sands that can be liquefied and used like oil.

Undernourished Those who receive less than 90 percent of the minimum dietary intake over a long-term time period; they lack energy for an active, productive life and are more susceptible to infectious diseases.

Understory That vegetation which characterizes the lower level of plants in a forest; the vegetation below the canopy.

Undiscovered resources Potential supplies of a mineral or other useful material believed to exist based on history, scientific theory, or general knowledge of geology, biology, or geography of an unexplored area.

Undulating membrane Name applied to two quite different structures in protozoa. In some Mastigophora it is a finlike ridge across the surface of a cell, with the axoneme of a flagellum near its surface. In some ciliates it is a line of cilia that are fused at their bases, usually beating to force food particles toward the gullet.

Undulating ridges Undulatory waves in the surface of some protozoa, probably aided by subpellicular microtubules; the means of locomotion in some species.

Unicellular The life-form in which all functions are carried out within the confines of a single plasma membrane; members of the kingdom Protista display unicellular organization; also called cytoplasmic organization.

Unified field theory The principle that all forces in the universe are part of a single concept.

Uniformitarianism Methodological assumptions that the laws of chemistry and physics have remained constant throughout the history of the earth, and that past geological events occurred by processes that can be observed today.

Uniformity principle The principle that the laws of physics here and now are not different, at least qualitatively, from the laws of physics in previous or future epochs of time, or elsewhere in the universe. This principle was scoffed at by the ancients who believed that the laws that governed the earth and those that governed the heavens were completely divorced; now it is used routinely in cosmology to describe the structure and evolution of the universe.

Uniramia The subphylum of arthropods whose members are characterized by a head with one pair of antennae and one pair of mandibles. All appendages are uniramous.

Uniramous appendage Arthropod appendage that is unbranched, characteristic of living arthropods other than Crustacea, although some crustacean appendages are uniramous.

Unisexual The condition of having either stamens or pistil, not both; an imperfect flower.

Units Labels which distinguish one type of measurable quantity from other types. Length, mass and time are distinctly different physical quantities, and therefore have different unit names, meters, kilograms and seconds. We use several systems of units, including the metric (SI) units, the English (or U.S. customary units), and a number of others of mainly historical interest. Some dimensionless quantities are assigned unit names, some are not. Specific gravity has no unit name, but density does. Angles are dimensionless, but have unit names: degree, radian, grad. Some quantities which are physically different, and have different unit names, may have the same dimensions, for example, torque and work.

Unit magnetic pole One that repels an exactly similar pole placed one centimeter away with a force of one dyne

Unit membrane The transmission electron microscopic interpretation of a biological membrane, consisting of two electron-dense lines separated by a translucent space.

Universal age paradox Two of the most straightforward methods of calculating the age of the universe — through redshift measurements, and through stellar evolution — yield incompatible results. Recent (mid 1990s) measurements of the distances of distant galaxies through the use of the hubble space telescope indicate an age much less than the ages of the oldest stars that we calculate through stellar evolution theory. At present there is no conclusion to this paradox; a cosmological constant would rectify the situation, but it's possible that the discrepancy will disappear with more accurate measurements of the age of the universe using both methods.

Universal constant of gravitation; g The constant of proportionality in newton's law of universal gravitation and which plays an analogous role in a. Einstein's general relativity. It is equal to 6.672 x 10^{-11} n m^2/kg^2.

Universal gas constant Constant of proportionality, R, in the ideal gas equation.

Universal veil A membrane that totally encloses some young mushrooms; after it breaks, its remnants appear as a volva at the base and scales on the cap.

Universalists Those who believe that some fundamental ethical principles are universal and unchanging. In this vision, these principles are valid regardless of the context or situation.

Unreactive Failing to react with a specified chemical species under specified conditions. The term should not be used in place of stable, since a relatively more stable species may nevertheless be more reactive than some reference species towards a given reaction partner.

Unsaturated fat A fat containing one or more double bonds between carbon atoms.

Unstable The opposite of stable, i.e. the chemical species concerned has a higher molar Gibbs energy than some assumed standard. The term should not be used in place of reactive or transient, although more reactive or transient species are frequently also more unstable.

(Very unstable chemical species tend to undergo exothermic unimolecular decompositions. Variations in the structure of the related chemical species of this kind generally affect the energy of the transition states for these decompositions less than they affect the stability of the decomposing chemical species. Low stability may therefore parallel a relatively high rate of unimolecular decomposition.)

Upwelling Convection currents within a body of water that carry nutrients from bottom sediments toward the surface.

Uracil A nitrogen base found in RNA.

Urban area An area in which a majority of the people are not directly dependent on natural resource-based occupations.

Urban Peculiar to the human environment, as contrasted with that found normally around wild animals.

Urban renewal Programs to revitalize old, blighted sections of inner cities.

Urbanization An increasing concentration of the population in cities and a transformation of land use to an urban pattern of organization.

Urea A nitrogen-containing waste product excreted in the urine of many vertebrates

Uredium The structure that produces uredospores in rust fungi; sometimes called a uredinium.

Uredospore A reddish, binucleate spore formed by rust fungi; often forms the repeating stage of the rust; also called a urediniospore.

Ureotelic excretion Having urea as the chief excretory product of nitrogen metabolism; occurs in mammals.

Ureter The tube which conveys urine from the kidney to the bladder.

Urethra The tube from the urinary bladder to the exterior in both sexes.

Uric acid the main nitrogenous excretory product in birds, reptiles, some invertebrates, and insects

Uricotelic excretion Having uric acid as the chief excretory product of nitrogen metabolism; occurs in reptiles and birds.

Urinary bladder The storage organ for urine.

Urn Region near the center of an infusoriform larva of a dicyemid mesozoan.

Urochordata The subphylum of chordates whose members have all four chordate characteristics as larvae. Adults are sessile or planktonic and enclosed in tunic that usually contains cellulose. Sea squirts or tunicates.

Urogenital opening The common opening for urine and gametes in bony fishes and other animals.

Urophysis A discrete structure in the spinal cord of the fish tail that produces neuropeptides that help control water and ion balance, blood pressure, and smooth muscle contractions. 586

Uropod Posteriormost appendage of many crustaceans.

Uropolar cells Somatoderm cells at the posterior end of the trunk of a dicyemid mesozoan.

Urosome Portion of the body posterior to the major point of body flexion in many copepods; usually includes one or more free thoracic segments and abdomen.

Urstigmata Sense organs between the coxae of the first and second pairs of legs on some mites. Apparently they are humidity receptors. Also called Claparedé organs.

Uterine bell Structure in female acanthocephalans that allows fully developed, shelled embryos to pass out of the body and that retains undeveloped ones.

Uterine tube The tube that leads from the ovary to the uterus; also called fallopian tube.

Uterus The portion of female mammals' reproductive tract in which the embryo develops.

Utilitarian conservation A philosophy that resources should be used for the greatest good for the greatest number for the longest time.

Utricle That part of the inner ear containing the receptors for dynamic body balance; the semicircular canals lead from and to the utricle.

V

Vaccine A preparation of either killed microorganisms; living, weakened (attenuated) microorganisms; or inactivated bacterial toxins (toxoids); administered to induce development of the immune response and protect the individual against a pathogen or toxin.

Vacuole A membrane-bounded, fluid-filled space in a cell.

Vagabond's disease Darkened, thickened skin caused by years of infestation with body lice, Pediculus humanus humanus.

Vagina Tubular organ that leads from the uterus to the vestibule of the female reproductive tract.

Valence Degree of combining power of an element as expressed by the number of atoms of hydrogen (or its equivalent) that the element can hold (if negative) or displace in a reaction (if positive). The oxidation state of an element in a compound. The number of electrons gained, shared, or lost by an atom when forming a bond with one or more other atoms.

Valence isomer A constitutional isomer interrelated with another by pericyclic reactions. For example, Dewar benzene, prismane and benzvalene are valence isomers of benzene.

Valence tautomerization The term describes simple reversible and generally rapid isomerizations or degenerate re-arrangements involving the formation and rupture of single and/ or double bonds, without migration of atoms or groups.

Value (rho-value) A measure of the susceptibility to the influence of substituent groups on the rate constant or equilibrium constant of a particular organic reaction involving a family of related substrates. Defined by Hammett for the effect of ring substituents in meta- and para-positions of aromatic side-chain reactions by the empirical "-equation" of the general form

lg(kx/kH) = X

in which x is a constant characteristic of the substituent X and of its position in the reactant molecule.

More generally, -values are used to designate the susceptibility of reaction series for families of various organic compounds to any substituent effects, as given by the modified set of -constants in an empirical -correlation.

Reactions with a positive -value are accelerated (or the equilibrium constants of analogous equilibria are increased) by substituents with positive -constants. Since the sign of was defined so that substituents with a positive increase the acidity of benzoic acid, such substituents are generally described as attracting electrons away from the aromatic ring. It follows that reactions with a positive -value are considered to involve a transition state (or reaction product) so that the difference in energy between this state and the reactants is decreased by a reduction in electron density at the reactive site of the substrate.

Valvifers Basal portions of the ovipositor in Hymenoptera, derived from coxae of segmental appendages.

Valvulae Processes from the valvifers to form the body of the ovipositor (terebra) and the ovipositor sheath (third valvulae) in Hymenoptera.

Van de Graaff generator A particle accelerator that transfers a charge from an electron source to an insulated sphere by means of a moving belt composed of an insulating material

Van der Waals forces The attractive or repulsive forces between molecular entities (or between groups within the same molecular entity) other than those due to bond formation or to the electrostatic interaction of ions or of ionic groups with one another or with neutral molecules. The term includes: dipole-dipole, dipole-induced dipole and London (instantaneous induced dipole-induced dipole) forces. The term is sometimes used loosely for the totality of nonspecific attractive or repulsive intermolecular forces.

Vapor The gaseous phase of a substance that exists as a liquid or solid under normal conditions.

Vaporization The change in phase from a solid or a liquid to a gas Variation.

Variability Generally refers to genetic differences between all organisms that results in uniqueness.

Variable A symbol representative of a set of numbers, points, values, etc. In science, variables represent values of measurements of quantities.

Much confusion exists about the meanings of dependent and independent variables. In one sense this distinction hinges on how you write the relation between variables. 1. If you write a function or relation in the form y = f(x), y is considered dependent on x and x is said to be the independent variable.

2. If one variable (say x) in a relation is experimentally set, fixed, or held to particular values while measuring corresponding values of y, we call x the independent variable. We could just as well (in some cases) set values of y and then determine corresponding values of x. In that case y would be the independent variable.

3. If the experimental uncertainties of one variable are smaller than the other, the one with the smallest uncertainty is often called the independent variable.

4. As a general rule independent variables are plotted on the horizontal axis of a graph, but this is not required if there's a good reason to do it otherwise.

In many cases these four different practical definitions do not conflict with each other and one may choose language, form of equation, and method of plotting graphs so that all definitions are satisfied. But not always. Let common sense and the need for clear communication decide how to deal with situations where there seems to be conflict.

Some common statistical packages for computers can only deal with situations where one variable is assumed error-free, and all the experimental error is in the other one. They cavalierly refer to the error-free variable as the independent variable. But in real science, there's always some experimental error in all values, including those we "set" in advance to particular values.

Variance A statistical measure of the amount of variation in a series of measurements.

Variant antigen type (VAT) Applied to certain trypanosomes, any one of numerous antigenic types expressed on the surface of the organisms and "seen" by the immune system of the host.

Variant-specific surface glycoprotein (VSG) Glycoprotein on the surface of certain trypanosomes recognized by the host's immune system. Each VSG is responsible for one VAT.

Variation Differences among individuals of a group or species that cannot be ascribed to age, sex, or position in the life cycle.

Vascular bundle A strand of tissue containing primary xylem and primary phloem, often surrounded by a bundle sheath.

Vascular cambium Meristematic tissue that gives rise to secondary xylem and secondary phloem.

Vascular cylinder The stele; vascular tissue making up the central cylinder of roots.

Vascular plant A general name for any plant that has xylem and phloem.

Vascular ray Sheet of parenchyma that extends radially through the wood, across the cambium and into the secondary phloem; rays are produced by the vascular cambium and function in lateral transport.

Vascular system The total of the tissues that function in the conduction of water and food by means of xylem and phloem in the plant body.

Vascular tissue Tissue that is specialized for the long-distance transport of water or photosynthetic products; xylem and phloem.

Vector quantity A quantity that is completely specified by a magnitude and a direction.

Vector The mathematical classification of a quantity which contains more than one piece of information which is organized in a specific manner. Vectors can always be described as provide information about how far and in which direction (from some point). For example, the position vector for a tree tells us that the tree in question is not just 5 meters from the house, but that it is five meters to the north of the house. The information provided by a vector can be denoted in a variety of ways, one of which is as an array (list) of numbers that denote the size of the components of the vector along specific axes.

Vegan A pure vegetarian consuming no animal products at all.

Vegetal pole The lower pole of an egg; usually more dense than the animal pole because it contains more yolk.

Vegetarian A person who does not consume animal flesh; some consume dairy products and eggs, while others are vegans.

Veins Blood vessels conducting blood toward the heart in any animal. Also more heavily sclerotized portions of wings of insects, which are remains of lacunae.

Velarium Shelf-like extension of the subumbrella edge in cubozoans (phylum Cnidaria).

Veliger Larval form of certain molluscs; develops from the trochophore and has the beginning of a foot, mantle, shell, and so on.

Veliger larva The second free-swimming larval stage of many molluscs; develops from the trochophore and forms rudiments of the shell, visceral mass, and head-foot before settling to the substrate and undergoing metamorphosis.

Velocity Speed in a particular direction.

Velum A membrane on the subumbrella surface of jellyfish of class Hydrozoa. Also, a ciliated swimming organ of the veliger larva.

Vena cava The major vein returning blood to the heart of vertebrates

Venation The pattern of vein development.

Veneer A thin sheet of wood, often with attractive grain, used to cover less expensive wood.

Venule A small blood vessel that collects blood from a capillary bed and joins a vein.

Vermicle Infective stage of Babesia in a tick.

Vermiculite Expanded mica used as a sterile medium for the rooting of cuttings; high water-holding capacity and relatively inert.

Vertebra Each of the bones that make up the backbone.

Vertebrata The subphylum of chordates whose members are characterized by cartilaginous or bony vertebrae surrounding a nerve cord. The skeleton is modified anteriorly into a skull for protection of the brain.

Vertical migration Daily or seasonal movement of small marine animals between the photic zone and midwater depths

Vertical stratification The vertical distribution of specific subcommunities within a community.

Vesicles Membrane-bound particles pinched off by constriction of a membrane, as in the Golgi apparatus.

Vesicular disease Any disease of the urinary bladder, such as vesicular schistosomiasis.

Vessel A tubelike column of vessel elements that are connected by open end walls and are specialized for the conduction of water and minerals.

Vessel element One of the cells forming a vessel and characterized by a perforation plate.

Vessel member One of the cells comprising a vessel, characterized by the possession of a lumen and perforation plate shared with other vessel members above and below in the series. Also termed vessel elements.

Vestibulum Cavity leading into another cavity or passage, such as in the ciliate order Vestibulifera.

Vestige A rudimentary organ that may have been well developed in some ancestor or in the embryo.

Vestigial structures Visible evidence of a structure that was present in an earlier stage in the evolution of an organism. One of the sources of evidence for evolution.

Vestimentiferans Pogonophorans that are common at hydrothermal vents.

Viability The period of time an organism remains alive; often used to describe the length of time before a seed will fail to germinate.

Vibrissa pl. Vibrissae Stiff hairs that grow from the nostrils or other parts of the face of many mammals and that serve as tactile organs; "whiskers."

Vicariance Geographical separation of populations, especially as imposed by discontinuities in the physical environment that fragmented populations that were formerly geographically continuous.

Village A collection of rural households linked by culture, custom, and association with the land.

Villus A small fingerlike, vascular process on the wall of the small intestine. Also one of the branching, vascular processes on the embryonic portion of the placenta.

Viroid An infectious agent consisting of a single-stranded RNA molecule with no protein coat; produces diseases in plants.

Virtual image The point(s) from which light rays converge as they emerge from a lens or mirror. The rays do not actually pass through each image point.

Virtual object The point(s) to which light rays converge as they enter a lens. The rays do not actually pass through each object point.

Virulence Degree of pathogenicity of an agent; how much damage the agent can cause.

Virus A submicroscopic noncellular particle composed of a nucleoprotein core and a protein shell; parasitic; will grow and reproduce in a host cell.

Viscera Internal organs in the body cavity.

Visceral mass The region of a mollusc's body that contains visceral organs.

Visceral organs of the thoracic, abdominal, and pelvic cavities.

Viscosity The ratio of shear stress to the rate of change of shear strain in a liquid or gas.

Visible light A portion of the electromagnetic spectrum that includes the wavelengths used for photosynthesis.

Visible spectrum That portion of all the sun's radiation that can be perceived as light by humans; between approximately 380 and 750 nm in wavelength.

Vitalism The view that natural processes are controlled by supernatural forces and cannot be explained through the laws of physics and chemistry alone, as opposed to mechanism.

Vitamin An organic substance other than a carbohydrate, lipid, or protein that is needed for normal metabolism but cannot be synthesized in adequate amounts by the body.

Vitellaria Structures in many flatworms that produce vitelline cells, that is, cells that provide eggshell material and nutrient for the embryo.

Vitelline membrane The noncellular membrane that encloses the egg cell.

Viviparity Reproduction in which eggs develop within the female body, with nutritional aid of maternal parent as in therian mammals, many reptiles, and some fishes; offspring are born as juveniles.

Viviparous An animal whose eggs develop inside the female while the embryo derives nutrition from the mother.

Volatile oils Terpenes composed of two to four isoprene units; also known as essential oils, such as lemon and peppermint.

Volatile organic compounds Organic chemicals that evaporate readily and exist as gases in the air.

Volatilization The process of evaporation; to pass into the atmosphere as a gas.

Volt The derived SI unit of electric potential, defined as the difference of potential between two points on a conductor carrying a constant current of 1 amp when the power dissipated between the points is 1 watt; it thus has units of watt/amp.

Voltage sensitivity Voltage per unit scale division of an electric instrument

Voltage-gated channels Ion-specific channels that open only in response to a specific polarity across the cell membrane.

Voltaic cell A device that changes chemical into electric energy by the action of two dissimilar metals immersed in an electrolyte.

Voltmeter An instrument used to measure the difference of potential between two points in an electric circuit.

Volume strain The ratio of the decrease in volume to the volume before stress is applied.

Voluntary (somatic) nervous system That part of the nervous system that relays commands to skeletal muscles.

Vulnerable species Naturally rare organisms or species whose numbers have been so reduced by human activities that they are susceptible to actions that could push them into threatened or endangered status.

Vulva The external genital organs in the female.

W

W particle The theoretical carrier for weak subatomic interactions.

Warm front A long, wedge-shaped boundary caused when a warmer advancing air mass slides over neighboring cooler air parcels.

Warning coloration Coloration that allows organisms to escape from predators by advertising something harmful or distasteful.

Waste stream The steady flow of varied wastes, from domestic garbage and yard wastes to industrial, commercial, and construction refuse.

Water column The vertical column of seawater that extends from the surface to the bottom.

Water cycle The recycling and reutilization of water on Earth, including atmospheric, surface, and underground phases and biological and nonbiological components.

Water droplet coalescence A mechanism of condensation that occurs in clouds too warm for ice crystal formation.

Water mass A body of water that can be identified by its temperature and salinity.

Water potential The relative ability of water molecules to do work by interacting with each other.

Water stress A shortage of water in plants resulting in decreased ability to function properly, because water affects all biochemical controls and is the medium for all chemical reactions.

Water table The top layer of the zone of saturation; undulates according to the surface topography and subsurface structure.

Water vascular system A network of water-filled canals in echinoderms used in locomotion and food-gathering.

Waterlogging Water saturation of soil that fills all air spaces and causes plant roots to die from lack of oxygen; a result of overirrigation.

Water-soluble vitamin A vitamin that is not readily stored in the body, with excess eliminated in the urine; includes B vitamins and vitamin C.

Water-vascular system. System of fluid-filled closed tubes and ducts peculiar to echinoderms; used to move tentacles and tube feet that serve variously for clinging, food handling, locomotion, and respiration.

Watt The derived SI unit of power, defined as a power of 1 J acting over a period of 1 s; it thus has units of J/s.

Wave The undulation that forms as a disturbance moves along the surface of the water. Waves can be described by their height (the vertical distance between crest and trough), wavelength (the horizontal distance between adjacent crests), and period (the time the wave takes to move past a given point).

Wave reinforcement The process by which two waves collide and add together to create a higher wave.

Wave shock The intensity of the impact of a wave.

Wave trough The lowest part of a wave.

Wavelength In a periodic wave, the distance between consecutive points of corresponding phase.

Wave-particle duality The principle of quantum mechanics which implies that light (and, indeed, all other subatomic particles) sometimes act like a wave, and sometime act like a particle, depending on the experiment you are performing. For instance, low frequency electromagnetic radiation tends to act more like a wave than a particle; high frequency electromagnetic radiation tends to act more like a particle than a wave.

Wax A lipid material with considerable oxygen inserted in the molecule; high melting point and relatively impermeable to water.

Weak equivalence principle The idea within general relativity that the worldline of a freefalling body is independent of its composition, structure, or state. This principle, embraced by newtonian mechanics and gravitation when newton set the inertial and gravitational masses equal to each other. This principle is incorporated into a stronger version with the equivalence principle.

Weak nuclear interaction The interaction that produces pairs of particles that have unusually long half-lives

Weather Description of the physical conditions of the atmosphere (moisture, temperature, pressure, and wind).

Weber The derived SI unit of magnetic flux equal to the flux that, linking a circuit of one turn, produces in it an electromotive force of 1 v as it is reduced to zero at a uniform rate in a period of 1 s; it thus has units of v s.

Weed A plant not valued for its use or beauty and not intentionally planted; a category of hay fever plants that includes nongrass and nontree species.

Weight The size of the external force required to keep a body at rest in its frame of reference.

Elementary textbooks almost universally define weight to be "the size of the gravitational force on a body." This would be fine if they would only consistently stick to that definition. But, no, they later speak of weightless astronauts, loss of weight of a body immersed in a liquid, etc. The student who is really thinking about this is confused. Some books then tie themselves in verbal knots trying to explain (and defend) why they use the word inconsistently. Our definition has the virtue of being consistent with all of these uses of the word.

In the special case of a body supported near the earth's surface, where the acceleration due to gravity is g, the weight happens to have size mg. So this definition gives the same size for the weight as the more common definition.

This definition is consistent with the statement: "The astronauts in the orbiting spacecraft were in a weightless condition." This is because they and their spacecraft have the same acceleration, and in their frame of reference (the spacecraft) no force is needed to keep them at the same position relative to their spacecraft. They and their spacecraft are both falling at the same rate. The gravitational force on the astronauts is still mg (though g is about 12% smaller at an altitude of 400 km than it is at the surface of the earth. It is not zero).

This definition is consistent with statements about the "loss of weight" of a body immersed in a liquid (due to the buoyant force). The "weight" meant here is the external force (not counting the buoyant force) required to support the body in equilibrium in the liquid.

Weir Filtration apparatus in flame-cell protonephridia of flatworms.

Weiss constant A characteristic constant dependent on the material, used in calculating the susceptibility of paramagnetic materials.

Westerlies The winds that blow from west to east at middle latitudes.

Wetlands Ecosystems of several types in which rooted vegetation is surrounded by standing water during part of the year.

Wheatstone bridge Instrument used for the measurement of electric resistance

Whirling disease Disease of fishes, caused by the protozoan Myxobolus cerebralis.

White blood cell (leukocyte) A type of blood cell involved with body defenses.

Whorled arrangement Three or more leaves per node.

Whorled Three or more leaves attached at a node.

Wiedemann-Franz law The ratio of the thermal conductivity of any pure metal to its electrical conductivity is approximately constant for any given temperature. This law holds fairly well except at low temperatures.

Wien displacement law For a blackbody, the product of the wavelength corresponding to the maximum radiancy and the thermodynamic temperature is a constant, the wien displacement law constant. As a result, as the temperature rises, the maximum of the radiant energy shifts toward the shorter wavelength (higher frequency and energy) end of the spectrum.

Wilderness An area of undeveloped land affected primarily by the forces of nature; an area where humans are visitors who do not remain.

Wilderness actlegislation of 1964 recognizing that leaving land in its natural state may be the highest and best use of some areas.

Wildlife Plants, animals, and microbes that live independently of humans; plants, animals, and microbes that are not domesticated.

Wildlife refuges Areas set aside to shelter, feed, and protect wildlife; due to political and economic pressures, refuges often allow hunting, trapping, mineral exploitation, and other activities that threaten wildlife.

Wind farms Large numbers of windmills concentrated in a single area; usually owned by a utility or large-scale energy producer.

Windbreak Rows of trees or shrubs planted to block wind flow, reduce soil erosion, and protect sensitive crops from high winds.

Windward The side toward which the wind is blowing.

Winnowing The process that separates the grain from the fragments of chaff.

Winter sleep A period of inactivity in which a mammal's body temperature remains near normal and the mammal is easily aroused.

Winterbottom's sign Swollen lymph nodes at the base of the skull, symptomatic of African sleeping sickness.

Wisconsin General Test Apparatus (WGTA) An apparatus designed originally at the University of Wisconsin Primate Center for use with various species of primates in a series of learning problems. Modifications of this apparatus have been used for a variety of animal species.

Wise Use Movement A coalition of ranchers, loggers, miners, industrialists, hunters, off-road vehicle users, land developers, and others who call for unrestricted access to natural resources and public lands.

Withdrawal A description of the total amount of water taken from a lake, river, or aquifer.

Wood pulp A watery suspension of pulverized wood used in the production of paper, cardboard, fiberboard, rayon, cellophane, and other products.

Wood Secondary xylem.

Woodland A forest where tree crowns cover less than 20 percent of the ground; also called open canopy.

Woodward-hoffmann rules Rules governing the formation of products during certain types of organic reactions.

Work function The minimum energy required to remove an electron from the surface of a material and send it into field-free space.

Work The amount of energy transferred to or from a body or system as a result of forces acting upon the body, causing displacement of the body or parts of it. More specifically the work done by a particular force is the product of the displacement of the body and the component of the force in the direction of the displacement. A force acting perpendicular to the body's displacement does no work on the body. A force acting upon a body which undergoes no displacement does no work on that body. Also, it follows that if there's no motion of a body or any part of the body, nothing did work on the body.

Working equation The equation derived from a basic equation as an expression of the unknown quantity in a problem directly in terms of the unknown quantities stated in the problem The working equation is the mathematical expression of the solution to the problem.

World conservation strategy A system of maintaining essential ecological processes, preserving genetic diversity, and ensuring that utilization of species and ecosystems is sustainable.

World Ocean The interconnected world seas and oceans.

X

X ray Very short wavelength in the electromagnetic spectrum; can penetrate soft tissue; although it is useful in medical diagnosis, it also damages tissue and causes mutations.

Xanthophore A chromatophore containing yellow pigment.

Xanthophyll a group of yellow or golden photosynthetic pigments

Xenodiagnosis Diagnosis of a disease by infecting a test animal.

Xenograft Graft of a piece of tissue or organ from one individual to another of a different species.

Xenosomes Body or organelle living within a cell that contains its own DNA and is capable of reproducing itself, once having functioned as a free-living organism. Examples are zooxanthellae and zoochlorellae.

Xeric Characterized by a very dry environment.

Xerophyte A plant adapted for growth under arid conditions.

Xiphidiocercaria Cercaria with a stylet in the anterior rim of its oral sucker.

X-organ. Neurosecretory organ in eyestalk of crustaceans that secretes molt-inhibiting hormone.

X-ray crystallography A technique for studying molecular and atomic structure of a substance.

Xylem In vascular plants, the tissue that conducts water and minerals; xylem consists, in various plants, of tracheids, vessel members, fibers, and parenchyma.

Xylem ray That part of a vascular ray which is located in the wood or secondary xylem. Consists of a radial file of parenchyma cells, Also termed wood ray.

Y

Yaws Bacterial disease caused by the spirochete Treponema pertenue, often transmitted by flies.

Yeasts One of several species of ascomycete fungi in which an extensive mycelium is not produced, so that the oganism is usually unicellular.

Yellow fever Virus disease transmitted by the mosquito Aedes aegypti.

Yellowcake The concentrate of 70 to 90 percent uranium oxide extracted from crushed ore.

Ylide A chemical species produced (actually or notationally) by loss of a hydron from an atom directly attached to the central atom of an "onium ion.

$$Ph_3P^+{-}CHRR' \longrightarrow \left(Ph_3P^+{-}C^-RR' \underset{\text{ylide}}{\longleftrightarrow} Ph_3P{=}CRR' \right)$$

Yolk plug The large yolk-filled cells that protrude from beneath the ectoderm at the blastopore of the amphibian gastrula.

Yolk sac The stored food reserve (yolk) and its surrounding membranes, which is found in embryonic reptiles, birds, and mammals.

Y-organ Gland in the antennal or maxillary segment of some crustaceans that secretes molting hormone.

Young's experiment; double-slit experiment A famous experiment which shows the wave nature of light (and indeed of other particles). Light is passed from a small source onto an opaque screen with two thin slits. The light is diffracted through these slits and develops an interference pattern on the other side of the screen.

Young's modulus The ratio of stress to strain in a solid.

Yukawa-Tsuno equation A multiparameter extension of the Hammett equation to quantify the role of enhanced resonance effects on the reactivity of meta- and para-substituted benzene derivatives.

Z

Zeeman effect The splitting of atomic energy levels into two or more sublevels by means of a magnetic field as seen in spectral lines under these conditions.

Zeitgeber Any entraining agent that plays a role in setting or resetting an internal biological clock. Examples include sunrise or sunset.

Zero population growth (ZPG) The number of births at which people are just replacing themselves; also called the replacement level of fertility.

Zeroth law of thermodynamics If body A is in thermal equilibrium with body B, and B is also in thermal equilibrium with C, then A is necessarily in thermal equilibrium with C.

This is equivalent to saying that thermal equilibrium obeys a transitive mathematical relation. Since we define equality of temperature as the condition of thermal equilibrium, then this law is necessary for the complete definition of temperature. It ensures that if a thermometer (body B) indicates that body A and C give the same thermometer reading, then they bodies A and C are at the same temperature.

Zoid Member of a colonial organism.

Zoitocyst Tissue phase in some of the coccidia of the Isospora group. They usually have internal septae and contain thousands of bradyzoites. Also called a sarcocyst or Miescher's tubule.

Zonation The presence of organisms within a particular range, as in the vertical zonation observed in the intertidal.

Zone of aeration Upper soil layers that hold both air and water.

Zone of differentiation An area within the plant, proximal to the apical meristems, where cells begin to assume their srtuctural and functional characters.

Zone of leaching The layer of soil just beneath the topsoil where water percolates, removing soluble nutrients that accumulate in the subsoil; may be very different in appearance and composition from the layers above and below it.

Zone of saturation Lower soil layers where all spaces are filled with water.

Zonite The individual body unit of a member of the phylum Kinorhyncha.

Zoochlorella Any of various minute green algae (usually Chlorella) that live symbiotically within the cytoplasm of some protozoa and other invertebrates.

Zooflagellates Members of the Zoomastigophora, the animal-like flagellates (phylum Sarcomastigophora).

Zooid An individual member of a colony of animals, such as colonial Cnidarians and ectoprocts, produced by incomplete budding.

Zoology The study of animals.

Zoomastigophorea The protozoan class where members lack chloroplasts; heterotrophic or saprozoic. Examples: Trypanosoma, Trichonympha, Trichomonas, Giardia.

Zoonosis Disease of animals that is transmissible to humans. Some authors subdivide the concept into zooanthroponosis, an infection humans can acquire from animals, and anthropozoonosis, a disease of humans transmissible to other animals.

Zooplankton The animal component of the plankton, having only limited locomotory powers.

Zoospore A motile asexual spore.

Zooxanthellae A group of dinoflagellates that live in mutualistic relationships with some Cnidarians. They promote high rates of calcium carbonate deposition in coral reefs.

Zosterophyllophyta An extinct division of early vascular plants characterized by lateral sporangia and commonly spine-like emergences. Believed by evolutionary biologists to be the ancestor of the lycopods.

Z-scheme A diagrammatic representation of electron flow through PS II and PS I.

Zugunruhe Restlessness that an animal displays at the time of its usual migration.

Z-value An index of the ionizing power of a solvent based on the frequency of the longest wavelength electronic absorption maximum of 1-ethyl-4-methoxycarbonylpyridinium iodide in the solvent.

Zwitterionic compound A neutral compound having electrical charges of opposite sign, delocalized or not on adjacent or nonadjacent atoms. Zwitterionic compounds have no uncharged canonical representations. Sometimes referred to as inner salts, ampholytes, dipolar ions (a misnomer).

Zygomorphic flower A bilaterally symmetrical flower, capable of being divided into two symmetrical halves only by a single longitudinal plane passing through the axis.

Zygomycetes Members of the fungal division Zygomycota, consisting of about 800 terrestrial species having primarily coenocytic hyphae. Bread mold is an example.

Zygosporangium (or zygospore) The thick-walled sexual spore formed by members of the zygomycetes.

Zygospore A thick-walled, resistant spore resulting from a zygote.

Zygote Diploid cell produced by the fusion of an egg and sperm; fertilized egg cell.

Zygotic meiosis. Meiosis that takes place within the first few divisions after zygote formation; thus all stages in the life cycle other than the zygote are haploid.